Special Functions:
Group Theoretical Aspects and Applications

Mathematics and Its Applications

Managing Editor:

M. HAZEWINKEL
Centre for Mathematics and Computer Science, Amsterdam, The Netherlands

Special Functions: Group Theoretical Aspects and Applications

edited by

R. A. Askey
Mathematics Department, University of Wisconsin, Madison, U.S.A.

T. H. Koornwinder
Centre for Mathematics and Computer Science, Amsterdam, The Netherlands

and

W. Schempp
Department of Mathematics, University of Siegen, F.R.G.

D. Reidel Publishing Company

A MEMBER OF THE KLUWER ACADEMIC PUBLISHERS GROUP

Dordrecht / Boston / Lancaster

Library of Congress Cataloging in Publication Data

Main entry under title:

Special functions: Group theoretical aspects and applications

        (Mathematics and its applications)
        Includes bibliographies and index.
        1. Functions, Special. I. Askey, Richard. II. Koornwinder,
· T. H., 1943-   III. Schempp, W. (Walter), 1938-   IV. Series:
Mathematics and its applications (D. Reidel Publishing Company)
QA351.S696   1984   515.9   84-15052
ISBN-13: 978-1-4020-0319-6   e-ISBN-13: 978-94-010-9787-1
DOI: 10.1007/978-94-010-9787-1

Published by D. Reidel Publishing Company
P.O. Box 17, 3300 AA Dordrecht, Holland

Sold and distributed in the U.S.A. and Canada
by Kluwer Academic Publishers,
190 Old Derby Street, Hingham, MA 02043, U.S.A.

In all other countries, sold and distributed
by Kluwer Academic Publishers Group,
P.O. Box 322, 3300 AH Dordrecht, Holland

TABLE OF CONTENTS

EDITOR'S PREFACE

Approach your problems from the right end and begin with the answers. Then one day, perhaps you will find the final question.

'The Hermit Clad in Crane Feathers' in R. van Gulik's The Chinese Maze Murders.

It isn't that they can't see the solution.
It is that they can't see the problem.

G.K. Chesterton. The Scandal of Father Brown 'The Point of a Pin'.

Growing specialization and diversification have brought a host of monographs and textbooks on increasingly specialized topics. However, the "tree" of knowledge of mathematics and related fields does not grow only by putting forth new branches. It also happens, quite often in fact, that branches which were thought to be completely disparate are suddenly seen to be related.

Further, the kind and level of sophistication of mathematics applied in various sciences has changed drastically in recent years: measure theory is used (non-trivially) in regional and theoretical economics; algebraic geometry interacts with physics; the Minkowsky lemma, coding theory and the structure of water meet one another in packing and covering theory; quantum fields, crystal defects and mathematical programming profit from homotopy theory; Lie algebras are relevant to filtering; and prediction and electrical engineering can use Stein spaces. And in addition to this there are such new emerging subdisciplines as "completely integrable systems", "chaos, synergetics and large-scale order", which are almost impossible to fit into the existing classification schemes. They draw upon widely different sections of mathematics.

This programme, Mathematics and Its Applications, is
devoted to such (new) interrelations as exampla gratia:

- a central concept which plays an important role in
  several different mathematical and/or scientific
  specialized areas;
- new applications of the results and ideas from one area
  of scientific endeavor into another;
- influences which the results, problems and concepts of
  one field of enquiry have and have had on the development
  of another.

The Mathematics and Its Applications programme tries to
make available a careful selection of books which fit the
philosophy outlined above. With such books, which are
stimulating rather than definitive, intriguing rather than
encyclopaedic, we hope to contribute something towards
better communication among the practitioners in diversified
fields.

Special functions as an area of research in mathematics
have had their ups and downs. There used to be a time when
every (applied) mathematician and physicist knew at least
a large chunk of special function theory and used this
knowledge regularly. On the other hand, Dick Askey, one of
the editors of the present volume, told me that when he
started work in the field some twenty years ago, this was
generally regarded as quixotic to a degree; the field being
considered practically dead.

As has happened before, it turned out far otherwise, and
this volume is fair testimony to that fact.

It may well be that in the mind of many mathematicians the
topic of special functions still carries some of the
following flavour: a nicely well-rounded topic, conveniently
classified under number 33 - classification schemes sometimes do a lot of harm this way -, useful when you need
them, but not all that exciting and certainly not one of
those cutting edges of modern mathematics where lots of
different specialisms come together. Perhaps I am over-
stating, but I certainly felt that way myself a number of
years ago, and now I wonder how in the world I made do with
such a poor understanding - perhaps here also phylogeny
repeats ontogeny.

Certainly a glance at the table of contents of the present volume will show how wrong these impressions are and perusing the volume itself may well emphasize to the attentive reader that special functions, instead of being relatively isolated, are in fact related to just about everything: from representation theory to radar and tomography and from combinatorics to gauge theory.

The unreasonable effectiveness of mathematics in science .....

Eugene Wigner

Well, if you knows of a better 'ole, go to it.

Bruce Bairnsfather

What is now proved was once only imagined.

William Blake

As long as algebra and geometry proceeded along separate paths, their advance was slow and their applications limited.
But when these sciences joined company they drew from each other fresh vitality and thenceforward marched on at a rapid pace towards perfection.

Joseph Louis Lagrange

Amsterdam, April 1984   Michiel Hazewinkel

Richard Askey

PREFACE

Certain functions appear so often that it is convenient to
give them names. These functions are collectively called
special functions. There are many examples and no single way
of looking at them can illuminate all examples or even all
the important properties of a single example of a special
function.

One of the early examples was called the Laplace
coefficient in the nineteenth century, and now is called
the <u>Legendre polynomial</u>. It is denoted by $P_n(x)$. It
arose from the generating function

(1) $$(1-2xr+r^2)^{-1/2} = \sum_{n=0}^{\infty} P_n(x) r^n$$

when Legendre and Laplace were studying problems in celestial
mechanics. They found the differential equation

(2) $$(1-x^2)y'' - 2xy'+n(n+1)y = 0, \quad y = P_n(x)$$

Earlier Lagrange had studied polynomials that are solutions
of the recurrence relation

(3) $$(2n+1) x R_n(x) = (n+1)R_{n+1}(x) + nR_{n-1}(x),$$

$R_{-1}(x) = 0$, $R_0(x) = 1$. This arose in his study of a
problem in probability theory. These polynomials are the
same as the Legendre polynomials.
    Legendre polynomials are a special case of <u>spherical
harmonics</u>. These can be defined as follows. A function
$f(x,y,z)$ is harmonic if it satisfies Laplace's equation

R. A. Askey et al. (eds.), Special Functions: Group Theoretical Aspects and Applications, xi-xxxiv.
© *1984 by D. Reidel Publishing Company.*

$$\frac{\partial^2 f}{\partial x^2} + \frac{\partial^2 f}{\partial y^2} + \frac{\partial^2 f}{\partial z^2} = 0.$$

It is homogeneous of degree   n   if

$$f(\lambda x, \lambda y, \lambda z) = \lambda^n f(x,y,z).$$

The homogeneous harmonic polynomials can be considered as functions on the unit sphere. They are called spherical harmonics.

The corresponding functions for R^2 are generated by

$$(x+iy)^n \quad \text{and} \quad (x-iy)^n$$

before restricting them to the unit circle, and by

$$\cos n\theta \quad \text{and} \quad \sin n\theta$$

or

$$e^{in\theta} \quad \text{and} \quad e^{-in\theta}$$

after restriction to the unit circle. In R^2 there are two linearly independent spherical harmonics of each degree except for those of degree zero, when there is only one. In R^3 the spherical harmonics of degree   n   form a (2n+1)   dimensional vector space. The Legendre polynomials play the same role here that $\cos n\theta$ plays in R^2.   A natural extension of the addition formula

$$\cos n(\theta_1 - \theta_2) = \cos n\theta_1 \; \cos n\theta_2 + \sin n\theta_1 \; \sin n\theta_2$$

was found by Legendre and Laplace. They showed that

(4) $P_n(\cos \theta_1 \cos \theta_2 + \sin \theta_1 \sin \theta_2 \cos \varphi)$

$$= P_n(\cos \theta_1) P_n(\cos \theta_2)$$

$$+ \sum_{k=1}^{n} a_{n,k} g_{k,n}(\theta_1) g_{k,n}(\theta_2) \; \cos k\varphi$$

for specific functions $g_{k,n}(\theta)$ and coefficients $a_{n,k}$
which do not concern us as this point. Their method was to
use rotations of the unit sphere and the fact that the class
of harmonic functions is invariant under rotation, as is the
class of homogeneous polynomials.

  Once attention is focused on the spherical harmonics
on the surface of the sphere it is natural to ask how to
determine them directly. Legendre polynomials can be
determined as solutions to the <u>functional equation</u>

$$(5) \quad g(\theta_1)g(\theta_2) = \frac{1}{\pi} \int_0^\pi f(\cos\theta_1 \cos\theta_2 + \sin\theta_1 \sin\theta_2 \cos\varphi) \, d\varphi$$

with $g(\theta) = f(\cos\theta)$. The fact that $f(\theta) = P_n(\cos\theta)$
satisfies this equation is immediate from (4), using

$$\int_0^\pi \cos n\theta \, d\theta = 0, \quad n = 1,2,\dots .$$

A basis for the remaining spherical harmonics of degree   n
can be found by differentiating $P_n(\cos\theta)$ and multiplying
by spherical harmonics on the circle.

  The functional equation (5) has been generalized to
many different settings. One is the following. Let   G
be a connected Lie group   and   K   a compact subgroup. A
nontrivial complex valued continuous function on   G   is a
<u>spherical function</u> if

$$(6) \qquad\qquad \int_K f(xky) \, dk = f(x)f(y)$$

for all $x,y \in G$,   where   dk   is invariant measure on
K   of mass one.

  The example above arises when $G = SO(3)$, $K = SO(2)$
and $G/K = S^2$.   The other compact rank one symmetric
spaces have spherical functions which are orthogonal polynomials in an appropriate variable. These polynomials are
special cases of a class called Jacobi polynomials. They
are orthogonal with respect to a beta distribution, and
can be given explicitly as hypergeometric functions.

  It is worthwhile looking at some of the objects mentioned in the last paragraph. The compact rank one symmetric

spaces are Riemannian manifolds M that are two point homogeneous. By this we mean there is a group G acting on M so that M is homogeneous under G, i.e. given x_1 and y_1 in M there is $g \in G$ so that $gx_1 = y_1$. There is a metric on M, say $d(x_1, x_2)$, if (x_1, x_2), (y_1, y_2) are given, $x_i \in G$, $y_i \in G$ and if $d(x_1, x_2) = d(y_1, y_2)$, then there is $g \in G$ so that $g(x_i) = y_i$, $i = 1, 2$.

Hypergeometric functions come in two types, the classical hypergeometric function, and generalized hypergeometric functions. Both are important. A hypergeometric series is a series

$$\sum c_n$$

with

$$\frac{c_{n+1}}{c_n} = \text{a rational function of} \quad n.$$

This is usually written as

$$(7) \qquad \frac{c_{n+1}}{c_n} = \frac{(n+a_1) \cdots (n+a_p)}{(n+b_1) \cdots (n+b_q)(n+1)} \, x$$

so

$$c_n = \frac{(a_1)_n \cdots (a_p)_n}{(b_1)_n \cdots (b_q)_n} \frac{x^n}{n!}$$

when $c_0 = 1$. Here the shifted factorial $(a)_n$ is defined by

$$(a)_n = a(a+1) \cdots (a+n-1) = \Gamma(n+a)/\Gamma(a), \quad n = 1, 2, \ldots,$$

$$= 1, \qquad\qquad\qquad\qquad\qquad\qquad n = 0.$$

Then the standard notation is

$$(8) \qquad {}_p F_q \left(\begin{matrix} a_1, \cdots, a_p \\ b_1, \cdots, b_q \end{matrix} ; x \right) = \sum_{n=0}^{\infty} \frac{(a_1)_n \cdots (a_p)_n}{(b_1)_n \cdots (b_q)_n} \frac{x^n}{n!}$$

If $p \leq q$ this is an entire function. If $p = q + 1$ this series gives an analytic function for $|x| < 1$, and it can be continued outside this circle to the complex plane

cut on $[1,\infty)$.   The resulting function is called a generalized hypergeometric function (generalized is often dropped).
The case $p = 2, q = 1$ is often called the hypergeometric
function. This special case arises in many other ways.
Euler found the integral representation

$$(9) \quad {}_2F_1(a,b;c;x) = \frac{\Gamma(c)}{\Gamma(b)\Gamma(c-b)} \int_0^1 (1-xt)^{-a}t^{b-1}(1-t)^{c-b-1}dt$$

and showed that $y = {}_2F_1(a,b;c;x)$ is a solution of the
differential equation

$$(10) \quad x(1-x)y'' + [c-(a+b+1)x]y' - aby = 0$$

This differential equation has regular singular points
at $x = 0,1,\infty$.   Riemann showed that the requirement that
a differential equation have regular singular points at
three given points and every other complex point is a regular
point is so strong a restriction that the differential equation is the hypergeometric equation with the singularities
moved to the three given points. Differential equations
with four or more singular points only infrequently have a
solution which can be given explicitly as a series whose
coefficients are known, or have an explicit integral representation. This partly explains why the classical hypergeometric function arises in many settings that seem to have
nothing to do with each other. The differential equation
they satisfy is the most general one of its kind that has
solutions with many nice properties.
    Jacobi polynomials can be given as follows.

$$(11) \quad P_n^{(\alpha,\beta)}(x) = \frac{(\alpha+1)_n}{n!} \, {}_2F_1\left(\begin{matrix}-n, & n+\alpha+\beta+1 \\ & \alpha+1\end{matrix}; \frac{1-x}{2}\right)$$

Their orthogonality relation is

$$(12) \quad \int_{-1}^1 P_n^{(\alpha,\beta)}(x)P_m^{(\alpha,\beta)}(x)(1-x)^{\alpha}(1+x)^{\beta}dx = 0, \quad m \neq n$$

when $\alpha,\beta > -1$.
    The Legendre polynomials mentioned above are the special
case $\alpha = \beta = 0$ of Jacobi polynomials. The sphere S^k
is a two point homogeneous space, and the spherical functions

are $P_n^{(\alpha,\alpha)}(\cos \theta)$ when $\alpha = (k-2)/2$. The other compact rank one symmetric spaces are real projective spaces, complex projective spaces, quaternionic projective spaces and a two dimensional projective space over the octonians. The spherical functions are $P_n^{(\alpha,\beta)}(\cos \theta)$ for $(\alpha,\beta) = ((k-2)/2, -1/2)$, $(k-2, 0)$, $(2k-1, 1)$ and $(7,3)$ respectively. See Gangolli [20].

There are other settings where similar structures exist. For noncompact Riemannian manifolds the spherical functions are <u>Bessel functions</u> in the case of Euclidean n-space and <u>Jacobi functions</u> for the hyperbolic manifolds. The last example is very important and various properties of the Jacobi functions get rediscovered periodically. This happens because they occur in a number of different settings in addition to the present one, and there has been no adequate survey of what was known about them. This lack has been removed in the present book by Koornwinder's extensive survey of Jacobi functions.

There are discrete two point homogeneous spaces that arise naturally, and a version of spherical harmonics on them is a powerful tool to study these spaces and the functions on them. One example is the set of N-tuples of zeros and ones with the Hamming distance. Geometrically this set is the vertices of the unit cube in R^N and the distance counts the minimal number of edges that need to be traversed to go from one point to another. The spherical functions are called <u>Krawtchouk polynomials</u>, and they can also be given as hypergeometric functions. They are

$$K_n(x,N) = {}_2F_1(-n, -x; -N; 2)$$

$$= \sum_{k=0}^{n} \frac{(-n)_k(-x)_k}{(-N)_k k!} 2^k$$

$n,x = 0,1,\ldots,N$. These polynomials are orthogonal with respect to the binomial distribution:

$$(13) \qquad \sum_{x=0}^{N} \binom{N}{x} K_n(x;N)K_m(x;N) = 0, \quad m \neq n \leq N.$$

There are a number of other discrete two point homogeneous spaces, and the spherical functions are orthogonal polynomials in an appropriate variable. Some of these

spherical functions can be given as hypergeometric series,
but others need a more general class of functions to represent them. Instead of taking a series $\sum c_n$ with
c_{n+1}/c_n a rational function of n, allow it to be a
rational function of q^n for a fixed parameter q. The
resulting series are called basic hypergeometric series.
Since these are almost never mentioned in undergraduate
courses, and most mathematicians have seen at most one or
two examples, a few examples will be given here. The first,
and probably the most famous, is the theta function

$$\sum c_n = \sum_{-\infty}^{\infty} q^{n^2} x^n$$

with

$$\frac{c_{n+1}}{c_n} = q^{2n+1} x$$

The terms are summed over positive and negative values of
n, and the reason is that there is no reason not to. This
will be explained below.

Another example is the q-binomial theorem which comes
from taking

(14)
$$\frac{c_{n+1}}{c_n} = \frac{1 - aq^n}{1 - q^{n+1}} x .$$

Then
$$c_n = \frac{(a;q)_n}{(q;q)_n} x^n$$

with $c_0 = 1$.

The notation above is defined by

$$(a;q)_n = (1-a)(1-aq) \cdots (1-aq^{n-1}), \quad n = 1,2,\ldots$$

$$=1, \qquad n = 0.$$

When $|q| < 1$ this can be redefined as

(15)
$$(a;q)_n = (a;q)_\infty / (aq^n;q)_\infty$$

with

$$(a;q)_\infty = \prod_{k=0}^{\infty} (1-aq^k)$$

From now on we will take $|q| < 1$, unless otherwise stated. A more general ratio than (14) is

$$\frac{c_{n+1}}{c_n} = \frac{(1-aq^n)}{(1-bq^n)} x$$

The resulting series was considered by Ramanujan [31, Chapter XVI, Entry 17, p. 196]. It is

$$\sum_{-\infty}^{\infty} \frac{(a;q)_n}{(b;q)_n} x^n$$

where $(a;q)_n$ is defined for negative values of n by (15). The series converges for $|b/a| < |x| < 1$. When $b = q$,

$$\frac{1}{(q;q)_n} = 0, \qquad n = -1, -2,\ldots$$

and the resulting series is

$$\sum_{n=0}^{\infty} \frac{(a;q)_n}{(q;q)_n} x^n$$

In the definition of the generalized hypergeometric series ${}_pF_q$ there was one factor that played a special role, $1/n!$. This is usually thought to be there because it occurs in the expansions of the functions that arise in elementary calculus such as $\exp(x)$ and $\sin x$, and the more complicated functions such as the Bessel function $J_\alpha(x)$. Thus to include it means that one less parameter needs to be mentioned explicitly. Another reason is that it occurs in Taylor series. However the real reason is different. It is there to terminate the series at $n = 0$. If the term ratio c_{n+1}/c_n is a rational function of n, so that

$$\frac{c_{n+1}}{c_n} = \frac{(n+a_1) \cdots (n+a_p)}{(n+b_1) \cdots (n+b_q)}$$

then there are three cases to consider. If $p > q$ the
series $\sum\limits_{n=0}^{\infty} c_n$ diverges for all x except $x = 0$. If
$p < q$ the series $\sum\limits_{-\infty}^{-1} c_n$ diverges for all finite x.
If $p = q$ the series $\sum\limits_{0}^{\infty} c_n$ diverges for $|x| > 1$
and $\sum\limits_{-\infty}^{-1} c_n$ diverges for $|x| < 1$. Thus the only case
where such a series can converge is when $p = q$ and then
it can only converge on the unit circle. There are some
important examples of these series, such as

$$\sum_{-\infty}^{\infty} \frac{(-1)^n}{n+x} = \frac{\pi}{\sin \pi x} .$$

However most of these series are of limited interest. This
is no longer the case for basic hypergeometric series. The
analogue of $1/n!$ is $1/(q;q)_n$, which terminates the
series at $n = 0$. However the more general series of
Ramanujan (18) converges for $|b|/|a| < |x| < 1$, and
represents an analytic function there.
     Each of the three basic hypergeometric series given
above can be summed. The resulting identities are

(16) $$\sum_{-\infty}^{\infty} q^{n^2} x^n = (q^2;q^2)_\infty (-qx;q^2)_\infty (-q/x;a^2)_\infty$$

(17) $$\sum_{n=0}^{\infty} \frac{(a;q)_n}{(q;q)_n} x^n = (ax;q)_\infty / (a;q)_\infty$$

(18) $$\sum_{-\infty}^{\infty} \frac{(a;q)_n}{(b;q)_n} x^n = \frac{(ax;q)_\infty (q/ax;q)_\infty (q;q)_\infty (b/a;q)_\infty}{(x;q)_\infty (b/ax;q)_\infty (b;q)_\infty (q/a;q)_\infty}$$

In the last identity of Ramanujan replace q by q^2,
x by $-xq/a$, set $b = 0$ and let $a \to \infty$. The
resulting identity is the <u>Jacobi triple product</u> (16).
Ramanujan sum (18) can be written as a series that extends
one of Euler's beta integrals (20). See [3]. Simple proofs
of (17) and (18) are given there as well.

There are two point homogeneous spaces over some finite Chevalley groups whose spherical functions are orthogonal polynomials that can be represented as basic hypergeometric series. Stanton's paper in the present volume surveys these examples.

There is a more general setting where much of this work can be done. This is an association scheme. Bannai and Ito [8] contains a treatment of association schemes as mathematical objects. Their second volume will give applications. Seidel's paper in the present volume contains a number of combinatorial applications of spherical harmonics and discrete versions of them.

Before going on to survey some of the classical orthogonal polynomials a little needs to be said about gamma and beta functions and integrals and their extensions. The gamma function has many properties, and a judicious choice of a few (or even one in some cases) will determine the function uniquely. One current mandarin, N. Bourbaki, defines $\Gamma(x)$ as the unique solution of the functional equation

$$f(x+1) = xf(x)$$

which is normalized by $f(1) = 1$ and whose logarithm is convex for positive x. My favorite definition is essentially Euler's original one:

$$\Gamma(x) = \lim_{k \to \infty} \frac{k!}{(x)_k} (k+1)^{x-1}$$

when $x \neq 0, -1, \ldots$. The real reason the gamma function is important is that it arises in the solution of many problems. One is the evaluation of many integrals, including the classical beta integral that was studied by Wallis and evaluated by Euler. It is

$$(19) \qquad \int_0^1 t^{x-1}(1-t)^{y-1}\, dt = \frac{\Gamma(x)\Gamma(y)}{\Gamma(x+y)} \quad \text{Re } x > 0, \quad \text{Re } y > 0.$$

There are three variants that should be mentioned. Setting $t = s(1+s)^{-1}$ gives

$$(20) \qquad \int_0^\infty \frac{s^{x-1}}{(1+s)^{x+y}}\, ds = \frac{\Gamma(x)\Gamma(y)}{\Gamma(x+y)}, \quad \text{Re } x > 0, \quad \text{Re } y > 0$$

Cauchy found two other beta integrals:

(21) $\qquad \dfrac{1}{2\pi} \displaystyle\int_{-\infty}^{\infty} \dfrac{dt}{(a+it)^x (b-it)^y} = \dfrac{\Gamma(x+y-1)}{\Gamma(x)\,\Gamma(y)} (a+b)^{1-x-y}$

when   Re a > 0,   Re b > 0,   Re (x+y) > 1,   and

(22) $\qquad \dfrac{1}{2\pi} \displaystyle\int_{-\pi}^{\pi} (1 - \cos\theta)^{(x+y-2)/2}\, e^{i(x-y)\theta/2}\, d\theta$

$$= \dfrac{\Gamma(x+y-1)}{\Gamma(x)\,\Gamma(y)}\, 2^{(2-x-y)/2}$$

The first three have the following form. Take two linear
functions, raise them to powers and integrate over an appropriate curve. In the first case connect the zeros of the
linear functions; in the second case connect a zero with
infinity, avoiding the second zero, in the third case separate the zeros by a curve that goes from infinity to infinity.
The fourth integral comes from the third by setting

$$\dfrac{2}{1+it} = 1 - z = 1 - e^{i\theta}$$

It has the same form with the two linear functions   1-z
and   -z,   for $(1 - e^{-i\theta})^y = (1-z)^y (-z)^{-y}$ and the
contour connects one zero to itself with a curve that has
the other zero inside it.
        Euler's integral for the classical hypergeometric
function (9) is just one of many such integrals. Take three
linear functions, raise them to powers and integrate over an
appropriate curve. The result is the hypergeometric function.
This is another of the many ways in which hypergeometric
functions arise.
        Two limits of the beta integral are Euler's integral for
the gamma function:

$$\Gamma(x) = \int_0^\infty t^{x-1} e^{-t}\, dt$$

and the normal integral

$$\sqrt{\pi} = \int_{-\infty}^\infty e^{-t^2}\, dt$$

Each of these measures is the weight function for an important set of orthogonal polynomials, Jacobi polynomials for the beta distribution, Laguerre polynomials for the gamma distribution, and Hermite polynomials for the normal distribution.

There are many extensions of these integrals. Some of the most important are sums. One is

$$(23) \qquad \sum_{k=0}^{N} \frac{(a)_k}{k!} \frac{(b)_{N-k}}{(N-k)!} = \frac{(a+b)_N}{N!}$$

There are polynomials orthogonal with respect to the terms in the series (23). These polynomials are called Hahn polynomials, although Tchebychef had an explicit representation, the orthogonality, and a discrete Rodriques type formula giving them as the n'th difference of a simple function over one hundred years ago. See [38]. As we should expect, the representation is as a hypergeometric function. One explicit formula is

$$(24) \qquad Q_n(x;\alpha,\beta,N) = \sum_{k=0}^{n} \frac{(-n)_k (n+\alpha+\beta+1)_k (-x)_k}{(\alpha+1)_k (-N)_k k!},$$

$$n,x = 0,1,\ldots,N$$

and the orthogonality is

$$(25) \qquad \sum_{k=0}^{N} Q_n(x) Q_m(x) \frac{(\alpha+1)_x}{x!} \frac{(\beta+1)_{N-x}}{(N-x)!} = 0, \qquad 0 \leq m \neq n \leq N.$$

$$= \frac{1}{h_n}, \qquad m = n \leq N$$

The norm is important, but it is a bit of a mess, so it will not be given here.

If $x \to N - x$ the weight function

$$w_{\alpha,\beta}(x) = \frac{(\alpha+1)_x (\beta+1)_{N-x}}{x! (N-x)!}$$

satisfies

$$w_{\alpha,\beta}(N-x) = w_{\beta,\alpha}(x)$$

This induces a transformation in the polynomials orthogonal
with respect to this measure, which in turn gives the transformation

(26) $$3^F2\left(\begin{matrix}-n,a,b\\c,d\end{matrix};1\right) = \frac{(c-a)_n}{(c)_n} \; 3^F2\left(\begin{matrix}-n,a,d-b\\a+1-n-c,d\end{matrix};1\right).$$

Other transformations can be obtained by iterating (26) and
by reading one of the series backwards. Using these transformations it is possible to show that the   3-j   symbols
of angular momentum theory (or the Clebsch-Gordan coefficients
for   SU(2)   or   SO(3))   are really hypergeometric functions, one of whose orthogonality relations is (25). See
[12] for a detailed treatment of   3-j   symbols.
     The information in the orthogonality relation (25) can
be given in matrix form. If the orthogonality relation is

(27) $$\sum_{x=0}^{N} P_n(x)P_m(x)w(x) = \delta_{m,n}/h_n , \qquad m,n = 0,1,\ldots,N$$

then

$$a(n,x) = [w(x)h_n]^{1/2}P_n(x), \qquad n,x = 0,1,\ldots,N$$

forms an orthogonal matrix, since the rows are orthogonal
and the inner product of a row with itself is one. Then the
columns have the same property, so

(29) $$\sum_{n=0}^{N} P_n(x)P_n(y)h_n = \delta_{x,y}/w(x), \qquad x,y = 0,1,\ldots,N.$$

In the case of the Hahn polynomials, their dual polynomials
formed by interchanging   n   and   x   are

$$3^F2\left(\begin{matrix}-n,-x,x+\alpha+\beta+1\\\alpha+1,N\end{matrix};1\right) = S_{2n}(x),$$

and so are polynomials in   x,   but this is a polynomial of
degree   2n   rather than   n.   However it is a polynomial
of degree   n   in the quadratic variable $\lambda(x) =$
$-x(x+\alpha+\beta+1)$.   One can ask what is the most general set of
orthogonal polynomials with the property? The answer is that
two more degrees of freedom can be added. First, there is a
more general set of orthogonal polynomials that can be repre-

sented by a hypergeometric series. The orthogonality relation comes in two forms depending on the parameters. The
discrete set is equivalent to the   6-j   symbols or Racah
coefficients of angular momentum theory. As orthogonal
polynomials they are

$$(30) \quad R_n(-x(x+\gamma+\delta+1)) = {}_4F_3\left(\begin{array}{c} -n,n+\alpha+\beta+1,-x,x+\gamma+\delta+1 \\ \alpha+1,\beta+\delta+1,\gamma+1 \end{array} ;1\right)$$

when one of $\alpha+1$, $\beta+\delta+1$ or $\gamma+1$ is $-N$.   Their
orthogonality relation is

$$(31) \quad \sum_{x=0}^{N} R_n(\lambda(x)) R_m(\lambda(x)) \frac{(\gamma+\delta+1)_x (\frac{\gamma+\delta+3}{2})_x (\alpha+1)_x (\beta+\delta+1)_x (\gamma+1)_x}{x!\,(\frac{\gamma+\delta+1}{2})_x (\gamma+\delta-\alpha+1)_x (\gamma-\beta+1)_x (\delta+1)_x}$$

$= 0$, $0 \le m \ne n \le N$,   where $\lambda(x) = -x(x+\gamma+\delta+1)$.
   The other form is more attractive.

$$(32) \quad W_n(x^2 ; a,b,c,d) = {}_4F_3\left(\begin{array}{c} -n, \ n+a+b+c+d-1, \ a+ix, \ a-ix \\ a+b, \ a+c, \ a+d \end{array} ;1\right)$$

When $a,b,c,d > 0$ the orthogonality relation is

$$(33) \quad \int_0^{\infty} W_n(x^2) W_m(x^2) \left| \frac{\Gamma(a+ix)\Gamma(b+ix)\Gamma(c+ix)\Gamma(d+ix)}{\Gamma(2ix)} \right|^2 dx = 0,$$
$$m \ne n.$$

See Wilson [40].
   The second parameter is the   q   that comes from the
change from hypergeometric to basic hypergeometric series.
These polynomials were introduced in [5], their discrete
orthogonality was given there, and the absolutely continuous
one was given in [6]. The theorem that this discrete set of
polynomials is the most general one whose duals are polynomials in an appropriate variable was found by Leonard [24].
   One special case of the   q-extension of the polynomials
in (32) was introduced by L. J. Rogers in 1894 [33], and used
to derive a pair of identities now known as the Rogers-
Ramanujan identities. One of these is

$$\frac{1}{(q;q^5)_\infty (q^4;q^5)_\infty} = \sum_{n=0}^{\infty} \frac{q^{n^2}}{(q;q)_n}$$

This identity was rediscovered three times; by
Ramanujan, by Schur and recently by R. J. Baxter. Baxter
discovered it while solving a two dimensional problem in
statistical mechanics, this one called the hard hexagon. His
recent book [9] is a very good introduction to some beautiful
work.

There is a different way to look at the polynomials
introduced by Rogers, and this leads to them as polynomials
considered by Szegö which are orthogonal on the unit circle.
Recall that

$$\int_0^{2\pi} (e^{i\theta})^n \overline{(e^{i\theta})^m}\, d\theta = 0, \qquad m \neq n$$

Szegö extended this as follows. A set of polynomials
$\{p_n(z)\}$ is <u>orthogonal on the unit circle</u> with respect to
a positive measure $d\alpha(\theta)$ if

(34)
$$\int_0^{2\pi} P_n(e^{i\theta}) \overline{P_m(e^{i\theta})}\, d\alpha(\theta) = 0, \qquad m \neq n.$$

The Rogers-Szegö polynomials at the end of the paper by
Andrews and Onofri are examples. The other orthogonal polynomials in this paper are discrete extensions of Jacobi
polynomials. These are basic hypergeometric functions, and
the weight function is a discrete extension of the beta
distribution with masses at q^k, $k = 0,1,\ldots$.   The mass
of the weight function and the moments can all be computed
using the   q-binomial theorem, which is identity (17).

I would like to emphasize one fact which I have tried
to illustrate above. The same few integrals and series which
can be explicitly evaluated keep coming up. Most integrals
cannot be evaluated exactly, and when a new one arises it is
potentially very important. The evaluation of the integral
in (33) when $m = n = 0$ is an instance of this, as is the
q-extension given in [6]. This integral is

(35)
$$\frac{1}{2\pi} \int_{-1}^1 \frac{h(x,1)h(x,q^{1/2})h(-x,1)h(-x,q^{1/2})}{h(x,a)h(x,b)h(x,c)h(x,d)} \frac{dx}{\sqrt{1-x^2}}$$

$$= \frac{(abcd;q)_\infty}{(q;q)_\infty (ab;q)_\infty (ac;q)_\infty (ad;q)_\infty (bc;q)_\infty (bd;q)_\infty (cd;q)_\infty}$$

when $-1 < a,b,c,d,q < 1$ and $h(x,a)$ is defined by

$$h(x,a) = \prod_{n=0}^{\infty} (1-2axq^n+a^2q^{2n}).$$

This even contains a useful integral when $q = 0$,   which is a surprise since most q identities become trivial when $q = 0$.   See Littlewood [25, Chapter XI] for an application of (35) when $q = 0$.

These two integrals will be useful, but a much better illustration is a <u>multidimensional beta integral</u> found by Selberg [36].   It is

(36)
$$\int_0^1 \cdots \int_0^1 [\prod_{1 \le i < j \le n} (t_i-t_j)^2]^z \prod_{i=1}^{n} t_i^{x-1}(1-t_i)^{y-1} dt_i$$

$$= \prod_{j=1}^{n} \frac{\Gamma(x+(j-1)z)\Gamma(y+(j-1)z)\Gamma(jz+1)}{\Gamma(x+y+(n+j-2)z)\Gamma(z+1)}$$

Onofri started his talk at the Oberwolfach meeting by saying he would talk about a contribution to Mehta mathematics.   The allusion is to a beautiful conjecture of Mehta and Dyson [28] which is nicely explained in Mehta's book <u>Random Matrices</u> [27].   This conjecture was

(37)
$$\frac{1}{(2\pi)^{n/2}} \int_{-\infty}^{\infty} \cdots \int_{-\infty}^{\infty} [\prod_{1 \le i < j \le n} (t_i-t_j)^2]^z \prod_{i=1}^{n} e^{-t_i^2/2} dt_i .$$

$$= \prod_{j=1}^{n} \frac{\Gamma(jz+1)}{\Gamma(z+1)}$$

The verb was put in the past tense since this is no longer a conjecture but a theorem.   It is an immediate consequence of Selberg's integral.   The only known proof of (36) is Selberg's original one, and the only known proof of (37) is via Selberg's integral.   I. G. Macdonald was in the process of rediscovering Selberg's integral and some extensions of it when Selberg's integral was brought to light by Bombieri asking Selberg how to evaluate another multivariate beta integral and Selberg showing him how to do it using (36).

Macdonald was looking at the constant term in certain Laurent
polynomials associated with the root systems of simple Lie
algebras and the associated affine root systems. He had
conjectured that

$$(38) \qquad \text{C.T.} \prod_{\rho \in R} (1-e^\rho)^k = \prod_{j=1}^{n} \binom{d_j k}{d_j}$$

where   R   is a root system, e^ρ is a formal exponential,
C.T.   means to take the constant term in the Laurent polynomial, and d_j are structure constants associated with
the root system (the degrees of the homogeneous polynomials
invariant under the Weyl group). Selberg's integral is
equivalent to a refinement of (38) for BC_n,   where the
integer   k   is replaced by three integers with the same
integer being used for roots of the same length. For BC_n
the result is

$$(39) \qquad \text{C.T.} \prod_{i=1}^{n} (1-t_i)^a (1-1/t_i)^a (1-t_i^2)^b (1-1/t_i^2)^b$$

$$\prod_{1 \le i < j \le n} (1-t_i t_j)^c (1-1/t_i t_j)^c (1-t_i t_j)^c (1-t_j/t_i)^c$$

$$= \prod_{j=0}^{n-1} \frac{(2a+2b+2jc)!\,(2b+2jc)!\,((j+1)c)!}{(a+2b+(n+j-1)c)!\,(a+b+jc)!\,(b+jc)!\,c!}$$

For A_n there is a more general identity where a new
parameter is added for each new dimension. The first conjecture here was formulated by Dyson [16]. It is

$$(40) \qquad \text{C.T.} \prod_{i \ne j} (1-t_i/t_j)^{a_i} = \frac{(a_1 + \cdots + a_n)!}{a_1! \cdots a_n!}$$

This was proven rapidly by Wilson [41] and Gunson [23] and a
beautiful proof was found by Good [22]. Andrews [1] gives
a conjectured extension of (41):

(41)   C.T. $\prod_{1 \le i < j \le n}$ $(t_i/t_j ; q)_{a_i} (qt_j/t_i ; q)_{a_j}$

$$= \frac{(q;q)_{a_1 + \cdots + a_n}}{(q;q)_{a_1} \cdots (q;q)_{a_n}}$$

Macdonald is the one who noticed the connection with A_n. The Dyson-Andrews conjecture has recently been proved by Zeilberger and Bressoud [42] by a very interesting method which has not been alluded to yet. They used a very ingenious counting argument involving tournaments.

Much of the recent interest in combinatorial problems comes from the development of computers, but there has always been an interest in counting things. Given k zeros and $n-k$ ones, the binomial coefficient $\binom{n}{k}$ counts the number of ways these can be arranged in n spots. MacMahon found a couple of ways of refining this counting, and a third way plays an essential role in the Zeilberger-Bressoud proof of (41). The easiest one to describe is one of MacMahon's statistics. Take k zeros and $n-k$ ones as above. This time count the number of times the zeros are to the right of the ones. For example, (0,1,0,1,0) has the label three, since the second zero is to the right of one one and the third zero is to the right of two ones. Then the number of ways of arranging k zeros and $n=k$ ones so that j zeros are to the right of ones (with the above counting) is the coefficient of q^j in the expansion of

$$\begin{bmatrix} n \\ k \end{bmatrix}_q = \frac{(q;q)_n}{(q;q)_k (q;q)_{n-k}}$$

Foata and others have given counting proofs of many interesting facts about the classical orthogonal polynomials. See [18], [19] for some examples.

One of the most promising places to look for uses of Selberg's integral is in doing detailed harmonic analysis on the higher rank symmetric spaces. See Vretare [39]. There are other extensions or analogues of Selberg's integral, but essentially nothing has been proven in three or more dimensions. See [4], [17], [29], [31]. This shows how little we really understand about explicit results in several variables. There are other promising ways to attack the problem of

finding important explicit results in several variables. One
way comes from angular momentum theory, and it has primarily
been developed by physicists. Two books by Biedenharn and
Louck [11], [12] give an extensive treatment of some of this
work. The paper by Biedenharn et al in the present volume
carries on this work using a more sophisticated setting and
vocabulary. My hope is that some explicit orthogonal polynomials in several variables will be found here, and they
will play the same central role in several variable hypergeometric series that orthogonal polynomials in one variable
play in one variable. Also we need to find out where the
basic hypergeometric orthogonal polynomials live. The
q-ultraspherical (symmetric Jacobi) polynomials of
L. J. Rogers [34] are a case in point. They have the
following generating function

(42)
$$\left| \frac{(\beta r e^{i\theta} ; q)_\infty}{(r e^{i\theta} ; q)_\infty} \right|^2 = \sum_{n=0}^{\infty} C_n(x;\beta|q) r^n$$

and the three term recurrence relation

(43) $2x(1-\beta q^n) C_n(x;\beta|q)$

$$= (1-q^{n+1}) C_{n+1}(x;\beta|q) + (1-\beta^2 q^{n-1}) C_{n-1}(x;\beta|q)$$

$$C_{-1}(x;\beta|q) = 0, \qquad C_0(x;\beta|q) = 1.$$

There are many incredible formulas that involve these
polynomials. For example, Rogers [34] showed that

(44) $C_n(x;\beta|q) C_m(x;\beta|q) = \sum_{k=0}^{\min(m,n)} a(k,m,n) C_{m+n-2x}(x;\beta|q)$

with

$a(k,m,n)$

$$= \frac{(q;q)_{m+n-2k} (\beta;q)_{n-k} (\beta;q)_{m-k} (\beta;q)_k (\beta^2;q)_{m+n-k} (1-\beta q^{m+n-2k})}{(\beta^2;q)_{m+n-2k} (q;q)_{n-k} (q;q)_{m-k} (q;q)_k (\beta q;q)_{m+n-k} (1-\beta)}$$

It has been my experience that such formulas are not acci-
dents, if they exist there is a good reason why. We can
prove (44), but we cannot explain it. One wants some
geometric setting where these polynomials arise. It exists
when $\beta = q^{\lambda}$, $q \to 1$ and λ is half an integer, for
the polynomials are then spherical functions on spheres. It
also exists when $q = 0$ for infinitely many values of
β.   The polynomials are spherical functions on two point
homogeneous infinite trees, or on a distance transitive
infinite graph over the   p-adics. See Cartier [13] and
Sawyer [35]. However there is no known setting for these
polynomials for other values of   q.

Very interesting uses of spheroidal wave functions were
discovered by some mathematicians at Bell Telephone Laboratory almost twenty-five years ago. See Slepian [37] for some
results and references to earlier work, and Grünbaum's paper
in this volume for new developments in this area. Polynomial
analogues of these functions have recently been constructed.
See Gilbert and Slepian [21] for one example and Perlstadt [30]
for many others. These polynomials seem to be very interesting
and should be studied carefully.

The remaining three papers in this book deal with
Heisenberg groups. Auslander's paper deals with the general
setting where a nilpotent group is constructed from a separable locally compact Abelian group. In earlier work he had
looked at special Abelian groups and showed how much of the
theory of theta functions can be done in this setting. See
the first three references in his paper. The other two
papers, by Schempp and by Beals, Greiner and Vauthier, take
specific groups and show how Laguerre and Hermite polynomials
arise naturally, and how the group theory can give new in-
sights on these functions.

There are many aspects of special functions which were
not mentioned above or treated in papers in this volume. One
is number theory. G. Chudnovsky [15] has obtained some very
impressive results on irrationality and transcendence using
hypergeometric functions. Apéry's proof of the irrationality
of $\zeta(3)$ used a recurrence relation satisfied by

$$\sum_{k=0}^{n} \binom{n}{k}^2 \binom{n+k}{2}^2 = {}_4F_3\left(\begin{matrix}-n,-n,n+1,n+1\\1,1,1\end{matrix};1\right)$$

This recurrence relation is an easy consequence of the three
term recurrence relation satisfied by the Racah or Wilson

polynomials, (30) or (32). See [7] for extensions of this
recurrence relation. Beukers [10] found a second proof of the
irrationality of $\zeta(3)$. He used Legendre polynomials. In
a different direction, Andrews has recently found a new proof
of a claim of Fermat that Gauss first proved. Fermat claimed
that every positive integer can be written as the sum of
three triangular numbers, numbers of the form $n(n+1)/2$.
Gauss proved this using ternary quadratic forms. Andrews [2]
showed that

$$[\sum_{n=0}^{\infty} q^{n(n+1)/2}]^3 = \sum_{n=0}^{\infty} \sum_{k=0}^{2n} q^{2n^2+2n-k(k+1)/2} \frac{1+q^{2n+1}}{1-q^{2n-1}},$$

and since $2n^2+2n-k(k+1)/2$ is n_j when $k = 2n$, it
is clear that the coefficient of q^j in the expansion is
positive; for it is at least one, and no negative terms
appear on the right hand side to lower this value.

I want to close by mentioning a very interesting set of
functions that I do not understand at all. These are solutions to some differential equations that arose from studying
black holes. S. Chandrasekhar obtained these differential
equations and solutions in a series of papers in the
Proceedings of the Royal Society. A detailed treatment of
them (6,000 pages or so) has been deposited at the University
of Chicago, and these calculations are partly redone and
summarized in his recent book [14, §79 to §96]. Here is
one quotation from this book [p. 497] that nicely "sums up"
the state of affairs. "Nevertheless, the emergence of the
various integral identities express relations whose origins
are shrouded in mystery." Chandrasekhar explained some of
this to me, and some of the numbers that arose suggested that
there were different groups that could be used to explain
some of these results. These equations and their solutions
should be looked at by a number of people, since something
very interesting is going on.

University of Wisconsin-Madison

REFERENCES

[1]   Andrews, G.E.: 1975, 'Problems and prospects for
      basic hypergeometric functions', in R. Askey (ed.),
      Theory and Application of Special Functions,
      Academic Press, New York, pp. 191-224.

[2]   Andrews, G.E.: 'The fifth and seventh order mock theta
        functions', to appear. A short version is tentatively
        titled 'ETPHKA! num = Δ + Δ + Δ'.

[3]   Askey, R.: 1980, 'Ramanujan's extensions of the gamma
        and beta functions', Amer. Math. Monthly 87, 346-59.

[4]   Askey, R.: 1980, 'Some basic hypergeometric extensions
        of integrals of Selberg and Andrews', SIAM J. Math.
        Anal. 11, 938-951.

[5]   Askey, R. and Wilson, J.: 1979, 'A set of orthogonal
        polynomials that generalize the Racah coefficients
        or   6-j   symbols', SIAM J. Math. Anal. 10, 1008-
        1016.

[6]   Askey, R. and Wilson, J.: 'Some basic hypergeometric
        orthogonal polynomials that generalize Jacobi poly-
        nomials', Memoirs Amer. Math. Soc., to appear.

[7]   Askey, R. and Wilson, J.: 1984, 'A recurrence relation
        generalizing those of Apéry', J. Austral. Math. Soc.
        (A) 36, 267-78.

[8]   Bannai, E. and Ito, T.: 1984, Algebraic Combinatorics
        I: Association Schemes, Benjamin/Cummins, Menlo
        Park, CA.

[9]   Baxter, R.J.: 1982, Exactly Solved Models in Statis-
        tical Mechanics, Academic Press, London, New York.

[10]   Beukers, F.: 1979, 'A note on the irrationality of
        $\zeta(2)$ and $\zeta(3)$', Bull. London Math. Soc. 11,
        268-272.

[11]   Biedenharn, L.C. and Louck, J.D.: 1981, Angular
        Momentum in Quantum Physics, Addison-Wesley, Reading,
        Mass.

[12]   Biedenharn, L.D. and Louck, J.D.: 1981, The Racah-
        Wigner Algebra in Quantum Theory, Addison-Wesley,
        Reading, Mass.

[13]   Cartier, P.: 1973, 'Harmonic analysis on homogeneous
        trees', in C. C. Moore (ed.), Harmonic Analysis on
        Homogeneous Spaces, Proc. Symp. Pure Math. 26, Amer.
        Math. Soc., Providence, R.I., pp. 419-424.

[14]   Chandrasekhar, S.: 1983, The Mathematical Theory of
        Black Holes, Oxford Univ. Press, Oxford, New York.

[15]   Chudnovsky, G.: 1979, 'Padé approximations to the
        generalized hypergeometric functions, I', J. math.
        pures et appl. 58, 445-476.

[16]   Dyson, F.J.: 1962, 'Statistical theory of the energy
        levels of complex systems: I', J. Math. Phys. 3,
        140-156.

[17] Evans, R.J.: 1981, 'Identities for products of Gauss
      sums over finite fields', Ensign. Math., sér II,
      27, 197-209.
[18] Foata, D. and Garsia, A.M.: 1979, 'A combinatorial
      approach to the Mehler formulas for Hermite polyno-
      mials, in D. K. Ray-Chaudhuri (ed.), Relations between
      combinatorics and other parts of mathematics, Proc.
      Symp. Pure Math. 34, Amer. Math. Soc., Providence,
      pp. 163-179.
[19] Foata, D., and Leroux, P.: 1983, 'Polynômes de Jacobi,
      interprétation combinatoire et fonction génératrice',
      Proc. Amer. Math. Soc. 87, 47-53.
[20] Gangolli, R.: 1967, 'Positive definite functions on
      certain homogeneous spaces, and certain stochastic
      processes related to Levy's Brownian motion of several
      parameters', Ann. Inst. H. Poincaré (B), 3, 121-226.
[21] Gilbert, E.N. and Slepian, D.: 1977, 'Doubly orthogo-
      nal concentrated polynomials', SIAM J. Math. Anal. 8,
      290-319.
[22] Good, I.J.: 1970, 'Short proof of a conjecture of
      Dyson', J. Math. Phys., 1884.

[23] Gunson, J.: 1962, 'Proof of a conjecture by Dyson in
      the statistical theory of energy levels', J. Math.
      Phys. 3, 752-753.
[24] Leonard, D.: 1982, 'Orthogonal polynomials, duality,
      and association schemes', SIAM J. Math. Anal. 13,
      656-663.
[25] Littlewood, D.E.: 1940, The Theory of Group Characters
      and Matrix Representations of Groups, Oxford Univ.
      Press, Oxford and New York.
[26] Macdonald, I.G.: 1982, 'Some conjectures for root
      systems', SIAM J. Math. Anal. 13, 988-1007.
[27] Mehta, M.L.: 1967, Random Matrices and the Statistical
      Theory of Energy Levels, Academic Press, New York.
[28] Mehta, M.L. and Dyson, F.J.: 1963, 'Statistical theory
      of the energy levels of complex systems: V;, J. Math.
      Phys. 4, 713-719.
[29] Morris, W.G., II: 1982, Constant Term Identities for
      Finite and Affine Root Systems: Conjectures and
      Theorems, Ph.D. Thesis, Univ. Wisconsin-Madison.
[30] Perlstadt, M.: Polynomial analogues of prolate sphe-
      roidal wave functions', to appear in SIAM J. Math.
      Anal.

[31] Rahman, M.: 'Another conjectured   q-Selberg integral'
      to appear.

[32] Ramanujan, S.: 1957, Notebooks of Srinivasa Ramanujan,
      vol. II, Tata Institute of Fundamental Research,
      Bombay.

[33] Rogers, L.J.: 1894, 'Second memoir on the expansion
      of certain infinite products', Proc. London Math.
      Soc. 25, 318-343.

[34] Rogers, L.J.: 1895, 'Third memoir on the expansion of
      certain infinite products', Proc. London Math. Soc.
      26, 15-32.

[35] Sawyer, S.: 1978, 'Isotropic random walks in a tree',
      Z. Wahrscheinlichkeits theorie und Verw. Gebiete 42,
      279-292.

[36] Selberg, A.: 'Bermerkninger om et Multipelt Integral',
      Norsk Mat. Tidsskr. 26, 71-78.

[37] Slepian, D.: 1978, 'Prolate spheroidal wave functions,
      Fourier analysis, and uncertainty - V: the discrete
      case', Bell System Tech. Jour. 57, 1371-1430.

[38] Tchebychef, P.L.: 1961, 'Sur l'interpolation de valeurs
      equidistants', Oeuvres de P. L. Tchebychef, vol. II,
      Chelsea, New York, pp. 217-242.

[39] Vretare, L.: 'Formulas for elementary spherical
      functions and generalized Jacobi polynomials;, to
      appear.

[40] Wilson, J.: 1980, 'Some hypergeometric orthogonal
      polynomials', SIAM J. Math. Anal. 11, 690-701.

[41] Wilson, K.: 1962, 'Proof of a conjecture of Dyson',
      J. Math. Phys. 3, 1040-1043.

[42] Zeilberger, D. and Bressoud, D.: 'A proof of Andrews'
      q-Dyson conjecture', to appear.

Tom H. Koornwinder

JACOBI FUNCTIONS AND ANALYSIS ON NONCOMPACT
SEMISIMPLE LIE GROUPS

CONTENTS

R. A. Askey et al. (eds.), Special Functions: Group Theoretical Aspects and Applications, 1–85.
© *1984 by D. Reidel Publishing Company.*

INTRODUCTION

A Jacobi function $\phi_\lambda^{(\alpha,\beta)}$ $(\alpha,\beta,\lambda\in\mathbb{C},\alpha\neq-1,-2,\ldots)$ is defined as the even C^∞-function on $\mathbb{R}$ which equals 1 at 0 and which satisfies the differential equation

$$(1.1) \qquad (d^2/dt^2+((2\alpha+1)\coth t+(2\beta+1)\th t)d/dt +$$
$$+\lambda^2+(\alpha+\beta+1)^2)\phi_\lambda^{(\alpha,\beta)}(t) = 0.$$

It can be expressed as an hypergeometric function (cf.(2.4), (2.7)). For $\alpha >-1, |\beta| < \alpha+1$ the Jacobi function occurs as a kernel in the Jacobi transform pair

$$(1.2) \qquad \hat{f}(\lambda) = \int_0^\infty f(t)\phi_\lambda^{(\alpha,\beta)}(t)(2\sh\ t)^{2\alpha+1}(2\ch\ t)^{2\beta+1}dt,$$

$$(1.3) \qquad f(t) = (2\pi)^{-1}\int_0^\infty \hat{f}(\lambda)\phi_\lambda^{(\alpha,\beta)}(t)|c_{\alpha,\beta}(\lambda)|^{-2}d\lambda,$$

where $c_{\alpha,\beta}(\lambda)$ is a certain quotient of products of gamma functions, cf. (2.18). This transform generalizes the Fourier cosine transform ($\alpha=\beta=-\frac{1}{2}$).

For special α,β there are many group theoretic interpretations of Jacobi functions (cf.§4), first of all as spherical functions on noncompact Riemannian symmetric spaces of rank one, but also as associated spherical functions, intertwining functions, matrix elements of $SL(2,\mathbb{R})$ and spherical functions on certain nonsymmetric Gelfand pairs. This makes Jacobi functions an ideal subject for a case study of the interactions between special functions and group theory: Harmonic analysis can be developed for the Jacobi transform, considered either as a generalization of the Fourier-cosine transform or as a specialization of a group Fourier transform. Both forms of harmonic analysis influence each other. The harmonic analysis applies some "hardware" (a set of meaningful explicit formulas for Jacobi functions), while conversely it raises the need of finding some further formulas. These formulas exist both in analytic and group theoretic form and the two versions of the formulas may be derived from each other.

The interactions already listed form the heart of this paper, but there are two kinds of side interactions which will sometimes be mentioned. First one may consider eigenfunctions of second order ordinary differential operators

more general than the one in (1.1) and study the associated
transform. For such analysis the Jacobi function case can
serve as a prototype, while conversely it makes clear what is
general and what is special in the Jacobi case. Second, there
is the analysis on noncompact Riemannian symmetric spaces of
higher rank, in particular regarding the spherical Fourier
transform. This is helpful for putting the rank one analysis
in a conceptual framework, while conversely rank one results
suggest open problems in the higher rank case.

In the scheme below I summarize these interactions:

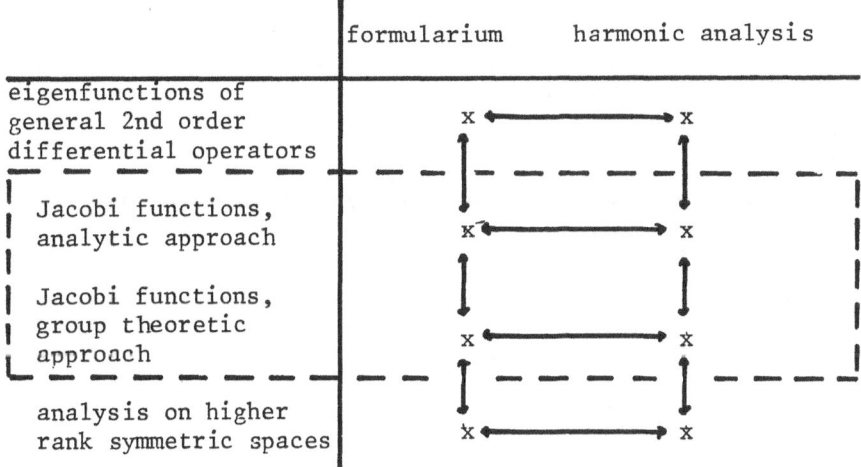

The analysis on semisimple symmetric pseudo-Riemannian
spaces is a particularly active area in analysis on Lie
groups at the moment. I will also discuss some recent developments in this area and the possible relevance of Jacobi
functions there.

Many people may have met Jacobi functions without being
aware of them, because they were written as hypergeometric
functions. Here I want to advertise the use of the Jacobi
function notation, because it enables one to make contact
with the existing literature on Jacobi functions and because
the arrangement of parameters in the Jacobi function notation is better adapted to harmonic analysis than in the hypergeometric notation.

The Contents will give the reader an impression of what
is treated inthe various sections. This paper is rather long
because it serves a multiple purpose: to provide an introduction to analysis on groups (notably the spherical Fourier
transform on rank one spaces), to develop the analytic

theory of the Jacobi transform and to discuss the interaction
between both. The backbone of this paper is material from a
sequence of papers by Flensted-Jensen and the author [41],
[48], [81], [49], [50] on the analytic approach to the Jacobi
transform with motivation from group theory. For the treatment of analysis on rank one Riemannian symmetric spaces the
presentation owes much to work by Helgason [64], [67], [70],
[71], [72] and Faraut [39]. The role of the Abel transform,
both in the analytic and group theoretic approach, will be
emphasized. Possibly new results are given in (5.22), (5.23),
§5.4 and §9. Complete proofs will usually not be given, but
many proofs are sketched in fairly large detail. Notes with
further results and references are often given at the end of
a section or subsection.

Acknowledgement. I thank Erich Badertscher, Bob Hoogenboom
and, in particular, Mogens Flenstad-Jensen for reading the
manuscript and suggesting improvements and corrections.

How to read this paper

Because of the organization of this paper, it is possible to
just read certain parts of it. Here are some suggestions:
(a) One or more subsections of §3 as a tutorial on analysis
    on groups, eventually to be used in the rest of the paper
    or elsewhere in this volume.
(b) Only section 2 for main results and history of the
    Jacobi transform.
(c) Only section 4 (with occasional reference to sections
    2,3) for the various group theoretic settings of the
    Jacobi transform.
(d) Sections 4.1, 5.1, 6, 7, 8.2, 8.3 (with occasional re-
    ference to sections 3.1, 3.2) for analysis in the sphe-
    rical rank one case.
(e) Sections 2, 5.3, 5.4, 6, 7, 8.1, 8.3, 9 for the analytic
    theory of the Jacobi transform.
(f) Sections 2, 4.1 and 5 for the Abel transform.

2. INVERSION OF THE JACOBI TRANSFORM, STATEMENT OF RESULTS

2.1. Definition of Jacobi functions

The Gaussian hypergeometric function is defined for $|z| < 1$
by the convergent power series

$$(2.1) \qquad {}_2F_1(a,b;c;z) := \sum_{k=0}^{\infty} \frac{(a)_k (b)_k}{(c)_k k!} z^k,$$

where a,b,c are complex, $c \neq 0,-1,-2,\ldots$ and the shifted factorial $(a)_k$ is defined by $(a)_0 := 1$, $(a)_k := a(a+1)\ldots$
$(a+k-1)$. It has an analytic continuation to a one-valued
analytic function on $\mathbb{C}\backslash[1,\infty)$. Numerous explicit formulas
exist for this function of four complex variables a,b,c,z
(cf.[33,Ch.2]), which often deter the uninitiated reader.
A fruitful way to discover more structure in this formula-
rium is to fix two of the four variables as parameters and
to consider the two other variables as the variable and the
dual variable in a discrete or continuous orthogonal system.
Thus one may consider various families of orthogonal polynomials (Jacobi, Krawtchouk, Meixner and Pollaczek polynomials, see [5, Lecture 2] and [34, 10.21]) and the continuous
orthogonal system of Jacobi functions, which will interest
us here.

Considered as a function of z, $_2F_1(a,b;c;z)$ is the
unique solution of the hypergeometric differential equation

(2.2) $z(1-z)u''(z)+(c-(a+b+1)z)u'(z)-abu(z) = 0$

which is regular at 0 and equal to 1 at 0. Now consider c
and $a+b$ (or equivalently $\alpha := c-1$ and $\beta := a+b-c$) as parameters and choose the eigenvalue ab in (2.2) as dual variable
to z. Depending on the way we restrict the z-variable to an
interval connecting two of the three singular points $0,1,\infty$
of (2.2) and on the boundary conditions we get interesting
discrete or continuous orthogonal systems. These are in particular the Jacobi   polynomials

(2.3) $R_n^{(\alpha,\beta)}(x) = \dfrac{n!}{(\alpha+1)_n} P_n^{(\alpha,\beta)}(x) :=$

$$:= {}_2F_1(-n,n+\alpha+\beta+1;\alpha+1;\tfrac{1}{2}(1-x)),$$

which, for $n = 0,1,2,\ldots$, are orthogonal polynomials on the
interval $(-1,1)$ with respect to the weight function
$(1-x)^\alpha(1+x)^\beta(\alpha,\beta>-1)$, and the Jacobi functions

(2.4) $\phi_\lambda^{(\alpha,\beta)}(t) := {}_2F_1(\tfrac{1}{2}(\alpha+\beta+1-i\lambda),\tfrac{1}{2}(\alpha+\beta+1+i\lambda);$

$$\alpha+1;-\text{sh}^2 t).$$

For $|\beta| < \alpha+1$ the system $\{\phi_\lambda^{(\alpha,\beta)}\}_{\lambda\geq 0}$ is a continuous or-
thogonal system on $\mathbb{R}_+$ with respect to the weight function

(2.5) $\Delta_{\alpha,\beta}(t) := (2\text{sh } t)^{2\alpha+1}(2\text{ch } t)^{2\beta+1}, \; t > 0,$

cf. Theorem 2.3. Jacobi functions are called so because of
their relationship with Jacobi polynomials:

(2.6) $\phi_\lambda^{(\alpha,\beta)}(t) = R_{\frac{1}{2}(i\lambda-\alpha-\beta-1)}^{(\alpha,\beta)}(\text{ch } 2t).$

From (2.4) and [33,2.1(22)] one can derive:

(2.7) $\phi_\lambda^{(\alpha,\beta)}(t) = (\text{ch } t)^{-\alpha-\beta-1-i\lambda}$

$\cdot \; {}_2F_1(\tfrac{1}{2}(\alpha+\beta+1+i\lambda),\tfrac{1}{2}(\alpha-\beta+1+i\lambda;\alpha+1;\text{th}^2 t).$

If no confusion is possible we will suppress the parameters
α,β in our notation and we will write

$\rho := \alpha+\beta+1.$

Assume that α,β,λ are complex, $\alpha \neq -1,-2,\ldots$. Let

(2.8) $L := d^2/dt^2+(\Delta'(t)/\Delta(t))d/dt, \quad t > 0,$

with Δ given by (2.5), so

(2.9) $L_{\alpha,\beta} = d^2/dt^2+((2\alpha+1)\coth t+(2\beta+1)\text{th } t)d/dt.$

Then rewriting of (2.2) shows that ϕ_λ is the unique even
C^∞-function v on $\mathbb{R}$ such that $v(0) = 1$ and

(2.10) $(L+\lambda^2+\rho^2)v = 0.$

It follows from this characterization of ϕ_λ that

(2.11) $\phi_\lambda^{(-\frac{1}{2},-\frac{1}{2})}(t) = \cos\lambda t, \phi_\lambda^{(\frac{1}{2},\frac{1}{2})}(t) = \dfrac{2\sin\lambda t}{\lambda\text{ sh}2t}.$

2.2. The Jacobi transform

Let us define the (Fourier-) Jacobi transform $f \mapsto \hat{f}$ by

(2.12) $\hat{f}(\lambda) := \displaystyle\int_0^\infty f(t)\phi_\lambda(t)\Delta(t)dt$

for all functions f on $\mathbb{R}_+$ and complex numbers λ for which
the right hand side is well-defined, possibly by analytic
continuation with respect to α. For instance, if

$f \in \mathcal{D}_{even}(\mathbb{R})$ (the space of even C^{∞}-functions with compact support on $\mathbb{R}$) then, for $n = 0,1,2,\ldots$, (2.12) has an analytic continuation from $\mathrm{Re}\ \alpha > -1$ to $\mathrm{Re}\ \alpha > -n-1$, $\alpha \neq -1,-2,\ldots,-n$, in the form

(2.13) $\hat{f}(\lambda) = \dfrac{(-1)^n}{2^{4n}(\alpha+1)_n} \displaystyle\int_0^{\infty} \left(\dfrac{1}{sh2t}\dfrac{d}{dt}\right)^n f(t) \cdot$

$\cdot\ \phi_{\lambda}^{(\alpha+n,\beta+n)}(t)\Delta_{\alpha+n,\beta+n}(t)dt,$

where we used $[33,2.8(27)]$. Thus, in all cases, if $f \in \mathcal{D}_{even}$ $(\mathbb{R})$ then $\hat{f}$ is an even entire analytic function and, by self-adjointness of L with respect to $\Delta(t)dt$, we have

(2.14) $(Lf)^{\wedge}(\lambda) = -(\lambda^2+\rho^2)\hat{f}(\lambda).$

Note that $f \mapsto \hat{f}$ reduces to the Fourier-cosine transform for $\alpha = \beta = -\frac{1}{2}$ and that it is immediately related to the Fourier-sine transform if $\alpha = \beta = \frac{1}{2}$. In order to give meaning to the notion "continuous orthogonal system" for the functions ϕ_{λ} we have to invert the transform $f \mapsto \hat{f}$. To formulate the inversion formula we will introduce the function $\lambda \mapsto c(\lambda)$ occuring in the asymptotics of $\phi_{\lambda}(t)$ as $t \to \infty$.

From $[33,\S2.9]$ or by straightforward computation we obtain, for $\lambda \neq -i,-2i,\ldots$, another solution Φ_{λ} of (2.10) on $(0,\infty)$ given by

(2.15) $\Phi_{\lambda}^{(\alpha,\beta)}(t) :=$

$:= (2ch\ t)^{i\lambda-\rho}\ {}_2F_1(\frac{1}{2}(\rho-i\lambda),\frac{1}{2}(\alpha-\beta+1-i\lambda);1-i\lambda;ch^{-2}t)=$

$= (2sh\ t)^{i\lambda-\rho}\ {}_2F_1(\frac{1}{2}(\rho-i\lambda),\frac{1}{2}(-\alpha+\beta+1-i\lambda);1-i\lambda;-sh^{-2}t).$

(The second equality is implied by $[33,2.1(22)]$.) Thus

(2.16) $\Phi_{\lambda}(t) = e^{(i\lambda-\rho)t}(1+o(1))$ as $t \to \infty.$

For $\lambda \notin i\mathbb{Z}$, Φ_{λ} and $\Phi_{-\lambda}$ are two linearly independent solutions of (2.10), so ϕ_{λ} is a linear combination of both. In fact, we get from $[33,2.10(2)$ and $2.10(5)]$ that, for $\lambda \notin i\mathbb{Z}$,

(2.17) $\phi_{\lambda} = c(\lambda)\Phi_{\lambda}+c(-\lambda)\Phi_{-\lambda},$

where

(2.18) $c(\lambda) = c_{\alpha,\beta}(\lambda) := \dfrac{2^{\rho-i\lambda}\Gamma(\alpha+1)\Gamma(i\lambda)}{\Gamma(\frac{1}{2}(i\lambda+\rho))\Gamma(\frac{1}{2}(i\lambda+\alpha-\beta+1))}$.

Hence

(2.19) $\phi_\lambda(t) = c(\lambda)e^{(i\lambda-\rho)t}(1+o(1))$ as $t \to \infty$ if $\mathrm{Im}\lambda < 0$.

This follows from (2.16) and (2.17) for $\lambda \neq -i,-2i,\ldots$, but it extends to $\mathrm{Im}\ \lambda < 0$. We can now formulate the main theorems about the Jacobi transform. Their proofs will be postponed until §6.

Theorem 2.1 (Paley-Wiener theorem). For all complex α,β with $\alpha \neq -1,-2,\ldots$ the Jacobi transform is a 1-1 map of $\mathcal{D}_{even}(\mathbb{R})$ onto the space of all even entire functions g for which there are positive constants A_g and $C_{g,n}$ $(n=0,1,2,\ldots)$ such that

(2.20) $|g(\lambda)| \leq C_{g,n}(1+|\lambda|)^{-n}e^{A_g|\mathrm{Im}\lambda|}$, $\lambda \in \mathbb{C}$, $n = 0,1,\ldots$

For $\alpha = \beta = -\frac{1}{2}$ this is the classical Paley-Wiener theorem (cf. Rudin [118, Theorem 7.22]). As a refinement of Theorem 2.1 we have that f has support in $[-a,a]$ iff $\hat{f}$ satisfies (2.20) with $A_{\hat{f}} := a$.

Theorem 2.2 (inversion formula, first form). If $\alpha,\beta \in \mathbb{C}$, $\alpha \neq -1,-2,\ldots$, $\mu \geq 0$, $\mu > -\mathrm{Re}(\alpha\pm\beta+1)$, $f \in \mathcal{D}_{even}(\mathbb{R})$, $t > 0$, then

(2.21) $f(t) = \dfrac{1}{2\pi}\displaystyle\int_{-\infty}^{\infty} \hat{f}(\lambda+i\mu)\Phi_{\lambda+i\mu}(t)(c(-\lambda-i\mu))^{-1}d\lambda.$

For convenience assume now that $\alpha > -1$, $\beta \in \mathbb{R}$. Then $\lambda \mapsto (c(\lambda))^{-1}$ has only simple poles for $\mathrm{Im}\ \lambda \geq 0$ which lie in the finite set

(2.22) $D_{\alpha,\beta} := \{i(|\beta|-\alpha-1-2m)\ |\ m = 0,1,2,\ldots;\ |\beta|-\alpha-1-2m>0\}.$

If $|\beta| \leq \alpha+1$ then $D_{\alpha,\beta}$ is empty. Put

(2.23) $d(\lambda) := -i\ \mathrm{Res}_{\mu=\lambda}(c(\mu)c(-\mu))^{-1}),\ \lambda \in D_{\alpha,\beta}.$

The next two theorems are versions of the Plancherel theorem for the Jacobi transform.

Theorem 2.3 (inversion formula, second form). If $\alpha > -1$, $\beta \in \mathbb{R}$, $f \in \mathcal{D}_{even}(\mathbb{R})$, $t \in \mathbb{R}$, then

$$(2.24) \qquad f(t) = \frac{1}{2\pi} \int_0^\infty \hat{f}(\lambda)\phi_\lambda(t)|c(\lambda)|^{-2}d\lambda +$$

$$+ \sum_{\lambda \in D_{\alpha,\beta}} \hat{f}(\lambda)\phi_\lambda(t)d(\lambda).$$

If, moreover, $\alpha \pm \beta + 1 \geq 0$ then

$$(2.25) \qquad f(t) = \frac{1}{2\pi} \int_0^\infty \hat{f}(\lambda)\phi_\lambda(t)|c(\lambda)|^{-2}d\lambda.$$

Define the measure ν on $\mathbb{R}_+ \cup D_{\alpha,\beta}$ by

$$(2.26) \qquad \int_{\mathbb{R}_+ \cup D_{\alpha,\beta}} g(\lambda)d\nu(\lambda) := \frac{1}{2\pi} \int_0^\infty g(\lambda)|c(\lambda)|^{-2}d\lambda +$$

$$+ \sum_{\lambda \in D_{\alpha,\beta}} g(\lambda)d(\lambda).$$

Theorem 2.4 (Parseval formula). If $\alpha > -1$, $\beta \in \mathbb{R}$ and $f,g \in \mathcal{D}_{even}(\mathbb{R})$ then

$$(2.27) \qquad \int_0^\infty f(t)\overline{g(t)}\Delta(t)dt = \int_{\mathbb{R}_+ \cup D_{\alpha,\beta}} \hat{f}(\lambda)\overline{\hat{g}(\lambda)}d\nu(\lambda).$$

The mapping $f \mapsto \hat{f}$ extends to an isometry of $L^2(\mathbb{R}_+,\Delta(t)dt)$ onto $L^2(\mathbb{R}_+ \cup D_{\alpha,\beta},\nu)$.

In this form these theorems are proved in [41], [44, Appendix 1] and [81], but special cases of Theorems 2.3, 2.4 have a long history. (We will deal with background to Theorem 2.1 in §6.) Mehler [103] (without proof), Heine [63], both in 1881, and Fock [52] in 1943 treated the case $\alpha = \beta = 0$. Consideration of a potential problem for the spherical lens (cf. Mehler [102], Lebedev [94]) brought them to the introduction of what is now known as the Mehler–Fock transform. This is usually formulated in terms of Legendre functions

$$(2.28) \qquad P_{i\mu-\frac{1}{2}}(ch2t) := \phi_{2\mu}^{(0.0)}(t)$$

as

$$f(\mu) = \mu \operatorname{th}\pi\mu \cdot \int_1^\infty P_{i\mu-\frac{1}{2}}(x)\psi(x)dx,$$

(2.29)

$$\psi(x) = \int_0^\infty P_{i\mu-\frac{1}{2}}(x)f(\mu)d\mu.$$

Fock [52] gave precise conditions on f or ψ (in terms of integrability of the function and some of its derivatives) in order that the one integral transform would invert the other.

In the literature one can find four main methods of proving Theorems 2.3, 2.4 or special cases of them, but interaction between the various approaches is rare:

a) Factorization of $f \mapsto \hat{f}$ as $f = F(F_f)$, where F is the classical Fourier transform and $f \mapsto F_f$ is the so-called Abel transform, which can be explicitly given and explicitly inverted. This fact is exploited in Heine [63,§75], Fock [52], Koornwinder [82] and, in group theoretic context, in Vilenkin [138], [139,Ch.10,§4] and Takahashi [129, 4.2]. We will return to this in section 5.

b) Use of spectral theory of second order o.d.e.'s with singularities at one or both endpoints of the interval. The Jacobi differential equation then usually occurs as an illustration of a quite general theory. This approach was started by Weyl [144] in 1910. He already has the example of the Jacobi o.d.e., including a remark about occurrence of the discrete spectrum for α, β. Later presentations are in Titchmarsh [136,§4.18-4.20], Dunford & Schwartz [28,Ch.13,§18,pp.1520-1526] (see also Flensted-Jensen [41,pp.155-156]) and Faraut [36], where [28],[36] emphasize the functional analytic aspects.

c) Use of function theory and asymptotics. Olevskiĭ [111], [112], van Nostrand [108] and Götze [58] proved that (2.25) implies (2.12) under certain conditions on $\hat{f}$ by use of

$$\int_0^M \phi_\lambda(t)\phi_\mu(t)\Delta(t)dt =$$

(2.30)

$$= (\mu^2-\lambda^2)^{-1}\Delta(M)(\phi_\lambda'(M)\phi_\mu(M)-\phi_\lambda(M)\phi_\mu'(M))$$

and the asymptotics of $\phi_\lambda(t), \phi_\lambda'(t)$ as $t \to \infty$. Inversion formulas in both directions were given in the very thorough paper by Braaksma & Meulenbeld [16].

d) As a special case of the inversion of the spherical Fourier transform on noncompact semisimple Lie groups. This was first done by Harish-Chandra [61]. A considerable simplification of his proof was given by Rosenberg [1977], see also Helgason [71,Ch.4,§7], [72, 4.2].

Yet order approaches occur in Lebedev [92], [93], Roehner & Valent [116], Stein & Wainger [128] and §9 of the present paper. In general we can say that approach a) connects the case (α,β) with the elementary case $(-\frac{1}{2},-\frac{1}{2})$ by the Abel transform $f \rightarrow F_f$, which has the transmutation property

$$(2.31) \qquad F_{Lf}(t) = \frac{d^2}{dt^2} F_f(t),$$

while the methods b), c), d) approximate the case (α,β) with the case $(-\frac{1}{2},-\frac{1}{2})$ by observing that, as $t \rightarrow \infty$, $L_{\alpha,\beta}$ and its eigenfunctions resemble $d^2/dt^2 + 2\rho d/dt$ and its eigenfunctions.

2.3. Generalizations of the Jacobi transform

Another limit case of the Jacobi o.d.e. (2.10) is the Bessel equation

$$(2.32) \qquad v''(t)+(2\alpha+1)t^{-1}v'(t)+\lambda^2 v(t) = 0.$$

Indeed, replace (t,λ) in (2.10) by $(\varepsilon t,\varepsilon^{-1}\lambda)$ and let $\varepsilon \downarrow 0$. The unique even C^∞-solution of (2.32) being equal to 1 at 0 equals

$$(2.33) \qquad \mathcal{J}_\alpha(\lambda t) := 2^\alpha \Gamma(\alpha+1)(\lambda t)^{-\alpha}J_\alpha(\lambda t),$$

where J_α is the usual Bessel function (cf.[34, (7.2(2))]).
Then

$$(2.34) \qquad \lim_{\varepsilon \downarrow 0} \phi_{\varepsilon^{-1}\lambda}^{(\alpha,\beta)}(\varepsilon t) = \mathcal{J}_\alpha(\lambda t).$$

A full asymptotic expansion of $\phi_\lambda^{(\alpha,\beta)}(t)$, $t \downarrow 0$, in terms of Bessel functions, which is, to a certain extent, uniform in λt, is given by Schindler [120] $(\alpha=\beta)$ and Stanton & Tomas [127]. (See Duistermaat [27] for a new approach in the group case.)

A lot of work (in particular by the Tunis school) has been done in extending the harmonic analysis for the Jacobi transform to the case of more general second order differential operators L of the form (2.8). It is then usually assumed that, for some $\alpha > -\frac{1}{2}$,

(2.35) $t \mapsto t^{-2\alpha-1}\Delta(t)$ is even, strictly positive C^∞-
             function,

while Trimèche [137] and Chébli [23] allow ρ^2 in (2.10) to
be replaced by some even C^∞-function on $\mathbb{R}$. Chébli [20],
[21] [22], [24] has the additional conditions

(2.36) $\Delta(t) \uparrow \infty$ as $t \to \infty$, $\dfrac{\Delta'(t)}{\Delta(t)} \downarrow 2\rho$ as $t \to \infty$.

The conditions (2.35), (2.36) are satisfied by the Bessel
equation (2.32) ($\alpha > -\frac{1}{2}$) and the Jacobi equation (2.10) with
(2.9) ($-\frac{1}{2} \neq \alpha \geq \beta \geq -\frac{1}{2}$). Under condition (2.35) ϕ_λ can again be
defined as the even C^∞-solution of (2.10) being equal to 1
at 0 and the transform $f \mapsto \hat{f}$ by (2.12). Under conditions
(2.34), (2.35) there are also solutions ϕ_λ and a function c
such that (2.16), (2.17) hold. Chébli [20], [21], [22], using
approach b), obtains the analogue of the inversion formula
(2.25). Trimèche [137], having only condition (2.35), uses
approach a) and obtains an inversion formula in the form

(2.37) $f(t) = \displaystyle\int_0^\infty \hat{f}(\lambda)\phi_\lambda(t)d\nu_1(\lambda) + \int_0^\infty \hat{f}(i\lambda)\phi_{i\lambda}(t)d\nu_2(\lambda),$

where ν_1 and ν_2 are positive measures, ν_1 being tempered
and ν_2 such that $\int_0^\infty e^{a\lambda}d\nu_2(\lambda) < \infty$ for all $a > 0$. Both Chébli
[24] and Trimèche [137] have asymptotic expansions of $\phi_\lambda(t)$
in terms of Bessel functions which are reminiscent to
Stanton & Tomas [127]. Chébli uses results of Langer [91],
Trimèche of Olver [113,Ch.12].

3. SOME GROUP THEORETIC PRELIMINARIES (A TUTORIAL)

In this section we put together some background material
from the general theory of analysis on Lie groups. The five
subsections (see contents) are rather disconnected. Each
subsection may be used as a reference in reading other parts
of this paper or as a tutorial to this whole volume or it
may just be read for its own interest.

3.1. Structure theory of noncompact semisimple Lie groups

References for this subsection are [70] and [86,Ch.1,2,3].
Let g be a real semisimple Lie algebra, i.e., g is a real
Lie algebra on which the Killing form $B(X,Y) := \mathrm{tr}(\mathrm{ad}X\,\mathrm{ad}Y)$
$(X,Y\epsilon g)$ is nondegenerate as bilinear form on g. Let θ be a

Cartan involution of g and $g = k+p$ the corresponding Cartan decomposition of g, i.e., θ is an automorphism of g, $\theta^2 = $ id, k and p are eigenspaces of θ for eigenvalues 1 and -1, respectively, and B is positive definite on p and negative definite on k. Such involutions do exist and they are all conjugate under inner automorphisms. Let G be a connected Lie group with Lie algebra g. Then G is called a semisimple Lie group. It can be shown that the Cartan involution θ is the differential of a unique involutive automorphism of G, also denoted by θ, and that the fixed point subgroup $K := \{g \in G \mid \theta(g) = g\}$ is precisely the connected Lie subgroup of G with Lie algebra k. The subgroup K is compact iff G has finite center; then K is also a maximal compact subgroup. Assume that G has finite center (such a choice of G is possible).

Let a be a maximal abelian subspace of p. All such subspaces are conjugate under Ad(K). The (real) rank of g is defined as the dimension of a. We will exclude the rank zero case, i.e., we assume that $\theta \neq$ id. For α in a^* (the real linear dual of a) put $g_\alpha := \{X \in g \mid [H,X] = \alpha(H)X$ for all H in $a\}$. Then $g_0 \cap p = a$. Put $m := g_0 \cap k$, $\Sigma := \{\alpha \in a^* \setminus \{0\} \mid \dim g_\alpha > 0\}$. Then g, as a linear space, has the direct sum decomposition $g = m + a + \Sigma_{\alpha \in \Sigma} g_\alpha$. Since B, restricted to a, is an inner product, it canonically gives rise to an inner product on a^*, denoted by $<.,.>$. The triple $\{a^*, <.,.>, \Sigma\}$ satisfies the axioms of a root system: Span $\Sigma = a^*$ and for all α,β in Σ, $\beta - 2 <\beta,\alpha>/<\alpha,\alpha>$ $\alpha \in \Sigma$ and $2<\beta,\alpha>/<\alpha,\alpha> \in \mathbb{Z}$. This root system is not necessarily reduced: if $\alpha, c\alpha \in \Sigma$ then $c = \pm\frac{1}{2}, \pm 1$ or ± 2; in a reduced root system only ± 1 would be possible. Introduce some linear vector space ordering $<$ on a^*. Put $\Sigma^+ := \{\alpha \in \Sigma \mid \alpha > 0\}$, $n := \Sigma_{\alpha \in \Sigma^+} g_\alpha$. Then n is a nilpotent subalgebra of g. Put $m_\alpha := \dim g_\alpha$, $\rho := \frac{1}{2} \Sigma_{\alpha \in \Sigma^+} m_\alpha \alpha$.

Let $N := \exp n$, $A := \exp a$. Then N,A are closed subgroups of G and diffeomorphic images of n,a under exp. Let $\bar{N} := \theta(N)$. Let $M := Z_K(a) = \{k \in K \mid Ad(k)H = H$ for all H in $a\}$, $M' := N_K(a) = \{k \in K \mid Ad(k)a = a\}$. Then M,M' are compact subgroups of K, both with Lie algebra m, M is a normal subgroup of M', the group $W := M'/M$ is finite. Via Ad the group M'/M acts as a group of orthogonal transformations on a, and hence on a^*. Under this identification W can be shown to be isomorphic to the Weyl group of the root system Σ, i.e. the group generated by the reflections $\lambda \to \lambda - 2<\lambda,\alpha>/<\alpha,\alpha> \alpha$ of a^* ($\alpha \in \Sigma$).

If g has rank one then, for some α in a^*, $\Sigma = \{\alpha, -\alpha\}$ or $\{\alpha, -\alpha, 2\alpha, -2\alpha\}$. Let H_1 in a be such that $\alpha(H_1) = 1$ and write $g_{\pm 1}$, $g_{\pm 2}$ instead of $g_{\pm \alpha}$, $g_{\pm 2\alpha}$ with dimension $m_{\pm 1}$, $m_{\pm 2}$, respectively. Choose the ordering on a^* such that α is positive. Then $n = g_1 + g_2$. Put $H_t := tH_1$, $a_t := \exp H_t$. We will always keep to these conventions in the rank one case.

We mention some decompositions of G:
(a) G = K exp p (<u>polar decomposition</u>), where $(k,X) \mapsto k \exp X$
    is an analytic diffeomorphism of K×p onto G.
(b) G = KAK (<u>Cartan decomposition</u>), where
    $KaK = KbK(a,b \in A)$ iff $b = kak^{-1}$ for some k in M'. In the
    rank one case: $Ka_sK = Ka_tK$ iff $s = \pm t$.
(c) G = KAN (<u>Iwasawa decomposition</u>), where $(k,a,n) \mapsto kan$ is
    an analytic diffeomorphism of K×A×N onto G. In the rank
    one case: if $g \in G$ and $g = ka_tn$ according to this decom-
    position then write u(g) := k, H(g) := t.
(d) NMAN is open and dense in G (part of <u>Bruhat decomposition</u>).

In the rank one case we finally need the result that the Ad(K)-orbits on p are the spheres $\{X \in p \mid B(X,X) = \text{const.}\}$. Up to local isomorphisms the different rank one cases are:

G	$SO_0(1,n)$	SU(1,n)	Sp(1,n)	$F_4(-20)$
K	SO(n)	S(U(1)×U(n))	Sp(1)×Sp(n)	Spin(9)
n	2,3,4,...	2,3,4,...	2,3,4,...	

Table 1

The first three columns in this Table can be treated in a uniform way as follows (cf.[39]). Let $\mathbb{F} := \mathbb{R}, \mathbb{C}$ or $\mathbb{H}$ ($\mathbb{H}$ denotes the skew field of quaternions) with real dimension d = 1, 2 or 4. Let U(p,q;$\mathbb{F}$) be the Lie group of (right) linear operators on $\mathbb{F}^{p+q}$ which leave invariant the hermitian form

$$\bar{y}_1 x_1 + \ldots + \bar{y}_p x_p - \bar{y}_{p+1} x_{p+1} - \ldots - \bar{y}_{p+q} x_{p+q}, x,y \in \mathbb{F}^{p+q}.$$

In particular, we will meet the groups U(n,$\mathbb{F}$) := U(n,o;$\mathbb{F}$) and U(1,n;$\mathbb{F}$). In case of the latter group we label coordinates on $\mathbb{F}^{n+1}$ such that the group elements leave invariant the form

$$\bar{y}_0 x_0 - \bar{y}_1 x_1 - \ldots - \bar{y}_n x_n, \quad x,y \in \mathbb{F}^{n+1}.$$

If $G := U(1,n;\mathbb{F})$ and $\theta g := (g^*)^{-1}$ ($g \in G$, g^* denoting $\mathbb{F}$- hermitian adjoint of g) then θ is an involutive automorphism of G and $K := U(1,\mathbb{F}) \times U(n,\mathbb{F}) := \{ (\begin{smallmatrix} u & 0 \\ 0 & V \end{smallmatrix}) \mid u \in U(1,\mathbb{F}), V \in U(n,\mathbb{F}) \}$ is the subgroup of elements fixed under θ. For $\mathbb{F} := \mathbb{H}$ this pair (G,K) matches the third column of Table 1, but for $\mathbb{F} := \mathbb{R}$ or $\mathbb{C}$ the groups $U(1,n;\mathbb{F})$ are reductive groups, bigger than $SO_0(1,n)$ and $SU(1,n)$, respectively, but with semisimple parts equal to the latter groups. This difference is rather harmless since G/K remains the same. All structure theory presented in this subsection will also hold for the groups $G = U(1,n;\mathbb{F})$. The remaining rank one group $F_{4(20)}$, which is related to the octonions, needs individual car. We will not treat it here, but refer to Takahashi [132]. From now on both the groups G from Table 1 and the groups $U(1,n;\mathbb{F})$ will be called <u>rank one groups</u>.

For $G := U(1,n;\mathbb{F})$ we already specified K and θ. Let us list some of the other structural elements of such G (cf. [39]):

$$p = \left\{ \begin{pmatrix} 0 & z^* \\ z & 0 \end{pmatrix} \;\middle|\; z \in \mathbb{F}^n \right\};$$

$$H_1 := \begin{pmatrix} 0 & \cdots & 0 & 1 \\ \vdots & & & 0 \\ \vdots & & 0 & \vdots \\ 0 & & & \\ 1 & 0 & & 0 \end{pmatrix}, \quad a_t = \begin{pmatrix} \mathrm{ch}\, t & 0 \ldots 0 & \mathrm{sh}\, t \\ 0 & & 0 \\ \vdots & I_{n-1} & \vdots \\ 0 & & 0 \\ \mathrm{sh}\, t & 0 \ldots 0 & \mathrm{ch}\, t \end{pmatrix};$$

$$M = \left\{ \begin{pmatrix} u & 0 & 0 \\ 0 & V & 0 \\ 0 & 0 & u \end{pmatrix} \;\middle|\; u \in U(1,\mathbb{F}) , V \in U(n-1,\mathbb{F}) \right\};$$

$$N = \{ n_{z,w} \mid z \in \mathbb{F}^{n-1}, w \in \mathrm{Im}\,\mathbb{F} \}, \quad \text{where}$$

$$(3.1) \, n_{z,w} := \begin{pmatrix} 1 + \frac{1}{2}|z|^2 + w & z^* & -\frac{1}{2}|z|^2 - w \\ z & I_{n-1} & -z \\ \frac{1}{2}|z|^2 + w & z^* & 1 - \frac{1}{2}|z|^2 - w \end{pmatrix}$$

and $\mathrm{Im}\mathbf{F} := \{w \in \mathbf{F} \mid w + \bar{w} = 0\};$

(3.2) $m_1 = d(n-1)$, $m_2 = d-1$,

(3.3) $\rho := \frac{1}{2}(m_1 + 2m_2) = \frac{1}{2}d(n+1) - 1.$

We will also use parameters

(3.4) $\alpha := \frac{1}{2}(m_1 + m_2 - 1) = \frac{1}{2}dn - 1$, $\beta := \frac{1}{2}(m_2 - 1) = \frac{1}{2}d - 1.$

3.2. Spherical functions on Gelfand pairs

In this subsection we summarize the facts needed from the general theory of spherical functions. References are Helgason [64,Ch.X], [72,Ch.IV], Faraut [39,Ch.I], Godement [57].

Let G be a locally compact group G with left Haar measure dg. The convolution product on G is defined by

$$(f_1 * f_2)(x) := \int_G f_1(y)f_2(y^{-1}x)dy, \quad f_1, f_2 \in L^1(G).$$

Let K be a compact subgroup of G with normalized Haar measure dk. The space $C_c(G//K)$ of K-biinvariant continuous functions on G with compact support is an algebra under convolution. The pair (G,K) is called a Gelfand pair if the algebra $C_c(G//K)$ is commutative. This property implies that the group G is unimodular. A sufficient condition in order that (G,K) is a Gelfand pair is the existence of a continuous involutive automorphism θ of G such that $\theta(KxK) = Kx^{-1}K$ for all x in G. In view of the polar decomposition $G = K \exp p$ (cf.§3.1) this criterium shows that for rank one groups G the pairs (G,K) are Gelfand pairs.

Let (G,K) be a Gelfand pair and provide $C_c(G//K)$ with the usual topology. Then each nonzero continuous character ω on the commutative topological algebra $C_c(G//K)$ determines a unique ϕ in $C(G//K)$ such that

(3.5) $\omega(f) = \int_G f(x)\phi(x)dx$

for all f in $C_c(G//K)$. Such functions ϕ are called spherical functions for the pair (G,K). They satisfy $\phi(e) = 1$. They can also be characterized as the nonzero functions ϕ in $C(G)$ which satisfy the functional equation (product formula)

(3.6) $\phi(x)\phi(y) = \int_K \phi(xky)dk$, $x, y \in G.$

The Banach algebra $L^1(G//K)$ is also commutative. Via (3.5)
its nonzero characters are in 1-1 correspondence with the
bounded spherical functions. Equip the set of bounded spherical functions with the Gelfand topology.

A continuous function ϕ on a locally compact group G is
called <u>positive definite</u> if for each finite subset
$\{x_1,\ldots,x_n\}$ of G and for all complex $c_1,\ldots,c_n$

$$\Sigma_{i,j=1}^n \; c_i \bar{c}_j \; \phi(x_i^{-1}x_j) \geq 0$$

or if, equivalently,

(3.7) $$\int_G \phi(x^{-1}y)f(x)\overline{f(y)} \; dxdy \geq 0$$

for each f in $C_c(G)$. If ϕ is positive definite then

(3.8) $$|\phi(x)| \leq \phi(e), \quad \phi(x^{-1}) = \overline{\phi(x)}$$

for all x in G. The formula

(3.9) $$\phi(x) = (\pi(x)e,e), \quad x \in G,$$

establishes a 1-1 correspondence between the nonzero positive definite functions ϕ on G and the equivalence classes
of pairs (π,e), where π is a unitary representation of G and
e is a cyclic vector in the representation space $H(\pi)$ of π.
By use of tensor products of representations it follows from
this correspondence that the product of two positive definite
functions on G is again positive definite.

Let (G,K) be a Gelfand pair. This property can be shown
to be equivalent to the fact that the representation 1 of K
occurs at most once in each irreducible unitary representa-
tion of G. Let $(G/K)^\wedge$ consist of all π in $\hat{G}$ in which the representation 1 of K occurs with multiplicity 1. Now (3.9)
gives in particular a 1-1 correspondence between the positive
definite spherical functions ϕ and the elements π of $(G/K)^\wedge$,
where e is chosen as a K-fixed unit vector in $H(\pi)$. Via this
correspondence and in view of (3.8), $(G/K)^\wedge$ is included in
the set of bounded spherical functions. Let $(G/K)^\wedge$ inherit
the Gelfand topology of this set.

If $f \in C_c(G//K)$ or $L^1(G//K)$ and ϕ is a spherical function or bounded spherical function, respectively, then write

(3.10) $$\hat{f}(\phi) := \int_G f(x)\phi(x)dx.$$

The transform $f \mapsto \hat{f}$ is called the <u>spherical Fourier transform</u> associated with the pair (G,K). The <u>Plancherel-Godement</u> theorem states that there is a unique positive measure ν on $\overline{(G/K)}^{\wedge}$ (the <u>Plancherel measure</u>) such that

$$(3.11) \qquad f(x) = \int_{(G/K)^{\wedge}} \hat{f}(\phi)\phi(x)d\nu(\phi), \qquad x \in G,$$

for all continuous and positive definite functions f in $L^1(G//K)$. Moreover, if $f \in L^1(G//K) \cap L^2(G//K)$ then the <u>Parseval formula</u>

$$(3.12) \qquad \int_G |f(x)|^2 dx = \int_{(G/K)^{\wedge}} |\hat{f}(\phi)|^2 d\nu(\phi)$$

holds and the mapping $f \mapsto \hat{f}$ extends to an isometry of $L^2(G//K)$ onto $L^2(\Omega,\nu)$.

Let (G,K) be a Gelfand pair such that G is a Lie group and G/K is connected. Identify functions on G/K with right-K-invariant functions on G in the obvious way. Then the algebra $\mathbb{D}(G/K)$ of G-invariant differential operators on G/K is commutative, the spherical functions ϕ are C^∞-functions and they are joint eigenfunctions of all D in $\mathbb{D}(G/K)$ for certain complex eigenvalues λ_D:

$$(3.13) \qquad D\phi = \lambda_D\phi, \qquad D \in \mathbb{D}(G/K).$$

Hence, since $\mathbb{D}(G/K)$ contains the Laplace-Beltrami operator Ω on G/K, which is elliptic, all spherical functions are analytic. Conversely, if $\phi \in C^\infty(G//K)$, $\phi(e) = 1$ and ϕ satisfies (3.13) for certain eigenvalues λ_D then ϕ is a spherical function uniquely determined by these eigenvalues.

As an application of this last property let G be a rank one group, cf.§3.1. Choose an orthonormal basis $X_1,\ldots,X_r$ of p. Since the $Ad(K)$-orbits on p are spheres (cf.§3.1), the $Ad(K)$-invariant polynomials on p are polynomial functions of the polynomial $x_1X_1 +\ldots+ x_rX_r \mapsto x_1^2 +\ldots+ x_r^2$ on p. Now, by identification of p with the tangent space to G/K at eK, B_θ induces a G-invariant Riemannian structure on G/K. Hence the corresponding <u>Laplace-Beltrami operator</u> Ω given by

$$(3.14) \quad (\Omega f)(x) := (\frac{\partial^2}{\partial t_1^2} +\ldots+ \frac{\partial^2}{\partial t_r^2})f(x \exp(t_1X_1+\ldots+t_rX_r))\Big|_{t_i=0}$$

$(f \in C^\infty(G/K))$ generates the algebra $\mathbb{D}(G/K)$, so to each complex eigenvalue of Ω there corresponds one and only one spherical function as an eigenfunction.

3.3. Associated spherical functions and addition formulas

Let (G,K) be a Gelfand pair. Let A be a subset of G such that
$G = KAK$. Let M be the centralizer of A in K, i.e.,
$M := \{k \in K \mid ka = ak \; \forall a \in A\}$. Suppose that (K,M) is also a
Gelfand pair. For instance, these conditions hold if
$G = U(1,n;\mathbb{F})$ and K,A,M are as in §3.1. Then (K,M) is a
Gelfand pair because the sufficient condition of §3.2 holds
with $\theta((g_{ij})) := (\bar{g}_{ij})((g_{ij}) \in U(1,\mathbb{F}) \times U(n,\mathbb{F}))$. (Kostant
[87] showed for all rank one cases that (K,M) is a Gelfand
pair.)

Let π be a strongly continuous, not necessarily unitary
representation of G on Hilbert space $H(\pi)$ such that $\pi|_K$ is a
unitary representation of K which is the direct sum of certain representations in $(K/M)^\wedge$, each occurring with multiplicity one, and which contains the trivial representation
of K, so $\pi|_K = \oplus_{\delta \in M(\pi)} \delta$ for some subset $M(\pi)$ of $(K/M)^\wedge$ and
$1 \in M(\pi)$. If $\delta \in M(\pi)$ then let e_δ be an M-fixed unit vector
in $H(\pi)$ behaving under K according to δ. Then

$$(3.15) \qquad \phi_\pi(x) := (\pi(x)e_1, e_1), \quad x \in G,$$

defines a spherical function ϕ_π for (G,K) and

$$(3.16) \qquad \psi_\delta(k) := (\pi(k)e_\delta, e_\delta), \quad k \in K, \; \delta \in M(\pi),$$

defines a spherical function ψ_δ for (K,M). Let d_δ be the
degree of δ. The function $\phi_{\pi,\delta}(\delta \in M(\pi))$ defined by

$$(3.17) \qquad \phi_{\pi,\delta}(x) := d_\delta^{-\frac{1}{2}}(\pi(x)e_1, e_\delta), \quad x \in G,$$

will be called an associated spherical function for (G,K).
Note that

$$(3.18) \qquad \phi_{\pi,\delta}(ka\ell) = \phi_{\pi,\delta}(a)\psi_\delta(k), \quad k,\ell \in K, \; a \in A.$$

Corresponding to (3.17) there is the expansion

$$(3.19) \qquad \pi(a)e_1 = \sum_{\delta \in M(\pi)} d_\delta^{\frac{1}{2}} \phi_{\pi,\delta}(a)e_\delta, \quad a \in A.$$

Let $\tilde{\pi}$ be the conjugate contragredient representation
to π, that is, the representation $\tilde{\pi}$ of G on $H(\pi)$ which
satisfies

$$(\pi(x)v, w) = (v, \tilde{\pi}(x^{-1})w), \quad v,w \in H(\pi), \; x \in G.$$

For a,b $\in$ A, k $\in$ K we can expand:

$$\phi(a^{-1}kb) = (\pi(a^{-1}kb)e_1,e_1) = (\pi(k)\pi(b)e_1,\tilde{\pi}(a)e_1) =$$

$$= \sum_{\gamma,\delta\in M(\pi)} (\pi(b)e_1,e_\gamma)(e_\delta,\tilde{\pi}(a)e_1)(\pi(k)e_\gamma,e_\delta).$$

Hence,

$$(3.20)\quad \phi(a^{-1}kb) = \sum_{\delta\in M(\pi)} d_\delta \phi_{\pi,\delta}(b)\overline{\phi_{\tilde{\pi},\delta}(a)}\psi_\delta(k), a,b\in A,k\in K.$$

The convergence in (3.20) is absolute, uniform for (a,b,k) in compact subsets of A×A×K. We call (3.20) the addition formula for the spherical function ϕ. The right hand side expands $\phi(a^{-1}kb)$ as a function of k on the compact group K. Integration of both sides with respect to k over K yields the product formula (3.6).

Now let G be a rank one group and K,A,M as in §3.1. For π we take the representation $\pi_\lambda(\lambda\in\mathbb{C})$ of G induced by the one-dimensional representation $ma_tn \mapsto e^{-i\lambda t}$ of the subgroup MAN (cf. Wallach [142,§8.3]). Then π_λ has a realization on $L^2(K/M)$ given by

$$(3.21)\quad (\pi_\lambda(x)f)(kM) = e^{(i\lambda-\rho)H(x^{-1}k)}f(u(x^{-1}k)M),$$

$$x \in G, \ k \in K, \ f \in L^2(k/M),$$

where u(x) and H(x) are as in §3.1(c). The series of representations π_λ is called the spherical principal series for G. If $\lambda \in \mathbb{R}$ then π_λ is unitary. Restriction of π_λ to K yields the regular representation of K on $L^2(K/M)$, which is unitary. By Frobenius reciprocity: $\pi_\lambda|_K = \oplus_{\delta\in(K/M)}\wedge\delta$. Thus all conditions are satisfied in order to have an addition formula of the form (3.20). Observe that, with ψ_δ defined by (3.16), we can now take for $e_\delta(\delta\in(K/M)^\wedge)$ the element

$$kM \mapsto d_\delta^{\frac{1}{2}} \overline{\psi_\delta(k)} \text{ of } L^2(K/M).$$

Thus (3.17) becomes

$$(3.22)\quad \phi_{\lambda,\delta}(a_t) = \int_K e^{(i\lambda-\rho)H(a_{-t}k)}\psi_\delta(k)dk,$$

where we replaced the subscript π_λ by λ (cf. Helgason's [69, 4] definition of generalized spherical function). In particular, for $\delta = 1$, we get for the spherical function ϕ_λ corresponding to π_λ:

$$(3.23) \qquad \phi_\lambda(a_t) = \int_K e^{(i\lambda-\rho)H(a_{-t}k)}dk.$$

Formules (3.19) and (3.20) can now be written as:

$$(3.24) \qquad e^{(-i\lambda-\rho)H(a_{-t}k)} = \sum_{\delta\in(K/M)^\wedge}d_\delta\overline{\phi^-_{\lambda,\delta}(a_t)}\psi_\delta(k),$$

$$(3.25) \qquad \phi_\lambda(a_{-s}ka_t) = \sum_{\delta\in(K/M)^\wedge}d_\delta\phi_{\lambda,\delta}(a_s)\overline{\phi^-_{\lambda,\delta}(a_t)}\psi_\delta(k).$$

For the derivation of (3.25) we used that $\tilde{\pi}_\lambda = \pi^-_\lambda$. See also
[50] for a treatment of addition formulas in group theoretic
form.

3.4. Generalized Gelfand pairs and spherical distributions

References for this subsection are Faraut [37,§I], Thomas
[135], van Dijk [30], Benoist[10],[11] and Flensted-Jensen
[47]. Let G be a unimodular Lie group. If $\phi \in \mathcal{D}(G)$, $x,y \in G$
then write $(\lambda(x)\phi)(y) := \phi(x^{-1}y)$, $(\rho(x)\phi)(y) := \phi(yx)$.
A distribution vector of a unitary representation π of G is
a continuous linear mapping u: $\mathcal{D}(G) \to H(\pi)$ such that
$\pi(x)(u(\phi)) = u(\lambda(x)\phi)$ $(x\in G, \phi\in\mathcal{D}(G))$. Let $H_{-\infty}(\pi)$ denote the
space of all distribution vectors of π. There is an embedding
$v \mapsto u$: $H(\pi) \hookrightarrow H_{-\infty}(\pi)$ defined by $u(\phi) := \pi(\phi) v$ $(\phi\in\mathcal{D}(G))$. The
representation π extends to a representation $\pi_{-\infty}$ of G on
$H_{-\infty}(\pi)$:

$$(\pi_{-\infty}(x)u)(\phi) := u(\rho(x^{-1})\phi), (\pi_{-\infty}(\psi)u)(\phi) := u(\phi*\psi)$$

$(x\in G, \phi, \psi\in\mathcal{D}(G), u\in H_{-\infty}(\pi))$. If $\psi \in \mathcal{D}(G)$, $u \in H_{-\infty}(\pi)$ then $\pi(\psi)u$
can be shown to lie in $H(\pi)$. A distribution vector u in
$H_{-\infty}(\pi)$ is called cyclic if $\pi(\mathcal{D}(G))u$ is dense in $H(\pi)$.
    A distribution T on G is called positive definite if
$T(\tilde{\phi}*\phi) \geq 0$ for all ϕ in $\mathcal{D}(G)$ ($\tilde{\phi}$ defined by (3.2.9)). The
formula

$$(3.26) \qquad T(\tilde{\psi}*\phi) = (\pi_{-\infty}(\phi)u, \pi_{-\infty}(\psi)u), \phi, \psi \in \mathcal{D}(G),$$

establishes a 1-1 correspondence between the nonzero positive definite distributions T on G and the equivalence classes of pairs (π,u), where π is a unitary representation of G
and u is a cyclic element of $H_{-\infty}(\pi)$ (cf. the corresponding
statement for positive definite functions in §3.2).
    Let H be a closed unimodular subgroup of G. The pair
(G,H) is called a generalized Gelfand pair if, for each

π in $\hat{G}$, the dimension of the space of H-invariant distribution vectors in $H_{-\infty}(\pi)$ is at most one-dimensional. (There are many equivalent definitions, cf. [30],[135].) Assume that $(G,\underline{H})$ has this property. Let $(G/H)^{\wedge}$ denote the set of all π in $\hat{G}$ for which there is a nonzero H-invariant distribution vector. A <u>positive definite spherical distribution</u> for (G,H) is a positive definite distribution T on G such that (3.26) holds for some π in $(G/H)^{\wedge}$ and some H-fixed nonzero u. Then $T \in \mathcal{D}'(G//H)$, i.e., T is H-biinvariant. If H is compact then the notions of generalized Gelfand pair, positive definite spherical distribution are equivalent to Gelfand pair, positive definite spherical function, respectively.

A potential source of generalized Gelfand pairs is given by the <u>symmetric pairs</u>, i.e. pairs (G,H) with an involutive automorphism σ of G such that the Lie algebra of H is precisely the 1-eigenspace of $d\sigma$. If H is compact then a symmetric pair is always a Gelfand pair, but, in general, symmetric pairs are not always generalized Gelfand pairs (cf. the end of this subsection). A particular class of symmetric pairs are the pairs $(G \times G, G^{\star})$, where G is a unimodular Lie group, $G^{\star}$ is the diagonal subgroup of $G \times G$ and $\sigma(x,y) := (y,x)$ $(x,y \in G)$. Then the homogeneous space $G \times G/G^{\star}$ can be identified with G, with the action of $G \times G$ on G given by $(x,y).z := xzy^{-1} (x,y,z \in G)$. If G is a type I group (for instance a semisimple Lie group) then $(G \times G, G^{\star})$ is a generalized Gelfand pair. Then $(G \times G/G^{\star})^{\wedge} = \{\pi \otimes \pi^{\star} | \pi \in \hat{G}\}$. Let T be a spherical distribution corresponding to $\pi \otimes \pi^{\star}(\pi \in \hat{G})$ and consider T as $G^{\star}$-invariant distribution on $G \times G/G^{\star}$, i.e. as central distribution on G. Then, up to a constant factor, $T(\phi) = tr(\phi)$, $\phi \in \mathcal{D}(G)$. $(\pi(\phi) := \int_G \phi(x)\pi(x)dx$ can be shown to be a trace class operator on $H(\pi)$.)

Next consider a symmetric pair (G,H) with G being a connected semisimple Lie group with finite center. Let $g = h + q$ be the corresponding Lie algebra decomposition with respect to $\sigma(=d\sigma)$. The <u>rank</u> of (G,H) is the dimension of a maximal abelian subspace of q consisting of semisimple elements. Infinitesimally, the rank one cases can be obtained from Berger's [14] classification. They fall into two classes depending on whether the space G/H is isotropic or not. The isotropic cases are (up to local isomorphisms) the pairs $(U(p,q;\mathbb{F}), U(1) \times U(p-1,q;\mathbb{F}))$ $(\mathbb{F}=\mathbb{R}, \mathbb{C}$ or $\mathbb{H})$ considered by Faraut [37] (and other authors mentioned in the references to [37]) and the pair $(F_{4(-20)}, Spin(1,8))$ considered by Kosters [88],[89,Ch.3]. (Note that, like in §3.1, $U(p,q;\mathbb{F})$ is taken reductive in order to get a

uniform presentation.) The non-isotropic cases are
$(SL(n,\mathbb{R})$, $GL(n-1,\mathbb{R}))$ (considered by Kosters [89,Ch.4]),
$(Sp(n,\mathbb{R})$, $Sp(1,\mathbb{R}) \times Sp(n-1,\mathbb{R}))$ and $(F_4(4)$, $Spin(4,5))$.

In the structure theory of these rank one cases it is
convenient to take a Cartan involution θ of G commuting with
σ. Let K be the fixed point group of θ (a maximal compact
subgroup of G) and let $g = k+p$ be the corresponding Lie algebra decomposition with respect to $\theta(=d\theta)$. Choose a one-
dimensional subspace a of $p \cap q$ and a non-zero element H_1 of
a. With g_λ defined as in the rank one case of §3.1, H_1 can be
chosen such that $g = g_0 + g_1 + g_2 + g_{-1} + g_{-2}$. Put
$A := \{a_t := \exp t H_1 \mid t \in \mathbb{R}\}$, $M := Z_H(A)$, $n := g_1 + g_2$,
$N := \exp n$. There is a generalized Cartan decomposition
$G = KAH$, where Ka_tH determines $|t|$ completely.

The isotropic rank one pairs (G,H) in the form given
above as well as the pairs $(SL(n,\mathbb{R})$,$GL(n-1,\mathbb{R}))$ $(n \geq 3)$ can be
shown to be generalized Gelfand pairs. But this is not true
for the pairs $(SL(2,\mathbb{R})$, $GL(1,\mathbb{R}))$ and $(O(1,n),O(1,n-1))$
(van Dijk, yet to be published).

If (G,H) is a semisimple symmetric pair then G/H
naturally becomes a pseudo-Riemannian symmetric space with
G-invariant metric. H-biinvariant distributions on G can be
identified with H-invariant distributions on G/H. For a
generalized Gelfand pair (G,H) of rank one all positive
definite spherical distributions are eigendistributions of
the Laplace-Beltrami operator Ω on G/H. More generally,
Faraut [37] defines a spherical distribution on a rank
one space G/H to be a H-biinvariant eigendistribution of Ω.

3.5. Plancherel theorems, general theory

Let (G,K) be a Gelfand pair and assume that G is a Lie group.
Then it can be shown that the spherical Plancherel theorem
(cf. (3.11), (3.12)) is equivalent to

$$(3.27) \qquad f(e) = \int_{(G/K)^\wedge} \hat{f}(\phi)d\nu(\phi), \quad f \in \mathcal{D}(G//K),$$

where $f \mapsto \hat{f}$ is the spherical Fourier transform defined by
(3.10), ν is the spherical Plancherel measure and $\mathcal{D}(G//K)$
is the space of K-biinvariant C^∞-functions on G with compact
support. For f in $\mathcal{D}(G)$ and for a spherical function ϕ put

$$(3.28) \qquad \phi(f) = \hat{f}(\phi) := \int_G f(x)\phi(x)dx.$$

Put

(3.29) $\tilde{f}(x) := \overline{f(x^{-1})}, \quad x \in G,$

for a function f on G. Now two versions of the spherical
Plancherel theorem equivalent to (3.27) are

(3.30) $\int_K f(k)dk = \int_{(G/K)^\wedge} \phi(f)d\nu(\phi), \quad f \in \mathcal{D}(G),$

(3.31) $\int_{G/K} |f(x)|^2 d(xK) = \int_{(G/K)^\wedge} \phi(\tilde{f}*f)d\nu(\phi), \quad f \in \mathcal{D}(G/K).$

The equivalence of (3.30), (3.31) follows from the fact that

(3.32) $\mathcal{D}(G) = \mathrm{span}(\mathcal{D}(G)*\mathcal{D}(G)),$

which was proved by Dixmier & Malliavin [26].
     Next, let G be a unimodular Lie group which is type I
(for instance a semisimple Lie group). Then the Plancherel
theorem for G (cf. Dixmier [25,§18]) states that there is a
unique measure ν on $\hat{G}$ such that the two following equivalent
statements hold:

(3.33) $f(e) = \int_{\hat{G}} \mathrm{tr}(\pi(f))d\nu(\pi), \quad f \in \mathcal{D}(G),$

(3.34) $\int_G |f(x)|^2 dx = \int_{\hat{G}} \mathrm{tr}(\pi(f)^*\pi(f))d\nu(\pi), \quad f \in \mathcal{D}(G).$

Equivalence of (3.33), (3.34) follows from (3.32). For f in
$\mathcal{D}(G/K)$ formulas (3.33), (3.34) imply their spherical analogues
(3.30), (3.31).
     Finally, let (G,H) be a generalized Gelfand pair (cf.
§3.4). Define

(3.35) $f^0(xH) := \int_H f(xh)dh, \quad f \in \mathcal{D}(G), x \in G.$

Then $f \mapsto f^0$ is a continuous surjection of $\mathcal{D}(G)$ onto $\mathcal{D}(G/H)$.
Attach to each π in $(G/H)^\wedge$ a corresponding spherical distribution T_π in a measurable way (with respect to the canonical
Borel structure of $(G/H)^\wedge$). Note that, in general, there is
no canonical normalization of spherical distributions. Now
there is a unique positive measure ν on $(G/H)^\wedge$ (cf. Thomas
[135, Theorem A]) such that the two following equivalent
statements hold:

$$(3.36) \qquad \int_H f(h)dh = \int_{(G/H)^\wedge} T_\pi(f)d\nu(\pi) \forall f \in \mathcal{D}(G),$$

$$(3.37) \qquad \int_{G/H} |f^0(xH)|^2 d(xH) = \int_{(G/H)^\wedge} T_\pi(\widetilde{f}*f)d\nu(\pi), \forall f \in \mathcal{D}(G).$$

If $(G,H) = (G_1*G_1, G_1^*)$ (cf.§3.4) then (3.36),(3.37) reduce to (3.33), (3.34) and if H is compact then (3.36), (3.37) imply (3.30),(3.31).

In the case that G is semisimple, K a maximal compact subgroup, the Plancherel measure in (3.27) was explicitly determined by work of Harish-Chandra [61] and Gindikin & Karpelevic [56]. The Plancherel measure in the group case (3.33) (G noncompact semisimple Lie group) was obtained in a number of papers by Harish-Chandra, cf. the survey paper by Schmid [121]. The Plancherel measure for G/H in (3.36) was only determined in a number of special cases, we mention Faraut [37], Kosters [88], [89,Ch.3], Benoist [10], [11].

We now want to emphasize one particular method to obtain the Plancherel measure, namely the method of K-finite functions. In special cases, to be treated in §4, this method allows reduction to the Jacobi transform.

Let G be a semisimple Lie group, K a maximal compact subgroup and, in the case (G,H), let K,H correspond to commuting involutions θ,σ. Let $\gamma,\delta \in \hat{K}$ and let χ_δ denote the character of δ in $\hat{K}$. We call f in $\mathcal{D}(G)$ a __K-finite function of double K-type γ,δ__ if

$$(3.38) \qquad f(x) = \int_K \int_K d_\gamma \chi_\gamma(k^{-1})f(kx\ell)d_\delta\chi_\delta(\ell^{-1})dkd\ell, x \in G.$$

Similarly, f in $\mathcal{D}(G/H)$ is a __K-finite function of K-type δ__ if

$$(3.39) \qquad f(\xi) = \int_K d_\delta\chi_\delta(k^{-1})f(k\xi)dk, \xi \in G/H.$$

Now, by density properties of K-finite functions, each of the Plancherel formulas (3.30), (3.31), (3.33), (3.34), (3.36), (3.37) will be valid for all C_c^∞-functions iff it is valid for all such functions which are K-finite. Observe that (3.30), (3.33), (3.36) become trivial for most K-types. More concretely, (3.30) only needs to be verified for double K-type (1,1), i.e., it is implied by (3.27). For (3.33) we can restrict ourselves to functions f of double K-type (δ,δ) ($\delta \in \hat{K}$) which are moreover K-central, i.e. $f(kxk^{-1}) = f(x)$ for

x in G, k in K. Finally, for (3.36) it is sufficient to con-
sider functions f in $\mathcal{D}(G)$ for which f^0 is K-finite in $\mathcal{D}(G/H)$
and moreover K∩H-invariant. On the other hand, the versions
(3.31), (3.34), (3.37) will imply Plancherel-type formulas
for functions of many more K-types than were needed in order
to verify (3.30), (3.33), (3.36). These things will become
more clear by examples in §4.

4. THE JACOBI TRANSFORM IN GROUP THEORY

In this section we treat a number of cases where Jacobi functions appear in the context of semisimple Lie groups: as
spherical or intertwining functions, as associated spherical
or intertwining functions and as matrix elements of irreducible representations. Correspondingly, the Jacobi differential operator arises by separation of variables of the
Casimir operator on the group and the Jacobi transform can
be interpreted as the group Fourier transform acting on certain function classes which possess special symmetries.

4.1. Jacobi functions as spherical functions

Let G be a rank one group and use the notation of §3.1.
The parameters α,β will be as in (3.4). In view of the
Iwasawa and Cartan decompositions (§3.1), the restriction
$f \mapsto f|_A$ identifies $C^\infty(N\backslash G/K)$, with $C^\infty(A)$ and $C^\infty(G//K)$ with
$C^\infty_{even}(A) := \{f \in C^\infty(A) | f(a_t) = f(a_{-t}), t \in \mathbb{R}\}$. Note also
that the Laplace-Beltrami operator Ω sends $C^\infty(N\backslash G/K)$ and
$C^\infty(G//K)$ into itself. It can be shown (cf. [72,Ch.2], [39,
Ch.3]) that Ω has the following A-radial parts with respect
to these two decompositions:

(4.1) $(\Omega f)(a_t) = (\dfrac{d^2}{dt^2} - 2\rho \dfrac{d}{dt})f(a_t)$, $f \in C^\infty(N\backslash G/K)$,

(4.2) $(\Omega f)(a_t) = (L_{\alpha,\beta}f)(a_t)$, $f \in C^\infty(G//K)$,

where $L_{\alpha,\beta}$ is the Jacobi differential operator (2.9).
      With H: G $\to$ $\mathbb{R}$ defined as in §3.1(c) the function
x $\mapsto$ exp$((i\lambda-\rho)H(x^{-1}))(\lambda\in\mathbb{C})$ is in $C^\infty(N\backslash G/K)$. Hence, by (4.1):

(4.3) $(\Omega+\lambda^2+\rho^2)\ e^{(i\lambda-\rho)H(x^{-1})} = 0$.

      Let

(4.4) $\phi_\lambda(x) := \displaystyle\int_K e^{(i\lambda-\rho)H(x^{-1}k)}dk$,   x $\in$ G, $\lambda \in \mathbb{C}$.

Then $\phi_\lambda \in C^\infty(G//K)$, $\phi_\lambda(e) = 1$ and, because of (4.2) and the G-invariance of Ω, ϕ_λ satisfies the differential equation

$$(4.5) \qquad (\Omega + \lambda^2 + \rho^2)\phi_\lambda = 0.$$

It follows from the results at the end of §3.2 that ϕ_λ is a spherical function, that the set $\{\phi_\lambda \mid \lambda \in \mathbb{C}\}$ equals the set of all spherical functions for (G,K) and that $\phi_\lambda = \phi_\mu$ iff $\lambda = \pm\mu$. In view of (3.23), (4.4), ϕ_λ equals the matrix element (3.15) of the principal series representation π_λ. In view of (4.5), (4.2),

$$(4.6) \qquad \phi_\lambda(a_t) = \phi_\lambda^{(\alpha,\beta)}(t),$$

where $\phi_\lambda^{(\alpha,\beta)}$ is the Jacobi function (2.4) and α,β are as in (3.4). The integral representation (4.4) is due to Harish-Chandra [61] (in the case of general rank).

For the groups G under consideration the spherical Fourier transform (3.10) can be rewritten as a transform $f \mapsto \hat{f}$ defined by

$$(4.7) \qquad \hat{f}(\lambda) := \int_G f(x)\phi_\lambda(x)dx,$$

where $f \in C_c(G//K)$, $\lambda \in \mathbb{C}$. It can be shown that, up to a positive constant factor, the Haar measure on G satisfies

$$(4.8) \qquad \int_G f(x)dx = \int_0^\infty f(a_t)\Delta(t)dt, \quad f \in C_c(G//K),$$

where Δ is given by (2.9). Normalize dx on G such that (4.8) holds exactly. Then combination of (4.7) and (4.8) shows that (2.12) holds with $f(t)$ replaced by $f(a_t)$, i.e., the spherical Fourier transform of f equals the Jacobi transform of $t \mapsto f(a_t)$.

By use of this identification we can now apply Theorems 2.3, 2.4 (which will be proved in section 6) in order to obtain the Plancherel measure ν (cf. (3.11), (3.12)) for the present groups. Note that α,β in (3.4) are such that $\alpha \geq \beta \geq -\frac{1}{2}$, so we get from Theorems 2.3, 2.4 and from (4.8) that

$$(4.9) \qquad f(x) = \frac{1}{2\pi}\int_0^\infty \hat{f}(\lambda)\phi_\lambda(x)|c(\lambda)|^{-2}d\lambda, \quad f\in\mathcal{D}(G//K), \; x\in G,$$

$$(4.10) \qquad \int_G f(x)\overline{g(x)}dx = \frac{1}{2\pi}\int_0^\infty \hat{f}(\lambda)\overline{\hat{g}(\lambda)}|c(\lambda)|^{-2}d\lambda, \quad f,g\in\mathcal{D}(G//K),$$

and (4.10) extends to an isometry of L^2-spaces. The spherical
functions $\phi_\lambda(\lambda\in\mathbb{R})$ occurring in (4.9), (4.10) are positive
definite because (3.15) holds with $\pi = \pi_\lambda(\lambda\in\mathbb{R})$ being a unitary principal series representation. Thus the Plancherel
measure ν (cf. (3.11), (3.12)) becomes the measure
$(2\pi)^{-1}|c(\lambda)|^{-2}d\lambda$ on $\mathbb{R}_+$ in the present case.

The above interpretation of the Jacobi transform as a
spherical Fourier transform was first observed by Olevskiĭ
[111] for the real hyperbolic spaces and by Harish-Chandra
[61,§13] in the general rank one case. Harish-Chandra recognized the radial part of (4.5) as a hypergeometric differential equation and he obtained the explicit value (2.18)
of $c(\lambda)$ by identifying (2.17) with an identity for hypergeometric functions. But he obtained the inversion formula
by specialization of his general rank result.

4.2. Jacobi functions as associated spherical functions

Let G be as in §4.1. Since (3.11) implies (3.31), (4.9) will
imply that, for F in $\mathcal{D}(G/K)$,

$$(4.11) \qquad \int_{G/K} |F(x)|^2 d(xK) = (2\pi)^{-1}\int_0^\infty (\tilde{F}*F)^\wedge(\lambda)|c(\lambda)|^{-2}d\lambda.$$

In particular, this identity will hold for F in $\mathcal{D}(G/K)$ of
the form

$$(4.12) \qquad F(ka_t K) = f(t)Y_\delta(k), \quad k \in K, \, t \in \mathbb{R},$$

where Y_δ is in $L^2(K/M)$ with norm 1 and is of K-type δ
$(\delta\in(K/M)^\wedge)$. (Any K-finite function is a finite sum of
functions of the form (4.12).) It follows from (4.7), (4.8),
(4.12), (3.25) that

$$(4.13) \qquad (\tilde{F}*F)^\wedge(\lambda) = \int_K\int_K\int_{\mathbb{R}_+}\int_{\mathbb{R}_+} \overline{F(\ell a_s)}F(ka_t)\cdot$$

$$\cdot \phi_\lambda(a_{-s}\ell^{-1}ka_t)\Delta(s)\Delta(t)ds\,dt\,dk\,d\ell =$$

$$= \int_{\mathbb{R}_+}\int_{\mathbb{R}_+} \overline{f(s)}f(t)\phi_{\lambda,\delta}(a_s)\overline{\phi_{\bar\lambda,\delta}(a_t)}\Delta(s)\Delta(t)ds\,dt.$$

Hence (4.11) becomes

(4.14) $\displaystyle\int_0^\infty |f(t)|^2 \Delta(t)dt =$

$$= (2\pi)^{-1} \int_0^\infty \left| \int_0^\infty f(t)\phi_{\lambda,\delta}(a_t)\Delta(t)dt \right|^2 |c(\lambda)|^{-2} d\lambda.$$

The set $(K/M)^\wedge$ was determined by Kostant [87], Johnson & Wallach [77], Johnson [76], while Helgason [68] computed the functions $\phi_{\lambda,\delta}$ in terms of hypergeometric functions. It turns out (see also §8.1) that the functions $t \mapsto \phi_{\lambda,\delta}(a_t)$ coincide with the <u>associated Jacobi function</u>

(4.15) $\phi_{\lambda,k,\ell}^{(\alpha,\beta)}(t) := (c_{\alpha,\beta}(-\lambda)/c_{\alpha+k+\ell,\beta+k-\ell}(-\lambda))\cdot$

$\cdot (2\text{sh } t)^{k-\ell}(2\text{ch } t)^{k+\ell}\phi_\lambda^{(\alpha+k+\ell,\beta+k-\ell)}(t),$

where the ϕ_λ-function and c-functions at the right hand side are defined by (2.4), (2.18), α,β are as in (3.4) and k,ℓ run over all integers with $k \geq \ell \geq 0$ if $\mathbb{F} = \mathbb{C}, \mathbb{H}$ or $\mathbb{O}$ (octonions) and $\ell = 0,1,2,\ldots,$ $k = \ell$ or $\ell-1$ if $\mathbb{F} = \mathbb{R}$. Thus, if we put

$$f(t) := (2\text{sh } t)^{k-\ell}(2\text{ch } t)^{k+\ell}g(t), \quad g \in \mathcal{D}_{\text{even}}(\mathbb{R})$$

in (4.14) then we have obtained a group theoretic interpretation of (2.27) with α,β replaced by $\alpha+k+\ell,\beta+k-\ell$ and α,β,k,ℓ having values as above. (Note that for these values $\alpha+k+\ell \geq \beta+k-\ell \geq -\tfrac{1}{2}$). An analogue of (4.14) in the case of general rank was proved by Helgason [69,Cor.10.2].

The explicit expressions for the associated spherical functions in the case of rank one were exploited by Helgason [66,pp.140,141], [68, sections 6,7], Lewis [95] and some other authors mentioned in [68,§1] in order to characterize the image of the <u>Poisson transform</u> on a rank one space G/K. This transform $T \mapsto f$, defined for λ in $\mathbb{C}$ by

(4.16) $f(x) = \displaystyle\int_{K/M} e^{(i\lambda-\rho)H(x^{-1}k)} dT(kM), \quad x \in G,$

maps the space of <u>analytic functionals</u> T on K/M into the space of all C^∞-functions f on G/K such that $\Omega f = -(\lambda^2+\rho^2)f$. This mapping is surjective, as Helgason showed in the case $\mathbb{F} = \mathbb{R}$ for Im $\lambda \geq 0$ and in the other rank one cases for $i\lambda \leq 0$. In the case of general rank one and general complex λ Helgason's elementary method using estimates in k,ℓ for the associated Jacobi functions (4.15) failed. Here the proof for general rank given by Kashiwara e.a. [78] and using the full machinery of hyperfunction theory is the only available proof until now. Lewis [95] shows for all rank one cases and for generic λ that the Poisson transform maps $\mathcal{D}'(K/M)$ onto the space of eigenfunctions of Ω of at most exponential growth.

See §8 for other applications of associated spherical functions.

In the following subsections we will consider some group theoretic interpretations of Jacobi transforms for which the Plancherel measure has a discrete part.

4.3. Jacobi functions as matrix elements of irreducible representations

The Plancherel measure in (3.33) with $G = SL(2,\mathbb{R})$ was first determined by Bargmann [8]. Here we will sketch an approach which uses K-finite functions. This approach was followed earlier by Takahashi [132] and (for the universal covering of $SL(2,\mathbb{R})$) by Flensted-Jensen [43].

For the following facts about $SL(2,\mathbb{R})$ the reader may consult, for instance, [85]. We will work with $G = SU(1,1)$, which is isomorphic to $SL(2,\mathbb{R})$. Consider an Iwasawa decomposition $G = KAN$ with

$$K := \{u_\theta := \operatorname{diag}(e^{\frac{1}{2}i\theta}, e^{-\frac{1}{2}i\theta}) \mid 0 \le \theta < 4\pi\},$$

$$A := \left\{a_t := \begin{pmatrix} \operatorname{ch} t & \operatorname{sh} t \\ \operatorname{sh} t & \operatorname{ch} t \end{pmatrix} \mid t \in \mathbb{R}\right\}.$$

Then $M = \{u_0, u_{2\pi}\}$. $\hat{K}$ consists of all δ_n, $n \in \frac{1}{2}\mathbb{Z}$, given by $\delta_n(u_\theta) := e^{in\theta}$ and $\hat{M}$ consist of $\delta_\xi|_M$, $\xi = 0$ or $\frac{1}{2}$. The principal series representation $\pi_{\xi,\lambda}$ ($\lambda \in \mathbb{C}, \xi = 0$ or $\frac{1}{2}$) of G is the representation of G induced by the representation $ma_t n \mapsto e^{-i\lambda t}\delta_\xi(m)$ of MAN. Then $\pi_{\xi,\lambda}|_K = \oplus_{n\in\mathbb{Z}} \delta_{n+\xi}$. In particular, each K-type occurs in $\pi_{\xi,\lambda}$ at most once. For a suitable orthonormal basis $\{e_n\}_{n\in\mathbb{Z}+\xi}$ of $H(\pi_{\xi,\lambda})$ with $\pi_{\xi,\lambda}(u_\theta)e_n = \delta_n(u_\theta)e_n$, the matrix elements

$$\pi_{\xi,\lambda;m,n}(x) := (\pi_{\xi,\lambda}(x)e_n, e_m), \quad x \in G, \ m,n \in \mathbb{Z}+\xi$$

can be expressed in terms of Jacobi functions (cf.[85, Theorem 2.1]):

(4.17) $\pi_{\xi,\lambda;m,n}(a_t) = (c_{\xi,\lambda;m,n}/(|m-n|)!) \cdot$

$\cdot (\text{sh } t)^{|m-n|}(\text{ch } t)^{m+n}\phi_\lambda^{(|m-n|,m+n)}(t),$

where

$$c_{\xi,\lambda;m,n} := \begin{cases} (-\tfrac{1}{2}i\lambda+n+\tfrac{1}{2})_{m-n} & \text{if } m \geq n, \\[2ex] (-\tfrac{1}{2}i\lambda-n+\tfrac{1}{2})_{n-m} & \text{if } n \geq m. \end{cases}$$

These explicit expressions (in terms of hypergeometric functions) were already obtained in [8].

The elements of $\hat{G}$ can all be obtained by unitarizing suitable subquotients of representations $\pi_{\xi,\lambda}$. They are (cf.[85]):

(a) <u>unitary principal series</u>: $\pi_{\xi,\lambda}(\lambda>0,\xi=0 \text{ or } \tfrac{1}{2})$,

$\pi_{0,0}, \; \pi_{\frac{1}{2},0}^{+}, \; \pi_{\frac{1}{2},0}^{-} \; (\pi_{\frac{1}{2},0} = \pi_{\frac{1}{2},0}^{+} \oplus \pi_{\frac{1}{2},0}^{-}).$

(b) <u>complementary series</u>: $\pi_{0,i\mu}(0<\mu<1)$ up to unitarization.

(c) <u>discrete series</u>: $\pi_{\xi,\lambda}^{+}, \pi_{\xi,\lambda}^{-}(\xi=0 \text{ or } \tfrac{1}{2}, \; \lambda = i(2\xi+1),$

$i(2\xi+3), \; i(2\xi+5),\ldots),$ where, up to unitarization,

$\pi_{\xi,\lambda;m,n}^{+} = \pi_{\xi,\lambda;m,n} (m,n = \tfrac{1}{2}|\lambda|+\tfrac{1}{2},\tfrac{1}{2}|\lambda|+ \dfrac{3}{2},\ldots),$

$\pi_{\xi,\lambda;m,n}^{-} = \pi_{\xi,\lambda;m,n} (m,n =-\tfrac{1}{2}|\lambda|-\tfrac{1}{2},-\tfrac{1}{2}|\lambda|-\dfrac{3}{2},\ldots).$

(d) <u>identity representation</u>.

Now we determine the Plancherel measure in (3.33) by the method of K-finite functions. It follows from (2.24), (2.18), (2.23) that, for f in $\mathcal{D}_{\text{even}}(\mathbb{R})$ and n = 0,1,2,...:

(4.18) $f(0) = \displaystyle\int_0^\infty \hat{f}_{0,2n}(\lambda)2^{-4n-2}\lambda \; \text{th}\tfrac{1}{2}\pi\lambda \; d\lambda \; +$

$+ \sum_{k=0}^{n-1} 2^{-4n-1}(2k+1)\hat{f}_{0,2n}((2k+1)i),$

(4.19) $f(0) = \displaystyle\int_0^\infty \hat{f}_{0,2n+1}(\lambda)2^{-4n-4}\lambda \; \text{coth}\tfrac{1}{2}\pi\lambda \; d\lambda \; +$

$+ \sum_{k=1}^{n} 2^{-4n-3}(2k) \; \hat{f}_{0,2n+1}(2ki),$

where $\hat{f}_{\alpha,\beta}(\lambda) = \hat{f}(\lambda)$ as defined by (2.12). Now let $F \in \mathcal{D}(G)$ such that

(4.20) $F(u_\theta a_t u_\eta) = (ch\ t)^{2n} f(t) e^{in(\theta+\eta)}$

with $f \in \mathcal{D}_{even}(\mathbb{R})$, $n \in \frac{1}{2}\mathbb{Z}$, i.e., F is K-central of K-type δ_n. Then (4.18), (4.19), (4.20), (2.12), (4.17), (4.8) imply that

$$F(e) = \int_0^\infty \left(\int_G F(x)\pi_{0,\lambda;n,n}(x)dx \right) \tfrac{1}{4}\lambda th\tfrac{1}{2}\pi\lambda\ d\lambda\ +$$

$$+ \sum_{k=0}^{|n|-1}(k+\tfrac{1}{2}) \int_G F(x)\pi_{0,(2k+1)i;n,n}(x)dx, n\in \mathbb{Z},$$

$$F(e) = \int_0^\infty \left(\int_G F(x)\pi_{\frac{1}{2},\lambda;n,n}(x)dx \right) \tfrac{1}{4}\lambda coth\tfrac{1}{2}\pi\lambda\ d\lambda\ +$$

$$+ \sum_{k=1}^{|n|} k \int_G F(x)\pi_{\frac{1}{2},2ki;n,n}(x)dx,\ n \in \mathbb{Z}+\tfrac{1}{2}.$$

Hence, for all F of the form (4.20) we have

(4.21) $$F(e) = \int_0^\infty tr\pi_{0,\lambda}(F)\tfrac{1}{4}\lambda th\tfrac{1}{2}\pi\lambda\ d\lambda\ +$$

$$+ \int_0^\infty tr\pi_{\frac{1}{2},\lambda}(F)\tfrac{1}{4}\lambda coth\tfrac{1}{2}\pi\lambda\ d\lambda\ +$$

$$\sum_{k=0}^\infty \sum_{\xi=0,\frac{1}{2}}(k+\xi+\tfrac{1}{2})tr(\pi^+_{\xi,i(2k+2\xi+1)}(F)+\pi^-_{\xi,i(2k+2\xi+1)}(F))$$

In view of §3.5, formula (4.21) is now valid for all F in $\mathcal{D}(G)$. This yields the Plancherel measure we looked for. Now (3.34) holds with the same measure ν. Specialization of this formula to functions f of arbitrary double K-type yields, in view of (4.17), Theorem 2.4 for all $\alpha, \beta \in \mathbb{Z}$ with $\alpha \geq 0$.

The explicit knowledge of the matrix elements of the principal series representations of $SL(2,\mathbb{R})$ in a K-basis, cf. (4.17), was exploited by Koornwinder [84], [85], Takahashi [133] in order to treat the representation theory of $SL(2,\mathbb{R})$ in a global, i.e. non-infinitesimal way.

For the universal covering group $\tilde{G}$ of $SL(2,\mathbb{R})$ the subgroup $\tilde{K}$ in the Iwasawa composition is no longer compact but isomorphic to $\mathbb{R}$. The principal series representations of $\tilde{G}$ restricted to $\tilde{K}$ still decompose as a multiplicity-free direct sum of irreducible $\tilde{K}$-types and the matrix-elements

of these representations can again be expressed in terms of
Jacobi functions. In particular, for the diagonal matrix
elements restricted to A we get the function
$a_t \mapsto (\mathrm{ch}\ t)^\beta \phi_\lambda^{(0,\beta)}(t)$ with β arbitrarily real. Flensted-
Jensen [43], [44] used (2.24) for these Jacobi functions in
order to obtain the Plancherel measure for $\tilde{G}$. This measure
was earlier determined by Pukanszky [114]. Another derivation
using K-finite functions is given by Matsushita [99].

4.4. Jacobi functions occurring in spherical functions on nonsymmetric Gelfand pairs

In §3.3 we pointed out that, for each rank one group G, (K,M)
is a Gelfand pair. Then the spherical functions for (K,M)
can be expressed in terms of Jacobi polynomials (cf. §8.1).
Case-by-case inspection shows that noncompact duals (in a
certain sense) of these pairs (K,M) are also Gelfand pairs
and that the corresponding spherical functions involve
Jacobi functions. If $\mathbf{F} = \mathbb{R}$ this yields nothing new, for
$\mathbf{F} = \mathbb{C}$ this has been considered by Flensted-Jensen [44],
while the quaternionic and octonionic cases were treated by
Takahashi [130], [131].
     In the complex case we take G = SU(1,n) or its univer-
sal covering group $\tilde{G}$, and K = SU(n) (n≥2). Then (G,K) and
$(\tilde{G},K)$ are Gelfand pairs. The K-biinvariant functions on G
are completely determined by their restriction to some
abelian subgroup L×A isomorphic to $\mathbf{T} \times \mathbb{R}$ or its universal
covering $\mathbb{R} \times \mathbb{R}$. The spherical functions for (G,K) restricted
to this subgroup are

$$(\ell_\theta, a_t) \mapsto e^{i\beta\theta}(\mathrm{ch}\ t)^\beta\ \phi_\lambda^{(n-1,\beta)}(t),$$

where ϕ_λ is a Jacobi function, $\lambda \in \mathbb{C}$ and β runs over $\mathbb{Z}$ or
$\mathbb{C}$, respectively. In [44] the spherical Plancherel measure
for this Gelfand pair is obtained from the Plancherel formula for the Jacobi functions involved. The associated
spherical functions can also be expressed in terms of Jacobi
functions. In the special case n = 1 we can work with
G = SU(1,1)×S(U(1)×U(1)) and K the diagonal of S(U(1)×U(1))
in this direct product. Then the spherical Plancherel formula for (G,K) yields the group Plancherel formula for
SU(1,1). A similar result holds for $\overline{SU(1,1)}$.
     In the quaternionic case we take G = Sp(1,n)×Sp(1),
K = Sp(n)×Sp(1)*, where Sp(1)* is the diagonal in the direct
product Sp(1,n)×Sp(1). Then Takahashi [130] shows that

(G,K) is a Gelfand pair and that the spherical functions, in suitable coordinates, have the form

$$(\theta,t) \mapsto (ch\ t)^k\ \phi_\lambda^{(2n-1,k+1)} R_k^{(\frac{1}{2},\frac{1}{2})}(\cos\theta),$$

where $\lambda \in \mathbb{C}$, $k = 0,1,2,\ldots,$ and $R_k^{(\frac{1}{2},\frac{1}{2})}$ is a Jacobi polynomial (cf.(2.3)). (Note that $R_k^{(\frac{1}{2},\frac{1}{2})}(\cos\theta) = \sin(k+1)\theta/((k+1)\sin\theta)$ and that it has an interpretation as a character on Sp(1).) For n = 1 this situation was earlier met by Takahashi [129] in the context of the representation theory of $SO_0(4,1)$.

Finally, in the octonionic case Takahashi [131] points out that $(Spin_0(1,8), Spin(7))$ is a Gelfand pair with spherical functions

$$(\theta,t) \mapsto (ch\ t)^k\ \phi_\lambda^{(3,k+3)}(t) R_k^{(3/2,3/2)}(\cos\theta).$$

4.5. Jacobi functions as K-finite functions on G/H

For preliminaries to this subsection we refer to §3.4, §3.5. We restrict ourselves to Faraut's [37] generalized Gelfand pairs $(G,H) = (U(p,q;\mathbb{F}), U(1,\mathbb{F}) \times U(p-1,q;\mathbb{F}))$.

$$\text{Let } B := \left\{ a_{i\theta} := \begin{pmatrix} \cos\theta & \sin\theta & 0 \\ -\sin\theta & \cos\theta & \\ 0 & & I_{p+q-2} \end{pmatrix} \right\}.$$

Then $HAH \cup HBH$ is dense in G and the H-biinvariant distributions on G are in 1-1 correspondence with a certain class of generalized functions on $\{a_t | t \geq 0\} \cup \{a_{i\theta} | 0 \leq \theta \leq \frac{1}{2}\pi\}$ (i.e. on $\mathbb{R}_+ \cup i[0,\frac{1}{2}\pi]$) with respect to a suitable class of test functions on $\mathbb{R}_+ \cup i[0,\frac{1}{2}\pi]$, having a certain singularity at 0. The "radial" part of the Laplace-Beltrami operator on G/H becomes, up to a constant factor, the Jacobi differential operator $L_{\alpha,\beta}$ (cf.(2.9)) on $\mathbb{R}_+ \cup i[0,\frac{1}{2}\pi]$ with $\alpha = \frac{1}{2}d(p+q-1)-1$, $\beta = \frac{1}{2}d-1$ ($d=\dim_\mathbb{R}\mathbb{F}$). Thus the spherical distributions are essentially distributional solutions T_λ of the Jacobi differential equation (2.10). Faraut shows that, for λ outside a specific discrete countable set, (2.10) has only a one-dimensional eigenspace, for the exceptional values λ the eigenspace is two-dimensional. The eigendistributions are regular on $(0,\infty)$ and on $i(0,\frac{1}{2}\pi)$, but they are not necessarily regular at 0. In particular, they do not necessarily coincide with the Jacobi functions $\phi_\lambda^{(\alpha,\beta)}$ on $(0,\infty)$, but for certain d,p,q they do, as follows from [37,§III]. We normalize the eigendistribution T_λ such that it depends analytically on λ and is nonzero for all λ.

In §3.4 we introduced subgroups K,M,N of G. (These M,N are different from M,N in §3.1.) For the present G we have $K = U(p,\mathbb{F}) \times U(q,\mathbb{F})$, $M=U(1,\mathbb{F}) \times U(p-1,\mathbb{F}) \times U(q-1,\mathbb{F})$. Let π_λ ($\lambda \in \mathbb{C}$) be the representation of G induced by the representation $ma_tn \mapsto e^{-i\lambda t}$ of the subgroup MAN (a maximal parabolic subgroup, while the subgroup MAN used for obtaining (3.21) is minimal parabolic). Then, as shown in [37], an H-invariant distribution vector u_λ and the spherical distribution T_λ can be associated with π_λ. The representation π_λ is unitary if $\lambda \geq 0$ (H-spherical unitary principal series) and $\pi_{i\mu}$ may contain unitarizable subquotients belonging to $(G/H)^\wedge$ if $\mu > 0$. This happens if $0 < \mu \leq \mu_0$, where μ_0 depending on p,q,d can be explicitly given (H-spherical complementary series) and, possibly, if $\mu > \mu_0$, $\mu - \rho \in \mathbb{Z}$ (H-spherical discrete series). It can only occur in the discrete series case that two distinct elements of $(G/H)^\wedge$ correspond to one π_λ. All of $(G/H)^\wedge$ can be obtained in the above way. Faraut obtains these results by using K-finite functions (cf. an analogous approach for the K-spherical case in §8.2).

We will now give some more details about the K-finite functions because they involve Jacobi functions and can be used for deriving the Plancherel formula for G/H. Since G = KAH, a K-finite function f on G/H can be written as a function $(kM,t) \mapsto f(ka_tH)$ on $K/M \times \mathbb{R}$. Futhermore, K/M can be identified with a space of orbits of $U(1,\mathbb{F})$ on $S(\mathbb{F}^p) \times S(\mathbb{F}^q)$ ($S(\mathbb{F}^p)$ is unit sphere in $\mathbb{F}^p$). Denote by $\mathcal{Y}_\ell^{dp}$ the space of spherical harmonics of degree ℓ on $S(\mathbb{F}^p)$. Then any K-finite function of certain K-type on K/M is in particular contained in the space $\mathcal{Y}_{\ell,m} := \mathcal{Y}_\ell^{dp} \times \mathcal{Y}_m^{dq}$ for certain ℓ,m in $\mathbb{Z}_+$. Faraut obtains the expression of the Laplace-Beltrami operator Ω as a differential operator on $K/M \times \mathbb{R}$. In this way it can be shown that the K-finite solutions f of $(\Omega+\lambda^2+\rho^2)f = 0$ which are of certain K-type are given by the functions

$$(4.22) \qquad ka_tH \mapsto (\text{sh } t)^m(\text{ch } t)^\ell \phi_\lambda^{(\frac{1}{2}dq-1+m, \frac{1}{2}dp-1+\ell)}(t)Y_{\ell,m}(kM),$$

where $Y_{\ell,m} \in \mathcal{Y}_{\ell,m}$. Now let f in $\mathcal{D}(G)$ be such that f^0 (cf. (3.35)) is of the form

$$(4.23) \qquad f^0(ka_tH) = (\text{sh } t)^m(\text{ch } t)^\ell F(t)Y_{\ell,m}(kM),$$

where $F \in \mathcal{D}_{even}(\mathbb{R})$, $Y_{\ell,m} \in \mathcal{Y}_{\ell,m}$. Then it follows that

$$(4.24) \qquad T_\lambda(f) = \delta_{m,0} b_{\lambda,\ell} Y_{\ell,0}(eK) \cdot$$

$$\cdot \int_0^\infty F(t) \phi_\lambda^{(\frac{1}{2}dq-1,\frac{1}{2}dp-1+\ell)}(t) \Delta_{\frac{1}{2}dq-1,\frac{1}{2}dp-1+\ell}(t)dt,$$

for certain constants $b_{\lambda,\ell}$ which are explicitly evaluated by
Faraut. A similar formula holds for $S_\lambda f$, where S_λ is a
spherical distribution for λ in the discrete spectrum of Ω.

Now, in view of (4.23), (4.24) the Plancherel measure ν
in (3.36) is obtained from inversion of the Jacobi transform
in the cases $(\alpha,\beta) = (\frac{1}{2}dq-1,\frac{1}{2}dp-1+\ell)$, where ℓ runs over a
certain subset of $\mathbb{Z}_+$. Thus Theorem 2.3 can be applied again.
In a similar way, the version (3.37) of the Plancherel theorem
can be reduced to Theorem 2.4, where now
$(\alpha,\beta) = (\frac{1}{2}dq-1+m,\frac{1}{2}dp-1+\ell)$. This is Faraut's second proof of
his Plancherel theorem in [37,§10]. (His first proof uses
direct spectral decomposition of $L_{\alpha,\beta}$ on $\mathbb{R}_+ \cup i[0,\frac{1}{2}\pi]$.) Observe that in Faraut's second proof more cases of the
Plancherel theorem for the Jacobi transform are used than is
strictly needed: m can be put zero.

Kosters [88], [89,Ch.3] derived the Plancherel formula
for $(F_{4(-20)}, \mathrm{Spin}(1,8))$ in a similar way, using Jacobi functions.

The fact that the K-invariant eigenfunctions of Ω on
G/H can be expressed as Jacobi functions holds for all semisimple symmetric pairs of rank one. More generally it holds
for semisimple symmetric pairs (G,H) where the maximal
abelian subspaces of $p \cap q$ (cf.§3.4) have dimension one.
This follows from the explicit expression for the radial
part of Ω with respect to the decomposition G = KAH (cf.[67],
[72,Ch.II], [45,(4.12)],[46,p.307], [75,Ch.10,11]). For the
cases (G,H) = (O(p,q),O(1)×O(p-1,q)) this was already observed [42]. Unfortunately, K-finite eigenfunctions of nontrivial K-type cannot always be expressed in terms of Jacobi
functions. For instance, on the space $SL(n,\mathbb{R})/GL(n-1,\mathbb{R})$
considered (for n≥3) by Kosters [89,Ch.4] and (for n=3) by
Molčanov [107] the K-finite eigenfunctions do not factorize,
in general, as $ka_tH \to Y(kM)F(t)$, but for the t-dependence
we get vector-valued functions satisfying a system of second
order o.d.e.'s. Thus many of the methods used in [37] fail
here, because the theory of these vector-valued special
functions is not yet developed.

It is still worthwhile to have knowledge about Jacobi
functions as K-invariant eigenfunctions of Ω on G/H (so-
called <u>intertwining functions</u>): in order to do harmonic

analysis for K-invariant functions on G/H, in order to get partial information about full harmonic analysis on G/H and in order to get new group theoretic interpretations of formulas for Jacobi functions. For the first and third objective see Flensted-Jensen [42], Badertscher [6], for the second objective see, for instance, Kosters [89,Ch.4], Sekiguchi [122,§7]. Sekiguchi [122] characterized the eigenhyperfunctions of Ω on $G/H = U(p,q;\mathbb{F})/U(1,\mathbb{F}) \times U(p-1,q;\mathbb{F})$ as Poisson transforms of the hyperfunctions on $K/K \cap M$ by using the full machinery of hyperfunction theory. W. Kosters (Leiden, to appear) did analogous work for $SL(n,\mathbb{R})/GL(n-1,\mathbb{R})$. It would be interesting to redo the results in [122] in the style of Helgason [68](cf.§4.2) by use of the explicit expressions (4.22) for the K-finite functions.

Badertscher [6] does harmonic analysis for K-invariant functions on G/H with $G = O(p,1)$, $K = O(p)\times O(1)$, $H = O(p-1,1)$ (so H is slightly smaller than in the case of [37]). The radial part of Ω with respect to the decomposition $G = KAH$ now becomes the Jacobi differential operator $L_{-\frac{1}{2},\frac{1}{2}p-1}$ on the full real axis. The eigenfunctions with eigenvalue $-\lambda^2 - \rho^2$ are the linear combinations of the even function $\phi_\lambda^{(-\frac{1}{2},\frac{1}{2}p-1)}$ and the odd function $t \to \text{sh } t \, \phi_\lambda^{(\frac{1}{2},\frac{1}{2}p-1)}(t)$. So, by decomposition into even and odd functions the spectral decomposition of $L_{-\frac{1}{2},\frac{1}{2}p-1}$ on $\mathbb{R}$ can be reduced to inversion of the Jacobi transform for $(\alpha,\beta) = (\pm\frac{1}{2},\frac{1}{2}p-1)$. The occurrence of multiplicity 2 for the K-invariant eigenfunctions of Ω already suggests that (G,H) is not a generalized Gelfand pair in this case.

Mizony [105], [106] and Faraut [38] point out that for $G = O(p,1)$, $H = O(p-1,1)$ and $A_+ := \{a_t | t > 0\}$ the subset HA_+H of G is a subsemigroup of G. Thus the continuous H-biinvariant functions on G with support included in HA_+H form a convolution algebra which turns out to be commutative. After restriction to a smaller algebra of C^∞-functions of at most exponential growth on A_+, the characters of this algebra can be determined. It turns out that they have the form

$$(4.25) \qquad f \mapsto \int_0^\infty f(a_t) \frac{2^{p-2}\Gamma(\frac{1}{2}p-\frac{1}{2})\Gamma(-\frac{1}{2}p+\frac{1}{2}-i\lambda)}{\Gamma(1-i\lambda)} \Phi_\lambda^{(\frac{1}{2}p-1,-\frac{1}{2})}(t) \cdot$$

$$\cdot (\text{sh}t)^{p-1}dt,$$

where Φ_λ is a Jacobi function of the second kind (cf.(2.15)) and $\text{Im}\lambda$ is sufficiently large. Thus we have a group theoretic interpretation of the <u>Laplace-Jacobi transform</u> if $\beta = -\frac{1}{2}$,

$\alpha = 0, \frac{1}{2}, 1, \ldots$. Mizony [105] and Carroll [19] consider the
Laplace-Jacobi transform also for more general α, β, without
group theoretic interpretation, and they obtain inversion
formulas. Mizony [106] points out that, in the case of group
theoretic interpretation, the functions $Ha_tH \to \Phi_\lambda(t)$ can be
considered as certain generalized matrix elements of "principal series" representations of the semigroup HA_+H. The interpretation as characters of a convolution algebra is interesting, since such an interpretation is not known for
spherical distributions.

5. THE ABEL TRANSFORM

As we already observed in §2, the Jacobi transform $f \mapsto \hat{f}$ has
a factorization

(5.1)

where F is the classical Fourier transform and $f \mapsto Ff$ is the
<u>Abel transform</u>. This last transform can be defined both in
a group theoretic (geometric) way and in a purely analytic
way. Since fairly much is known about the properties of F,
a study of the Abel transform will teach us a lot about the
Jacobi transform. Moreover, the Abel transform is an interesting object in its own right. Roughly the following
aspects of the Abel transform will be discussed:
(a) the homomorphism property with respect to suitable con-
    volution algebras;
(b) the transmutation property with respect to suitable dif-
    ferential operators;
(c) the bijection property with respect to suitable function
    spaces;
(d) the inversion of the Abel transform;
(e) the images of certain special functions;
(f) the dual Abel transform.
    One can start reading this section either in §5.1, where
the Abel transform is treated in the spherical rank one case,
or in §5.3, where an analytic treatment of the Abel transform
is presented. In §5.2 the transposition of formulas from
group theoretic into analytic form is discussed. Finally,
§5.4 contains a generalization of the Abel transform and
§5.5 discusses results and references.

5.1. The spherical rank one case

The main reference for this subsection is [39]. Assume that
G is a rank one group, use the notation of §3.1 and use the
results and conventions of §4.1. The Haar measure dn on N can
be normalized such that the Haar measure on G, normalized by
(4.8) has the following expression with respect to the
Iwasawa decomposition:

$$(5.2) \qquad \int_G f(x)\,dx = \int_K \int_{\mathbb{R}} \int_N f(ka_t n) e^{2\rho t}\,dk\,dt\,dn, f \in C_c(G).$$

For f in $C_c(G//K)$ define the Abel transform $f \mapsto F_f$ by

$$(5.3) \qquad F_f(t) := e^{\rho t} \int_N f(a_t n)\,dn, \quad t \in \mathbb{R}.$$

Combination of (4.7), (4.4), (5.2), (5.3) shows that, for f
in $C_c(G//K)$,

$$(5.4) \qquad \hat{f}(\lambda) = \int_{\mathbb{R}} F_f(t) e^{i\lambda t}\,dt.$$

Thus the spherical Fourier transform is the composition of
the Abel transform and the classical Fourier transform (cf.
(5.1)). It can be shown that $f \mapsto F_f$ is an homomorphism of
the convolution algebra $C_c(G//K)$ (or $\mathcal{D}(G//K)$) into the convolution algebra $C_{c,even}(\mathbb{R})$ (or $\mathcal{D}_{even}(\mathbb{R})$) :

$$(5.4) \qquad F_{f*g} = F_f * F_g, \quad f,g \in C_c(G//K),$$

and that the mapping has the transmutation property (use(4.1)):

$$(5.5) \qquad F_{\Omega f}(t) = \left(\frac{d^2}{dt^2} - \rho^2\right) F_f(t),$$

where $f \in \mathcal{D}(G//K)$.

Let the dual Abel transform $g \mapsto E_g$ be the linear mapping
of $C(\mathbb{R})$ into $\overline{C(G//K)}$ which satisfies

$$(5.6) \qquad \int_G f(x) E_g(x)\,dx = \int_{\mathbb{R}} F_f(t) g(t)\,dt$$

for all f in $C_c(G//K)$. Then (use (5.2))

$$(5.7) \qquad E_g(x) = \int_K g(H(x^{-1}k)) e^{-\rho H(x^{-1}k)}\,dk, \quad x \in G, g \in C(\mathbb{R}),$$

and there is the transmutation property

$$(5.8) \qquad \Omega E_g = E_{g''-\rho^2 g}, \quad g \in E(\mathbb{R}) \ (=C^\infty(\mathbb{R})).$$

If $g(t) := e^{i\lambda t}$ then $E_g = \phi_\lambda$ (cf.(4.4)).

Now make the further assumption that $G = U(1,n;\mathbb{F})$ (cf. §3.1). Then we can rewrite (4.4) and (5.3) in a more concrete form. If x in $U(1,n;\mathbb{F})$ has matrix $(x_{ij})_{i,j=0,\ldots,n}$ then it can be shown that

$$(5.9) \qquad H(x) = \log(|x_{00}+x_{0n}|), \quad x \in G.$$

Let $k = \begin{pmatrix} u & 0 \\ 0 & v \end{pmatrix}$ be an element of $K = U(1,\mathbb{F}) \times U(n,\mathbb{F})$. Then it follows from (5.9) that

$$H(a_{-t}k) = \log(|u\,\mathrm{ch}\,t - V_{nn}\mathrm{sh}\,t|).$$

Thus (4.4) can be rewritten as

$$(5.10) \qquad \phi_\lambda(a_t) = \int_{S(\mathbb{F}^n)} |\mathrm{ch}\,t + y_n\mathrm{sh}\,t|^{i\lambda-\rho}dy$$

where $S(\mathbb{F}^n)$ is the unit sphere in $\mathbb{F}^n$, $y = (y_1,\ldots,y_n) \in S(\mathbb{F}^n)$ and dy is the normalized $U(n,\mathbb{F})$-invariant measure on $S(\mathbb{F}^n)$. Note that $S(\mathbb{F}^n)$ is the homogeneous space K/M.

Next we rewrite (5.3). In terms of the elements $n_{z,w}$ (cf.(3.1)) the Haar measure on N equals $dn_{z,w} = c_0\,dz\,dw$, where dz and dw are Lebesgue measures on $\mathbb{F}^{n-1} = \mathbb{R}^{d(n-1)}$ and $\mathrm{Im}\mathbb{F} = \mathbb{R}^{d-1}$, respectively, and the positive constant c_0 has yet to be determined. For a K-biinvariant function f on G write

$$(5.11) \qquad f[\mathrm{cht}] := f(a_t), \quad t \in \mathbb{R}.$$

Then

$$(5.12) \qquad f(x) = f[|x_{00}|], \quad x \in G.$$

For an even function g on $\mathbb{R}$ also write

$$(5.13) \qquad g[\mathrm{cht}] := g(t), \quad t \in \mathbb{R}.$$

Now we can rewrite (5.3) (using (3.1) and (5.12)) as

$$F_f(t) = c_0 e^{\rho t}\int_{\mathbb{F}^{n-1}\times\mathrm{Im}\,\mathbb{F}} f[|\mathrm{cht}+e^t(\tfrac{1}{2}|z|^2+w)|]dz\,dw,$$

hence

$$(5.14) \qquad F_f[x] = c_0(2x)^\rho\int_{\mathbb{F}^{n-1}\times\mathrm{Im}\,\mathbb{F}} f[x|1+|z|^2+2w|]dz\,dw, x \geq 1.$$

It was observed by Godement [57] that (5.14) reduces for
$\mathbb{F} = \mathbb{R}$, $n = 2$ to

$$F_f[x] = c_0(2x)^{\frac{1}{2}} \int_{\mathbb{R}} f[x(1+z^2)]dz =$$

$$= c_0 2^{-\frac{1}{2}} \int_x^\infty f[y](y-x)^{-\frac{1}{2}}dy,$$

which is a version of the classical Abel transform (cf.
Abel [1]). This explains the name of the transform $f \mapsto F_f$.
    If we substitute

(5.15) $f[x] := (2x)^{-i\lambda-\rho}$, $\text{Im}\lambda < 0$,

in (5.14) then

(5.16) $F_f[x] = c_0(2x)^{-i\lambda} \int_{\mathbb{F}^{n-1} x \text{Im } \mathbb{F}} |1+|z|^2+2w|^{-i\lambda-\rho}dz\ dw.$

Here $f[\cdot]$ determines a function $f(\cdot)$ in $C_c(G//K)$ and F_f is
well-defined although f does not have compact support. It is
possible to determine c_0 from (5.16). Observe that, for K-
biinvariant f,

$$\int_0^\infty f(a_t)\Delta(t)dt = \int_{\mathbb{R}} F_f(t)e^{\rho t}dt.$$

Substitute (5.15), (5.16) in this identity, put $\lambda := i\nu$ and
let $\nu \uparrow -\rho$. Then we obtain

(5.17) $c_0^{-1} = \int_{\mathbb{F}^{n-1} x \text{Im } \mathbb{F}} |1+|z|^2+2w|^{-2\rho}dz\ dw.$

    Remember the c-function, cf. (2.17), (2.18), (2.19). By
application of (2.19) to (5.10) we can derive an integral
representation for the c-function. Rewrite (5.10) as

$$\phi_\lambda(a_t) = e^{(i\lambda-\rho)t} \int_{S(\mathbb{F}^n)} \left|1+e^{-2t}\frac{1-y_n}{1+y_n}\right|^{i\lambda-\rho} |\tfrac{1}{2}(1+y_n)|^{i\lambda-\rho}dy.$$

Let $t > 0$. Then the integrand is dominated by
$|\tfrac{1}{2}(1+y_n)|^{-\text{Im}\lambda-\rho}$ if $-\rho \leq \text{Im}\lambda < 0$ and by 1 if $\text{Im}\lambda \leq -\rho$. Hence,
an application of the dominated convergence theorem shows
that $\phi_\lambda(a_t)$ satisfies (2.19), where

(5.18) $c(\lambda) = \int_{S(\mathbb{F}^n)} |\tfrac{1}{2}(1+y_n)|^{i\lambda-\rho}dy$, $\text{Im}\lambda < 0$.

Next we will express the constant factor in (5.16) in terms of the c-function. By the Bruhat and Iwasawa decompositions (cf. §3.1) the mapping $\bar{n} \mapsto u(\bar{n})M$ is a diffeomorphism of $\bar{N}$ onto an open dense subset of K/M. Here $u(\bar{n})$ is as in §3.1(c). The corresponding Jacobian occurs in the formula

$$(5.19) \qquad \int_{K/M} h(kM)d(kM) = \int_{\bar{N}} h(u(\bar{n})M)e^{-2\rho H(\bar{n})} d\bar{n}, h \in C(K/M).$$

This formula can be rewritten as

$$(5.20) \qquad \int_{S(\mathbb{F}^n)} h((y_1,\ldots,y_n))dy =$$

$$= c_0 \int_{\mathbb{F}^{n-1} \times \text{Im } \mathbb{F}} h\left(\frac{(-2z_1,\ldots,-2z_{n-1},1-|z|^2-2w)}{1+|z|^2+2w}\right)$$

$$|1+|z|^2+2w|^{-2\rho} dz\, dw,$$

where $h \in C(S(\mathbb{F}^n))$. Formula (5.20) can also be derived by straightforward computation. Now put $h(y) := |\tfrac{1}{2}(1+y_n)|^{i\lambda-\rho}$ in (5.20) and combine with (5.18). Then we obtain

$$(5.21) \qquad c(\lambda) = c_0 \int_{\mathbb{F}^{n-1} \times \text{Im } \mathbb{F}} |1+|z|^2+2w|^{-i\lambda-\rho} dz\, dw, \quad \text{Im}\lambda < 0,$$

and, in view of (5.15), (5.16):

$$(5.22) \qquad F_f[x] = (2x)^{-i\lambda} \text{ if } f[x] := (c(\lambda))^{-1}(2x)^{-i\lambda-\rho}, \quad \text{Im}\lambda<0.$$

Formula (5.22) has an analogue for Jacobi functions of the second kind defined by (2.15):

$$(5.23) \qquad F_f(t) = e^{-i\lambda t}(t>0) \text{ if } f(k_1 a_t k_2) := (c(\lambda))^{-1}\Phi_{-\lambda}(t)$$

$$(t>0, k_1, k_2 \in K), \quad \text{Im}\lambda < 0.$$

Here the function f is well-defined on $G\backslash\{e\}$ and so is its Abel transform defined by the right hand side of (5.14). It follows from (2.9) and (5.5) that $(d^2/dt^2+\lambda^2)F_f(t) = 0$ and (5.22), (5.14) and (2.15) show that $F_f(t) = e^{-i\lambda t}(1+o(1))$ as $t \to \infty$. Thus (5.23) is proved.

The results (5.22), (5.23) seem to be unobserved in literature until now. It would be of interest to find an higher rank analogue of (5.22).

5.2. Elimination of group variables from the integration formulas

In order to pass smoothly to the analytic treatment in §5.3, which does not use group theory, we will rewrite some of the previous integrals like (5.10), (5.14) in a form which does not involve group variables, thus allowing generalization to other values of α, β. The key observation is that, for $\mathbb{F} = \mathbb{C}$ or $\mathbb{H}$,

$$(5.24) \qquad \int_{S(\mathbb{F}^n)} f(\text{Rey}_n + i|\text{Imy}_n|)dy = \int_0^1 \int_0^\pi f(r\, e^{i\psi})dm_{\alpha,\beta}(r,\psi),$$

where f is a function on the upper half unit disk and

$$(5.25) \qquad dm_{\alpha,\beta}(r,\psi) :=$$

$$= \frac{2\Gamma(\alpha+1)}{\Gamma(\frac{1}{2})\Gamma(\alpha-\beta)\Gamma(p+\frac{1}{2})} (1-r^2)^{\alpha-\beta-1}(r\,\sin\psi)^{2\beta}rdrd\psi,$$

and that

$$(5.26) \qquad c_0 \int_{\mathbb{F}^{n-1}x\text{Im}\,\mathbb{F}} f(|1+|z|^2+2w|)\, dz\, dw =$$

$$= \int_0^\infty \int_0^\infty f(((1+s^2)^2+4t^2)^{\frac{1}{2}})\, dn_{\alpha,\beta}(s,t),$$

where f is a function on $(1,\infty)$ and

$$(5.27) \qquad dn_{\alpha,\beta}(s,t) := \frac{2^{2\rho+1}\Gamma(\alpha+1)}{\Gamma(\frac{1}{2})\Gamma(\alpha-\beta)\Gamma(\beta+\frac{1}{2})} s^{2\alpha-2\beta-1}t^{2\beta}ds\, dt.$$

These formulas remain valid for $\mathbb{F} = \mathbb{R}$ ($\beta = -\frac{1}{2}$), but the measures $m_{\alpha,-\frac{1}{2}}$, $n_{\alpha,-\frac{1}{2}}$ degenerate to measures with one-dimensional support: the weak limits of $m_{\alpha,\beta}$, $n_{\alpha,\beta}$ as $\beta \downarrow -\frac{1}{2}$. Throughout the rest of the paper we will keep to this convention, so we will not give the formulas for $\beta = -\frac{1}{2}$ (or $\alpha = \beta$) separately.

Formulas (5.10), (5.14), (5.18), (5.21) now can be rewritten as

$$(5.28) \qquad \phi_\lambda(a_t) = \int_0^1 \int_0^\pi |\text{ch } t + r\, e^{i\lambda}\text{sh } t|^{i\lambda-\rho}dm(r,\psi).$$

$$(5.29) \qquad F_f[x] = (2x)^\rho \int_0^\infty \int_0^\infty f[x((1+s^2)^2+4t^2)^{\frac{1}{2}}]dn(s,t),$$

$$(5.30) \qquad c(\lambda) = \int_0^1 \int_0^\pi |\tfrac{1}{2}(1+r\ e^{i\psi})|^{i\lambda-\rho} dm(r,\psi) =$$

$$= \int_0^\infty \int_0^\infty ((1+s^2)^2+4t^2)^{-\frac{1}{2}(i\lambda+\rho)} dn(s,t), \quad \mathrm{Im}\lambda < 0.$$

The equality of the two integrals in (5.30) also follows by the transformation of integration variables $\tfrac{1}{2}(1+r\ e^{i\psi}) =$ $= (1+s^2-2it)^{-1}$. This is seen by straightforward computation or by use of (5.20). The explicit expression (2.18) of $c(\lambda)$ can also be obtained by evaluation of one of the integrals in (5.30).

5.3. The analytic case

A reference for this subsection is [81]. We will obtain a pair of dual integral transforms $f \mapsto F_f$, $g \mapsto E_g$ such that the transmutation properties (5.5), (5.8) hold for more general α,β. These transforms will be built up from two fractional integrals and a quadratic transformation. So let us first introduce these building blocks.

Let $L_{\alpha,\beta}$ be defined by (2.9). Then there is the quadratic transformation (QT)

$$(5.31) \qquad (L_{\alpha,\alpha}f)(t) = 4(L_{\alpha,-\frac{1}{2}}g)(2t) \quad \text{if } f(t) = g(2t),$$

$$(5.32) \qquad \phi_{2\lambda}^{(\alpha,\alpha)}(t) = \phi_\lambda^{(\alpha,-\frac{1}{2})}(2t), \Phi_{2\lambda}^{(\alpha,\alpha)}(t) = \Phi_\lambda^{(\alpha,-\frac{1}{2})}(2t).$$

For $\mathrm{Re}\ \mu > 0$ define the fractional integral operators R_μ of Riemann-Liouville type and W_μ of Weyl type by

$$(5.33) \qquad (R_\mu f)(x) := \frac{1}{\Gamma(\mu)} \int_1^x f(y)(x-y)^{\mu-1} dy,$$

$$(5.34) \qquad (W_\mu f)(x) := \frac{1}{\Gamma(\mu)} \int_x^\infty f(y)(y-x)^{\mu-1} dy,$$

where $f \in L^1([1,\infty))$ and, in (5.34), $f \to 0$ sufficiently fast as $x \to \infty$ (cf. [35,Ch.13]). Some simple properties are:

$$(5.35) \qquad R_\mu \circ R_\nu = R_{\mu+\nu}, \quad DR_\mu = R_\mu D = R_{\mu-1}, \quad D^n R_n = \mathrm{id},$$

$$(5.36) \qquad W_\mu \circ W_\nu = W_{\mu+\nu}, \quad DW_\mu = W_\mu D = -W_{\mu-1}, \quad D^n W_n = (-1)^n \mathrm{id},$$

where $D := d/dx$ and $n \in \mathbb{N}$. Define the class $H_\sigma (\sigma > 0)$ by

(5.37) $H_\sigma := \{f \in C^\infty ((1,\infty)) | \forall n \in \mathbb{Z}_+ \ f^{(n)}(x) = O(x^{-\sigma - n}), \ x \to \infty\}$.

The following two mappings are bijections:

$$R_\mu : (x-1)^\alpha C^\infty ([1,\infty)) \to (x-1)^{\alpha + \mu} C^\infty ([1,\infty)), \ \text{Re } \alpha > -1,$$

$$W_\mu : H_\sigma \to H_{\sigma - \text{Re}\mu}, \quad \text{Re } \mu < \sigma.$$

A generalized integration-by-parts formula is given by

(5.38) $\displaystyle \int_1^\infty f(x)(W_\mu g)(x)(x-1)^\alpha (x+1)^\beta dx =$

$$= \int_1^\infty (R_\mu^{(\alpha, \beta)} f)(x) g(x) (x-1)^{\alpha + \mu} (x+1)^{\beta + \mu} dx,$$

where $f \in C^\infty ([1,\infty))$, $g \in C_c^\infty ([1,\infty))$ and

(5.39) $\displaystyle (R_\mu^{(\alpha, \beta)} f)(x) := (x-1)^{-\alpha - \mu}(x+1)^{-\beta - \mu} \cdot$

$$\cdot R_\mu (y \mapsto (y-1)^\alpha (y+1)^\beta f(y))(x).$$

$R_\mu^{(\alpha, \beta)}$ is a bijection of $C^\infty ([1,\infty))$ onto itself and of
$C^\infty ([1,\infty)) \cap H_\sigma$ onto $C^\infty ([1,\infty)) \cap H_{\sigma - \text{Re}\mu}$.
     Let $L_{\alpha, \beta}$ be the differential operator on $(1,\infty)$ obtained
from $L_{\alpha, \beta}$ by making the transformation $x = \text{ch } 2t$:

(5.40) $(L_{\alpha, \beta} g)(x) := 4(x^2 - 1)g''(x) + 4((\alpha + \beta + 2)x + \alpha - \beta)g'(x)$.

A straightforward computation yields the transmutation
formula

(5.41) $(L_{\alpha, \beta} + (\alpha + \beta + 1)^2)W_\mu f = W_\mu (L_{\alpha + \mu, \beta + \mu} + (\alpha + \beta + 2\mu + 1)^2)f$,

where $f \in H_\sigma$, $\text{Re } \mu < \sigma$. By using (5.41), (5.38) and the self-
adjointness of $L_{\alpha, \beta}$ with respect to the weight function
$(x-1)^\alpha (x+1)^\beta$ we obtain another transmutation formula

(5.42) $(L_{\alpha + \mu, \beta + \mu} + (\alpha + \beta + 2\mu + 1)^2)R_\mu^{(\alpha, \beta)} f = R_\mu^{(\alpha, \beta)}(L_{\alpha, \beta} + (\alpha + \beta + 1)^2)f$,

where $f \in C^\infty ([1,\infty))$. Three applications of the beta integral
yield:

$$(5.43) \qquad W_\mu((x-a)^{-\sigma}) = \frac{\Gamma(\sigma-\mu)}{\Gamma(\sigma)} \, (x-a)^{-\sigma+\mu}, \; 0 < \mathrm{Re}\,\mu < \mathrm{Re}\,\sigma,$$

$$(5.44) \qquad (R_\mu^{(\alpha,\beta)} f)(1) = \frac{2^{-\mu}\Gamma(\alpha+1)}{\Gamma(\alpha+\mu+1)} \, f(1), \; f \in C^\infty([1,\infty)),$$

$$(5.45) \qquad R_\mu^{(\alpha,\beta)}(x^\sigma) = \frac{\Gamma(\alpha+\beta+\sigma+1)}{\Gamma(\alpha+\beta+\sigma+\mu+1)} \, x^{\sigma-\mu} \, (\mathrm{mod}\,H_{\mu-\sigma-1}).$$

Now we consider maps $g \mapsto E_g^{(\alpha,\beta)}$ and $f \mapsto F_f^{(\alpha,\beta)}$ which are, schematically, compositions of the following maps:

$$(5.46) \qquad E^{(\alpha,\beta)} : (-\tfrac{1}{2},-\tfrac{1}{2}) \xrightarrow{\;R_{\alpha-\beta}^{(-\frac{1}{2},-\frac{1}{2})}\;} (\alpha-\beta-\tfrac{1}{2},\alpha-\beta-\tfrac{1}{2}) \xrightarrow{\;QT\;}$$

$$\longrightarrow (\alpha-\beta-\tfrac{1}{2},-\tfrac{1}{2}) \xrightarrow{\;R_{\beta+\frac{1}{2}}^{(\alpha-\beta-\frac{1}{2},-\frac{1}{2})}\;} (\alpha,\beta),$$

$$(5.47) \qquad F^{(\alpha,\beta)} : (\alpha,\beta) \xrightarrow{\;W_{\beta+\frac{1}{2}}\;} (\alpha-\beta-\tfrac{1}{2},-\tfrac{1}{2}) \xrightarrow{\;QT\;}$$

$$\longrightarrow (\alpha-\beta-\tfrac{1}{2},\alpha-\beta-\tfrac{1}{2}) \xrightarrow{\;W_{\alpha-\beta}\;} (-\tfrac{1}{2},-\tfrac{1}{2}).$$

We will work in the t-variable and we will normalize $E^{(\alpha,\beta)}$ and $F^{(\alpha,\beta)}$ such that $E_g^{(\alpha,\beta)}(0) = g(0)$ and $E^{(\alpha,\beta)}$ and $F^{(\alpha,\beta)}$ are adjoint to each other in a suitable sense. More concretely, we define

$$(5.48) \qquad E_g^{(\alpha,\beta)}(t) := (\Delta_{\alpha,\beta}(t))^{-1} \int_0^t g(s) A_{\alpha,\beta}(s,t)\,ds,$$

$$(5.49) \qquad F_f^{(\alpha,\beta)}(s) := \int_s^\infty f(t) A_{\alpha,\beta}(s,t)\,dt,$$

where

$$(5.50) \qquad A_{\alpha,\beta}(s,t) := \frac{2^{3\alpha+3/2}\Gamma(\alpha+1)\,\mathrm{sh}\,2t}{\Gamma(\tfrac{1}{2})\Gamma(\alpha-\beta)\Gamma(\beta+\tfrac{1}{2})} \int_s^t (\mathrm{ch}\,2t-\mathrm{ch}\,2w)^{\beta-\frac{1}{2}} \cdot$$

$$\cdot \; (\mathrm{ch}\,w-\mathrm{ch}\,s)^{\alpha-\beta-1} \mathrm{sh}\,w\,dw, \; 0 < s < t, \alpha > \beta > -\tfrac{1}{2},$$

with degenerate cases

$$A_{\alpha,-\frac{1}{2}}(s,t) = \tfrac{1}{2} A_{\alpha,\alpha}(\tfrac{1}{2}s,\tfrac{1}{2}t) =$$

$$= \frac{2^{3\alpha+\frac{1}{2}}\Gamma(\alpha+1)}{\Gamma(\alpha+\frac{1}{2})\Gamma(\frac{1}{2})} \; \mathrm{sh}\,t(\mathrm{ch}\,t-\mathrm{ch}\,s)^{\alpha-\frac{1}{2}}, \; \alpha > -\tfrac{1}{2}.$$

Again we call the transform $f \mapsto F_f$ defined by (5.49) an <u>Abel</u> <u>transform</u>. It equals the classical Abel transform if $\alpha = \beta = 0$.

It follows from (5.42), (5.31), (5.44) that

(5.51) $g \mapsto E_g^{(\alpha,\beta)} : C_{even}^{\infty}(\mathbb{R}) \to C_{even}^{\infty}(\mathbb{R})$,

(5.52) $E_g^{(\alpha,\beta)}(0) = g(0)$,

(5.53) $E_{g''}^{(\alpha,\beta)} = (L_{\alpha,\beta} + (\alpha+\beta+1)^2 E_g^{(\alpha,\beta)}$.

In order to describe the mapping properties of $F^{(\alpha,\beta)}$ let us introduce, for σ in $\mathbb{R}$, the class

$$H_\sigma := \{f \in C^{\infty}((0,\infty)) | \forall n \in \mathbb{Z}_+, f^{(n)}(t) = \mathcal{O}(e^{-\sigma t}), t \to \infty\}.$$

Then it follows from (5.41), (5.31) that

(5.54) $f \mapsto F_f^{(\alpha,\beta)} : H_\sigma \to H_{\sigma-\alpha-\beta}$, $\sigma > \alpha + \beta + 1$,

(5.55) $(F_f^{(\alpha,\beta)})'' = F_{(L_{\alpha,\beta} + (\alpha+\beta+1)^2)f}$.

It follows from (5.43) that

(5.56) $F_f^{(\alpha,\beta)}(s) = c_{\alpha,\beta}(\lambda)(2\cosh s)^{-i\lambda}$

           if $f(t) = (2\cosh t)^{-i\lambda-\alpha-\beta-1}$,

where $c(\lambda)$ is given by (2.18) and $Im\lambda < 0$.

By combination of the above results about $f \mapsto F_f$ with (2.11) and the characterization of ϕ_λ as special solution of (2.10) we obtain the integral representation

(5.57) $\phi_\lambda(t) = 2(\Delta(t))^{-1} \int_0^t \cos\lambda s \, A(s,t)ds$,

which, in the case $\alpha = \beta = 0$, goes back to Mehler [103]. Similarly, if we combine the above results about $f \mapsto F_f$ with the characterization of Φ_λ as solution of (2.10) satisfying

(5.58) $\Phi_\lambda^{(\alpha,\beta)}(t) = e^{(i\lambda-\rho)t}(mod H_{Im\lambda+\rho+2})$,

(in view of (2.15)) then we obtain

$$(5.59) \qquad e^{i\lambda s} = \int_s^\infty \frac{\phi_\lambda(t)}{c(-\lambda)} A(s,t)dt, \quad \text{Im}\lambda > 0.$$

Different proofs of (5.57), (5.59) were given in [81,(2.16), (2.17)] by the use of fractional integrals for hypergeometric functions.

Formula (5.57) together with (5.45) yields again (2.19) for $\alpha \geq \beta \geq -\frac{1}{2}$. We now give another proof of (2.17). For $\lambda \notin i\mathbb{Z}$ we have $\phi_\lambda = a(\lambda)\Phi_\lambda + b(\lambda)\Phi_{-\lambda}$ for certain coefficients $a(\lambda)$, $b(\lambda)$. From (2.19) and (2.16) we find $a(\lambda) = c(\lambda)$ if $\text{Im}\lambda < 0$. By analyticity in λ and since $\phi_\lambda = \phi_{-\lambda}$ we have $b(\lambda) = a(-\lambda)$. This proves (2.17).

The kernel $A_{\alpha,\beta}(s,t)$ can be written as a hypergeometric function by making the substitution $\tau = (\text{ch } t - \text{ch } w)/(\text{ch } t - \text{ch } s)$ in (5.50) and by using Euler's integral [33,2.1(10)]:

$$(5.60) \qquad A_{\alpha,\beta}(s,t) = \frac{2^{3\alpha+2\beta+\frac{1}{2}}\Gamma(\alpha+1)}{\Gamma(\alpha+\frac{1}{2})\Gamma(\frac{1}{2})} \text{ sh}2t(\text{ch } t)^{\beta-\frac{1}{2}} \cdot$$

$$\cdot (\text{ch } t - \text{ch } s)^{\alpha-\frac{1}{2}} {}_2F_1(\tfrac{1}{2}+\beta, \tfrac{1}{2}-\beta; \alpha+\tfrac{1}{2}; \frac{\text{ch } t - \text{ch } s}{2\text{ch } t}).$$

Instead of in kernel form we can write the formulas (5.57), (5.49) also in a more group-like way. In (5.57), (5.50) make the two successive transformations of variables $(s,w) \mapsto (s,\chi) \mapsto (r,\psi)$ given by ch w = cosχ ch t and ch t + sh t r $e^{i\psi}$ = $e^{s+i\chi}$. The resulting formula is (5.28) (with left hand side replaced by $\phi_\lambda(t)$), now proved for $\alpha \geq \beta \geq -\frac{1}{2}$. A different proof was given in [41,p.150]. In the case of Legendre polynomials ($\alpha=\beta=0$) this integral representation goes back to Laplace. Next consider (5.49), (5.50). With the convention (5.13) formula (5.49) can be written as

$$F_f[x] = \frac{2^{3\alpha+\beta+1}\Gamma(\alpha+1)}{\Gamma(\frac{1}{2})\Gamma(\alpha-\beta)\Gamma(\beta+\frac{1}{2})} \cdot$$

$$\cdot \int_{z=x}^\infty \int_{y=z}^\infty f[y](y^2-z^2)^{\beta-\frac{1}{2}}(z-x)^{\alpha-\beta-1}y \, dy \, dz.$$

By making the transformation of variables $y = x((s^2+1)^2+t^2)^{\frac{1}{2}}$, $z = x(s^2+1)$ we obtain (5.29) for general $\alpha \geq \beta \geq -\frac{1}{2}$. Finally (5.30) follows from (5.28), (2.19), (5.29), (5.56).

In order to invert the Abel transform we introduce a version W_μ^τ (Reμ>0,τ>0) of Weyl's fractional integral transform:

(5.61) $(W_\mu^\tau f)(s) := \dfrac{1}{\Gamma(\mu)} \displaystyle\int_s^\infty f(t)(\mathrm{ch}\tau t - \mathrm{ch}\tau s)^{\mu-1} \, d\mathrm{ch}\tau t,$

where, for convenience, we assume that $f \in \mathcal{D}_{even}(\mathbb{R})$. Then it follows from (5.36) that $W_\mu^\tau f$ has an analytic continuation to all complex μ: if $n = 0,1,2,\ldots$ and $\mathrm{Re}\,\mu > -n$ then

(5.62) $(W_\mu^\tau f)(s) = ((-1)^n / \Gamma(\mu+n)) \cdot$

$\cdot \displaystyle\int_s^\infty \dfrac{d^n f(t)}{d(\mathrm{ch}\tau t)^n} (\mathrm{ch}\tau t - \mathrm{ch}\tau s)^{\mu+n-1} \, d\mathrm{ch}\tau t.$

It follows (again using (5.36)) that W_μ^τ has inverse $W_{-\mu}^\tau$ and that it is a bijection of $\mathcal{D}_{even}(\mathbb{R})$ onto itself.

For f in $\mathcal{D}_{even}(\mathbb{R})$ formula (5.49) can be rewritten as

(5.63) $F_f^{(\alpha,\beta)} = 2^{3\alpha+\frac{1}{2}} \pi^{-\frac{1}{2}} \Gamma(\alpha+1) W_{\alpha-\beta}^1 \circ W_{\beta+\frac{1}{2}}^2 (f),$

this formula has an analytic continuation to all complex α, β with $\alpha \neq -1,-2,\ldots$ and it can be inverted as

(5.64) $f = 2^{-3\alpha-\frac{1}{2}} \pi^{\frac{1}{2}} (\Gamma(\alpha+1))^{-1} W_{-\beta-\frac{1}{2}}^2 \circ W_{\beta-\alpha}^1 (F_f^{(\alpha,\beta)}).$

Let us summarize the various expressions for the Abel transform obtained until now:
(a) kernel form (5.49) with kernel given by integral representation (5.50) or by hypergeometric function (5.60);
(b) composition (5.63) of two fractional integrals;
(c) group-like form (5.29) with only essential integration variables being preserved;
(d) group form (5.14) using special structure of N;
(e) group form (5.3) which generalizes to arbitrary semi-simple G.
We might add:
(f) geometric form (cf. [66]) involving integration over horospheres: the so-called Radon transform.
A similar list can be made for the integral representation for ϕ_λ.

5.4. A generalization of the Abel transform

Let

(5.65) $(A_{\alpha,\beta;\gamma,\delta} f)(s) := \displaystyle\int_s^\infty f(t) A_{\alpha,\beta;\gamma,\delta}(s,t) \, dt, \quad s > 0,$

where

(5.66) $\quad A_{\alpha,\beta;\gamma,\delta}(s,t) := \dfrac{2^{2(\alpha+\beta-\gamma-\delta)}\Gamma(\alpha+1)}{\Gamma(\gamma+1)\Gamma(\alpha-\gamma)} \cdot$

$\quad\quad\quad \cdot \; \text{sh}2t(\text{ch }t)^{\gamma-\delta-\alpha+\beta}(\text{ch}^2 t - \text{ch}^2 s)^{\alpha-\gamma-1} \cdot$

$\quad\quad\quad \cdot \; {}_2F_1\left(\tfrac{1}{2}(\alpha+\beta-\gamma+\delta),\tfrac{1}{2}(\alpha-\beta-\gamma+\delta);\alpha-\gamma;1-\dfrac{\text{ch}^2 s}{\text{ch}^2 t}\right),$

$\alpha > \gamma$, $f \in C^\infty((0,\infty))$ and sufficiently rapidly decreasing.
Then it can be shown (unpublished work of the author) that

(5.67) $\quad (L_{\gamma,\delta}+(\gamma+\delta+1)^2)A_{\alpha,\beta;\gamma,\delta} = A_{\alpha,\beta;\gamma,\delta}(L_{\alpha,\beta}+(\alpha+\beta+1)^2),$

(5.68) $\quad (A_{\alpha,\beta;\gamma,\delta}f)(s) = \dfrac{(2\text{chs})^{i\lambda-\gamma-\delta-1}}{c_{\gamma,\delta}(-\lambda)}$ if

$\quad\quad\quad\quad\quad f(t) := \dfrac{(2\text{cht})^{i\lambda-\alpha-\beta-1}}{c_{\alpha,\beta}(-\lambda)} \quad,$

(5.69) $\quad A_{\alpha,\beta;\gamma,\delta}\left(\dfrac{\phi_\lambda^{(\alpha,\beta)}}{c_{\alpha,\beta}(-\lambda)}\right) = \dfrac{\phi_\lambda^{(\gamma,\delta)}}{c_{\gamma,\delta}(-\lambda)} \quad,$

(5.70) $\quad \phi_\lambda^{(\alpha,\beta)}(t)\Delta_{\alpha,\beta}(t) = \displaystyle\int_0^t \phi_\lambda^{(\gamma,\delta)}(s)\Delta_{\gamma,\delta}(s)A_{\alpha,\beta;\gamma,\delta}(s,t)ds.$

Formulas (5.68), (5.69) are valid for $\text{Im}\lambda > -\gamma + |\delta|-1$.
Formula (5.69) may be derived from (5.68) by using that

(5.71) $\quad \dfrac{\phi_\lambda^{(\alpha,\beta)}(t)}{c_{\alpha,\beta}(-\lambda)} = \displaystyle\sum_{k=0}^\infty \dfrac{(-i\lambda)_{2k}}{(-i\lambda+1)_k}\dfrac{(2\text{cht})^{i\lambda-2k-\alpha-\beta-1}}{c_{\alpha,\beta}(-\lambda-2ki)}$

(this follows from (2.15), (2.18)).

In view of [33,2.11(22)], (5.66) and (5.60) we have
$A_{\alpha,\beta;-\frac{1}{2},-\frac{1}{2}} = A_{\alpha,\beta}$, hence $A_{\alpha,\beta;-\frac{1}{2},-\frac{1}{2}} f = F_f$ and formulas
(5.67)-(5.70) generalize (5.55), (5.56), (5.59) and (5.57).

Sprinkhuizen-Kuyper [125,(3.1)] defines a generalized
fractional integral operator $I_\nu^{\mu,\lambda}$ which operates on $C((0,1])$,
but which can immediately be extended to an action on sufficiently rapidly decreasing continuous functions on $(0,\infty)$.
Then

$$A_{\alpha,\beta;\gamma,\delta}(f\circ ch) = 2^{2(\beta-\delta)+3(\alpha-\gamma)}\frac{\Gamma(\alpha+1)}{\Gamma(\gamma+1)} \cdot$$

$$\cdot\ (I_{2\delta+1}^{\alpha-\beta-\gamma+\delta,\beta-\delta}f)\circ ch$$

and the composition property [125,(3.4)] can be translated as

(5.72) $$A_{\alpha_2,\beta_2;\alpha_3,\beta_3}\ A_{\alpha_1,\beta_1;\alpha_2,\beta_2} = A_{\alpha_1,\beta_1;\alpha_3,\beta_3}\ .$$

Like in [125,§3], analytic continuation of the operator (5.65) with respect to $\alpha,\beta,\gamma,\delta$ is possible by use of (5.72) and

(5.73) $$A_{\alpha,\beta;\alpha+1,\beta+1} = -2^{-5}(\alpha+1)^{-1}(ch\ t)^{-1}d/d(ch\ t),$$

(5.74) $$A_{\alpha,\beta;\alpha+2,\beta} = \frac{1}{2^6(\alpha+1)(\alpha+2)}\left(\left(\frac{d}{d(cht)}\right)^2 + \frac{2\beta+1}{cht}\frac{d}{d(cht)}\right).$$

The kernel (5.66) simplifies if $\alpha - \gamma = \beta + \delta$, $\beta - \delta$, $-\beta + \delta$ or $-\beta -\delta$. The kernel degenerates completely if $\alpha = \gamma$, $\beta = - \delta$ and then (5.69), (5.70) give rise to the symmetries

(5.75) $$\phi_\lambda^{(\alpha,\beta)}(t) = (cht)^{-2\beta}\phi_\lambda^{(\alpha,-\beta)}(t),$$

(5.76) $$\Phi_\lambda^{(\alpha,\beta)}(t) = (2cht)^{-2\beta}\ \Phi_\lambda^{(\alpha,-\beta)}(t),$$

which can also be derived from (2.4), (2.15).

It would be of interest to find integral transforms on groups which give interpretations of (5.65). One possibility would be to consider (5.3) with N replaced by some suitable subgroup of N. Another possibility is in Badertscher [6,§5], where formulas (3) and (14) give interpretations of our formulas (5.70) and (5.65) as a passage from K-biinvariant functions to left-K, right-H-invariant functions on $G(G = O(1,n), K = O(1)\times O(n), H = O(1,n-1)\times O(1))$.

5.5. Notes

The Abel transform (5.3) can also be considered in the case of higher rank. It is an interesting open problem to find analogues of (5.63) in those cases, maybe related to fractional integrals in several variables, and to find an explicit inversion formula. Partial answers to the inversion

problem in higher rank are given by Gindikin & Karpelevic
[56], Helgason [66, Theorem 2.6] and Aomoto [4]. Flensted-
Jensen & Ragozin [51] were motivated by the structure of
(5.57) to prove that also in the higher rank case spherical
functions $\phi_\lambda(a)$, considered as function of λ, are Fourier
transforms of L^1-functions.

Flensted-Jensen [45] considers integral transforms between function spaces on a complex semisimple Lie group and
on its normal real form. The special case $SL(2,\mathbb{C})$ yields a
pair of integral transforms connecting Jacobi functions of
order $(\frac{1}{2},\frac{1}{2})$ and $(0,0)$ (cf.[45,(10.4),(10.5)]).

Flensted-Jensen [42,§3] gives an analogue of the Abel
transform (5.3) for left-K, right-H invariant functions,
where (G,K,H) are certain triples as in §3.4, 4.5. In particular, this gives an interpretation in the Jacobi cases of
order (α,β), $\alpha \geq \beta \geq -\frac{1}{2}$, $\alpha,\beta \in \frac{1}{2}\mathbb{Z}$.

Lions [96,Ch.12] finds an analogue of the dual Abel
transform (5.48) in the case of a differential operator
$L+q$, $q \in E_{even}(\mathbb{R})$, L given by (2.8), $t \mapsto t^{-2\alpha-1}\Delta(t)$ in
$E_{even}(\mathbb{R})$ but not necessarily positive, $\alpha \in \mathbb{C}$ but $-\alpha \notin \mathbb{N}$.
He proves the existence of a unique continuous bijection E
of $E_{even}(\mathbb{R})$ such that $(Ef)(0) = f(0)$, $Ef'' = (L+q)(Ef)$.
For the proof he considers the hyperbolic p.d.e.

$$(5.77) \qquad (L_t+q(t)-\partial^2/\partial s^2)v(s,t) = 0.$$

Chébli [21], [22], [24] and Trimèche [137] have analogues
of the integral representation (5.57) under their conditions
(2.35) and eventually (2.36). Chébli [21], [22] uses (5.77)
for its derivation. Moreover, in view of (2.36), he can
apply a maximum principle and he thus obtains the positivity
of the kernel. Chébli [24] and Trimèche [137] obtain their
integral representation from asymptotics of ϕ_λ in terms of
Bessel functions and from properties of the Fourier transform. (Conversely, estimates for ϕ_λ can be derived from its
integral representation, cf.§6.) Trimèche [137] obtains
analogues of the Abel transform (5.49) and its dual (5.48)
from the integral representation.

Carroll [18] uses the Jacobi function example as a
model for a general theory of transmutation operators.

6. PROOF OF THE PALEY-WIENER AND PLANCHEREL THEOREM

This section contains proofs of the two above-mentioned
Theorems 2.1 and 2.3 by use of the Abel transform. We start

with some estimates which will be needed in the proofs and
we end with some notes. The papers [41] and [81] can be used
as a reference for this section

Assume that $\alpha \geq \beta \geq -\frac{1}{2}$. From (4.4), (5.10) or (5.28) it
follows that $|\phi_{\mu+i\nu}(t)| \leq |\phi_{i\nu}(t)| (\mu,\nu\in\mathbb{R})$, $\phi_{-i\rho}(t) \equiv 1$,
$\phi_{i\rho}(t) \equiv 1$ (since $\phi_\lambda = \phi_{-\lambda}$) and that $\nu \mapsto \phi_{i\nu}$ is a convex
function on $\mathbb{R}$ for all real t. It follows that

(6.1) $|\phi_\lambda(t) \leq 1$ if $|\mathrm{Im}\lambda| \leq \rho$, $t \in \mathbb{R}$.

By combination of this result with (2.19) we obtain that
ϕ_λ is bounded iff $|\mathrm{Im}\lambda| \leq \rho$. From (5.28) we see that

(6.2) $|\phi_\lambda(t)| \leq e^{t|\mathrm{Im}\lambda|}\phi_0(t)$, $t \geq 0$.

Now $\phi_0(t)$ can be written as a linear combination of two solu-
tions of $(L+\rho^2)u = 0$ behaving like $e^{-\rho t}$ and $te^{-\rho t}$ as $t \to \infty$
(cf.[33,Ch.2]). Hence, for some $C > 0$:

(6.3) $|\phi_\lambda(t)| \leq C(1+t)e^{t(|\mathrm{Im}\lambda|-\rho)}$for all $t \geq 0$, $\lambda \in \mathbb{C}$.

See [41, Lemmas 14,15], [81, Lemma 2.3], [101,(2.8)] for
estimates of derivatives of $\phi_\lambda(t)$.

Let now α,β be arbitrarily complex $(\alpha\neq-1,-2,\ldots)$. Observe that $(c(-\lambda))^{-1}$ has only finitely many poles for
$\mathrm{Im}\lambda \geq 0$ (none if $\mathrm{Im}\lambda \geq 0$ and $\mathrm{Re}(\alpha\pm\beta+1)>0$). Then an application of Stirling's formula [33,1.18(2)] shows that for each
$r > 0$ there is $C_r > 0$ such that

(6.4) $|c(-\lambda)|^{-1} \leq C_r(1+|\lambda|)^{\mathrm{Re}\alpha+\frac{1}{2}}$ if $\mathrm{Im}\lambda \geq 0$ and

           $c(-\mu) \neq 0$ for $|\mu-\lambda| \leq r$.

Finally we need the following estimate for ϕ_λ: For each
$\delta > 0$ there is $C_\delta > 0$ such that

(6.5) $|\phi_\lambda(t)| \leq C_\delta e^{-(\mathrm{Im}\lambda+\mathrm{Re}\rho)t}$ if $t \geq \delta \geq 0$, $\mathrm{Im}\lambda \geq 0$.

See Flensted-Jensen [41,pp.150-152] for a proof, which is
analogous to Harish-Chandra's [61] proof in the group case
(general rank). It proceeds by deriving a recurrence relation for the coefficients $\Gamma_m(\lambda)$ in the expansion

(6.6) $\phi_\lambda(t) = e^{(i\lambda-\rho)t} \sum_{m=0}^{\infty} \Gamma_m(\lambda)e^{-mt}$

and proving that $\Gamma_m(\lambda)$ is of at most polynomial growth in m, uniformly in λ. (Stanton & Thomas [127] give more precise estimates for $\Gamma_m(\lambda)$.)

Let $\alpha \geq \beta \geq -\frac{1}{2}$. In view of (5.57), (5.49) the Jacobi transform $f \mapsto \hat{f}$ defined by (2.12) factorizes as in (5.1) if, for instance, $f \in C_c(\mathbb{R})$. In the spherical rank one case this was already observed in (5.4).

We now prove the Paley-Wiener Theorem 2.1. The operator W_μ^τ defined by (5.61) can be shown to be an isomorphism of topological vector spaces in the two following cases:

$$W_\mu^\tau: \mathcal{D}_{even}(\mathbb{R}) \to \mathcal{D}_{even}(\mathbb{R}),$$

$$W_\mu^\tau: (ch\ t)^{-\sigma}S_{even}(\mathbb{R}) \to (ch\ t)^{-\sigma+\tau\mu}S_{even}(\mathbb{R})$$

$$(\sigma \geq \tau\mu \geq 0).$$

Thus, in view of (5.63), we have isomorphisms

(6.7) $f \mapsto F_f: \mathcal{D}_{even}(\mathbb{R}) \to \mathcal{D}_{even}(\mathbb{R}),$

(6.8) $f \mapsto F_f: (cht)^{-\sigma}S_{even}(\mathbb{R}) \to (cht)^{-\sigma+\rho}S_{even}(\mathbb{R})$

($\sigma \geq \rho$ in (6.8)). It follows from (6.7), (6.8), (5.1) and standard mapping properties of F that the Paley-Wiener Theorem 2.1 holds for $\alpha \geq \beta \geq -\frac{1}{2}$ and also

Theorem 6.1. For $\alpha \geq \beta \geq -\frac{1}{2}$, $\sigma \geq \rho$ the Jacobi transform is a 1-1 map of $(ch\ t)^{-\sigma}S_{even}(\mathbb{R})$ onto the space of even C^∞-functions g on $\{\lambda \in C \mid |Im\lambda| \leq \sigma-\rho\}$, holomorphic in its interior and satisfying

$$\sup_{|Im\lambda| \leq \sigma-\rho} (1+|\lambda|)^n |g^{(m)}(\lambda)| < \infty, \quad n,m \in \mathbb{Z}_+.$$

Flensted-Jensen [41, Theorem 4] shows that the bijections in Theorems 2.1,6.1 are homeomorphisms and that, in Theorem 2.1, there is also a bijection between the dual spaces.

Theorem 2.1 can similarly be proved for general complex $\alpha,\beta(\alpha \neq -1,-2,..)$ by use of the analytic continuation of (5.1) with respect to α,β (use (5.63), (5.62), (2.13)). Note that Theorem 2.1 both generalizes and is derived from the classical Paley-Wiener theorem, cf. for instance Rudin [118, Theor.7.22].

Next let us prove the inversion formulas (2.21), (2.25) for the Jacobi transform in the case $\alpha \geq \beta \geq -\frac{1}{2}$. Let $g(t)$ denote the right hand side of (2.25). Because of Theorem 2.1, (6.3), (6.4) g is well-defined and continuous. Observe that $\overline{c(\lambda)} = c(-\lambda)$ if $\lambda \in \mathbb{R}$ and use (2.17). Then, for $t > 0$,

$$g(t) = \frac{1}{2\pi} \int_{-\infty}^{\infty} \hat{f}(\lambda) \frac{\Phi_\lambda(t)}{c(-\lambda)} \, d\lambda$$

$$= \frac{1}{2\pi} \int_{-\infty}^{\infty} \hat{f}(\lambda+i\mu) \frac{\Phi_{\lambda+i\mu}(t)}{c(-\lambda-i\mu)} \, d\lambda, \quad \mu \geq 0,$$

because of the estimate

$$\left| \hat{f}(\lambda+i\mu) \frac{\Phi_{\lambda+i\mu}(t)}{c(-\lambda-i\mu)} \right| \leq CC_{\hat{f},n} \, C_\delta (1+|\lambda+i\mu|)^{-n+\alpha+\frac{1}{2}} .$$

$$\cdot \, e^{(A_{\hat{f}}-t)\mu-\rho t}, \quad t \geq \delta > 0, \lambda \in \mathbb{R}, \mu \geq 0, n \in \mathbb{Z}_+$$

(use Theorem 2.1, (6.4), (6.5)). Hence $g(t) = 0$ if $t > A_{\hat{f}}$ so $g \in C_{c,\text{even}}(\mathbb{R})$. Now, by injectivity, $f - g$ will follow from $F_f = F_g$. We have

$$F_g(s) = \frac{1}{2\pi} \int_s^{\infty} \left(\int_{-\infty}^{\infty} \hat{f}(\lambda+i\mu) \frac{\Phi_{\lambda+i\mu}(t)}{c(-\lambda-i\mu)} \, d\lambda \right) A(s,t) dt, \mu > 0, \; s > 0.$$

The above estimate together with an estimate for $A(s,t)$ following from (5.50) allows us to apply Fubini's theorem. Combination with (5.59) yields

$$F_g(s) = \frac{1}{2\pi} \int_{-\infty}^{\infty} \hat{f}(\lambda+i\mu) e^{(i\lambda-\mu)s} d\lambda = \frac{1}{2\pi} \int_{-\infty}^{\infty} \hat{f}(\lambda) e^{i\lambda s} d\lambda .$$

Thus, because of (5.1) and the inversion formula to the classical Fourier transform we obtain $F_f = F_g$ and we have proved (2.21) and (2.25).

By a slight adaptation of the above reasoning (cf. [82,§4]) Theorem 2.2 can be shown to hold for general α, β. Then Theorem 2.3 follows from Theorem 2.2 by moving μ to 0 in (2.21) and by taking account of the poles of the integrand which are passed. Formula (2.27) follows from (2.24) and (2.12), while the extension of (2.27) to an isometry of L^2-spaces is unique because of Theorem 2.1.

There is a close relationship between the Paley-Wiener and the Plancherel theorem, but there are many possibilities

for the order of their proofs. For instance, Flensted-Jensen
[41] first proves the Plancherel theorem and then uses it to
prove the Paley-Wiener theorem, but in the proofs given in
[81] and in the present section we have the converse order,
while in Rosenberg's [117] proof for the spherical case of
general rank (see also Helgason [72,Ch.4,§7], [71,§4.2])
ingredients of proofs of both theorems follow each other in
logical order.

The Paley-Wiener theorem was first proved by Ehrenpreis
& Mautner [31], [32] for $SL(2,\mathbb{R})$, respectively in the
spherical case and in the case of arbitrary double K-type,
by Helgason [65] in the spherical rank one case, by Gangolli
[53] in the spherical case of general rank, by Helgason
[69, Cor.10.2] for arbitrary K-types on rank one spaces G/K.
Chébli [21], [22], [23] and Trimèche [137] obtain a Paley-
Wiener theorem under their more general conditions (2.35),
(2.36). Nussbaum [110] announces without proof that for
positive Δ satisfying (2.36) but not necessarily (2.35) still
a Paley-Wiener theorem can be proved.

7. CONVOLUTION

In this section the convolution structure associated with the
Jacobi transform is discussed. A reference is [48]. We start
in §7.1 with the hardware, i.e. explicit formulas for the
product of two Jacobi functions and for the generalized
translation, both in the spherical rank one case and in the
analytic case. In §7.2 the corresponding harmonic analysis
is treated: the Jacobi transform of LP-functions, the convolution of two LP-functions and the Kunze-Stein phenomenon.
This subsection ends with some notes.

7.1. Product formulas and generalized translation

Let $G = U(1,n;\mathbb{F})$ as in §3.1. Remember that the spherical
function ϕ_λ on G of argument a_t equals the Jacobi function
$\phi_\lambda = \phi_\lambda^{(\alpha,\beta)}$ of argument t, for suitable order (α,β). Now
apply the product formula (3.6) to the spherical function ϕ_λ
with $x = a_s$, $y = a_t$ and use (5.12). Then for $k = \begin{pmatrix} u & 0 \\ 0 & V \end{pmatrix}$ in
$K = U(1,\mathbb{F}) \times U(n,\mathbb{F})$ the integrand of (3.6) becomes

$$\phi_\lambda[|u \text{ chs ch } t + V_{nn} \text{ sh } s \text{ sh } t|].$$

Hence, just as we got (5.10) and (5.28) we can rewrite (3.6)
in terms of Jacobi functions as

(7.1) $\phi_\lambda(s)\phi_\lambda(t) = \int_{S(\mathbb{F}^n)} \phi_\lambda[|ch\ s\ ch\ t + y_n\ sh\ s\ sh\ t|]dy$

(7.2) $= \int_0^1 \int_0^\pi \phi_\lambda[|ch\ s\ ch\ t + r\ e^{i\psi}sh\ s\ sh\ t|]dm(r,\psi).$

If $f_1, f_2 \in \mathcal{D}(G//K)$ then

(7.3) $(f_1 * f_2)(a_t) = \int_G f_1(y)f_2(y^{-1}a_t)dy$

$= \int_G f_1(y)\left(\int_K f_2(y^{-1}ka_t)dk\right)dy$

$= \int_0^\infty f_1(a_s)(T_s f_2)(a_t)\Delta(s)ds,$

where, for $f \in C^\infty(G//K)$,

(7.4) $(T_s f)(a_t) := \int_K f(a_s ka_t)dk$

$= \int_{S(\mathbb{F}^n)} f[|ch\ s\ ch\ t + y_n sh\ s\ sh\ t|]dy$

$= \int_0^1 \int_0^\pi f[|ch\ s\ ch\ t + r\ e^{i\psi}sh\ s\ sh\ t|]dm(r,\psi)$

$T_s f$ is called the generalized translate of f. Easy group theoretical arguments show that $(T_s f)(a_t)$ is symmetric in s and t and that

(7.5) $\begin{cases} (L_{(s)} - L_{(t)})(T_s f)(a_t) = 0, \\[2mm] T_0 f = f, \quad \frac{\partial}{\partial s}(T_s f)(a_t)|_{s=0} = 0. \end{cases}$

Furthermore, from the definition of spherical functions (§3.2) together with (4.7) we obtain

(7.6) $(f_1 * f_2)^\wedge(\lambda) = \hat{f}_1(\lambda)\hat{f}_2(\lambda).$

Also remember the homomorphism property (5.4) of the Abel transform.

Let us now extend the above results to other values of

α,β $(\alpha\geq\beta\geq-\frac{1}{2})$, without using group theory. First we sketch a proof of (7.2) (cf.[48,§4]). Observe that (5.28) implies that

$$(7.7) \qquad (\tfrac{1}{2}(\mathrm{ch}2s+\mathrm{ch}2t))^{i\lambda-\rho}\phi_\lambda\left[\frac{2^{\frac{1}{2}}\mathrm{ch\ s\ ch\ t}}{(\mathrm{ch}2s+\mathrm{ch}2t)^{\frac{1}{2}}}\right] =$$

$$= \int_0^1\int_0^\pi |\mathrm{ch\ s\ ch\ t} + r\ e^{i\psi}\mathrm{sh\ s\ sh\ t}|^{i\lambda-\rho}\,dm\ (r,\psi).$$

Next it can be shown (cf.[48, Theorem 4.2]) that

$$(7.8) \qquad \Phi_\lambda(s)\phi_\lambda(t) = \sum_{n=0}^\infty A_n(\tfrac{1}{2}(\mathrm{ch}2s+\mathrm{ch}2t))^{i\lambda-\rho-n}.$$

$$\cdot\phi_{\lambda+in}\left[\frac{2^{\frac{1}{2}}\mathrm{ch\ s\ ch\ t}}{(\mathrm{ch}2s+\mathrm{ch}2t)^{\frac{1}{2}}}\right], \quad s > t \geq 0,$$

where the coefficients A_n are the ones occurring in

$$(7.9) \qquad \Phi_\lambda(s) = \sum_{n=0}^\infty A_n(\tfrac{1}{2}(\mathrm{ch}2s+1))^{i\lambda-\rho-n}, \quad s > 0.$$

A similar expansion was independently obtained by Bellandi Fo & Capelus de Oliveira [9]. Combination of (7.7), (7.8), (7.9) yields

$$(7.10) \quad \Phi_\lambda(s)\phi_\lambda(t) = \int_0^1\int_0^\pi \Phi_\lambda[|\mathrm{ch\ s\ ch\ t} + r\ e^{i\psi}\mathrm{sh\ s\ sh\ t}|]dm(r,\psi).$$

Finally, (7.10) together with (2.17) yields (7.2).

In §5.2 we described how to pass from the kernel form (5.57) of the integral reprsentation for ϕ_λ to the group-like form (5.28). We can use a similar change of integration variables in converse direction, namely $e^{i\chi}\mathrm{ch\ u} = \mathrm{ch\ s\ ch\ t} + r\ e^{i\psi}\mathrm{sh\ s\ sh\ t}$, in order to derive from the product formula (7.2) the kernel form

$$(7.11) \qquad \phi_\lambda(s)\phi_\lambda(t) = \int_0^\infty \phi_\lambda(u)K(s,t,u)\Delta(u)\,du,$$

where

$$(7.12) \qquad K(s,t,u) := \frac{2^{1-2\rho}\Gamma(\alpha+1)}{\pi^{\frac{1}{2}}\Gamma(\alpha-\beta)\Gamma(\beta+\frac{1}{2})}\ (\mathrm{sh\ s\ sh\ t\ sh\ u})^{-2\alpha}\cdot$$

$$\cdot\int_0^\pi (1-\mathrm{ch}^2s-\mathrm{ch}^2t-\mathrm{ch}^2u+2\mathrm{ch\ s\ ch\ t\ chu\ cos}\chi)_+^{\alpha-\beta-1}\cdot$$

$$\cdot(\sin\chi)^{2\beta}d\chi$$

if $|s-t| < u < s+t$ and $K(s,t,u) := 0$ otherwise, and
$x_+^\gamma := x^\gamma$ if $x > 0$ and 0 otherwise. By the use of Euler's integral representation 2.1(10), formula (7.12) can be re-
written as

(7.13) $K(s,t,u) = \dfrac{2^{-2\rho}\Gamma(\alpha+1)(\text{ch } s \text{ ch } t \text{ ch } u)^{\alpha-\beta-1}}{\pi^{\frac{1}{2}}\Gamma(\alpha+\frac{1}{2})(\text{sh } s \text{ sh } t \text{ sh } u)^{2\alpha}}$.

$\cdot (1-B^2)^{\alpha-\frac{1}{2}} \; _2F_1(\alpha+\beta,\alpha-\beta;\alpha+\frac{1}{2};\frac{1}{2}(1-B))$, $|s-t| < u < s+t$,

where

(7.14) $B := \dfrac{\text{ch}^2 s + \text{ch}^2 t + \text{ch}^2 u - 1}{2\text{ch } s \text{ ch } t \text{ ch } u}$.

Observe that K is nonnegative and symmetric in its three
variables. From (7.11) with $\lambda = i\rho$ we obtain that

(7.15) $\displaystyle\int_0^\infty K(s,t,u)\Delta(u)\,du = 1$.

　　　　The generalized translate $T_s f$ of a function f (in $\mathcal{D}_{even}$
(ℝ) , for convenience), is defined by

(7.16) $(T_s f)(t) := \displaystyle\int_0^\infty f(u)K(s,t,u)\Delta(u)\,du$

$= \displaystyle\int_0^1\int_0^\pi f[|\text{ch } s \text{ ch } t + r e^{i\psi}\text{sh } s \text{ sh } t|]\,dm(r,\psi)$.

Then, obviously, $T_s f \in \mathcal{D}_{even}(\mathbb{R})$, $T_s f \geq 0$ if $f \geq 0$ and, by
use of (7.11),

(7.17) $(T_s f)^\wedge(\lambda) = \phi_\lambda(s)\hat{f}(\lambda)$.

Also $(T_s f)(t) = (T_t f)(s)$ is C^∞ in (s,t), $T_0 f = f$, $T_s f = T_{-s} f$
and, by (7.17),

(7.18) $(L_{(s)} - L_{(t)})(T_s f)(t) = 0$.

For $f,g \in \mathcal{D}_{even}(\mathbb{R})$ define the convolution product $f*g$ by

(7.19) $(f*g)(t) := \displaystyle\int_0^\infty (T_t f)(s)g(s)\Delta(s)\,ds =$

$= \displaystyle\int_0^\infty\int_0^\infty f(r)g(s)K(r,s,t)\Delta(r)\Delta(s)\,dr\,ds$.

Then $f*g = g*f \in \mathcal{D}_{even}(\mathbb{R})$, $f*g \geq 0$ if $f,g \geq 0$ and, by (7.17),

$$(7.20) \qquad (f*g)^\wedge(\lambda) = \hat{f}(\lambda)\hat{g}(\lambda).$$

From (7.20) we conclude that the convolution product is associative and that

$$(7.21) \qquad F_{f*g} = F_f*F_g.$$

7.2. Harmonic analysis

Let $\alpha \geq \beta \geq -\frac{1}{2}$, $(\alpha,\beta) \neq (-\frac{1}{2},-\frac{1}{2})$. Write L^p for $L^p(\mathbb{R}_+ ,\Lambda(t)dt)$ and $\|f\|_p$ for the L^p-norm of f. We will first discuss the mapping properties of the Jacobi transform on L^p, cf. [48, §3]. Let $1 \leq p < 2$ and take q such that $p^{-1} + q^{-1} = 1$. Let

$$(7.22) \qquad D_p := \{\lambda \in \mathbb{C} \mid |Im\lambda| < (2p^{-1}-1)\rho\}.$$

It follows from (6.3) and (6.1) that

$$(7.23) \qquad \|\phi_\lambda\|_q < \infty \quad \text{if} \quad \lambda \in D_p, \quad 1 < p \leq 2$$

$$(7.24) \qquad \|\phi_\lambda\|_\infty < \infty \quad \text{if} \quad \lambda \in \bar{D}_1.$$

Thus, by Hölder's inequality we obtain from (2.12) that

$$(7.25) \qquad |\hat{f}(\lambda)| \leq \|f\|_p\|\phi_\lambda\|_q \quad \text{if } f \in L^p, \lambda \in D_p, \; 1 < p \leq 2,$$

$$(7.26) \qquad |\hat{f}(\lambda)| \leq \|f\|_1, \quad f \in L^1, \lambda \in \bar{D}_1.$$

By the analyticity of $\phi_\lambda(t)$ in λ we conclude that f is holomorphic in the strip D_p if $f \in L^p$. Moreover, if f is in L^1 then f is continuous on $\bar{D}_1$ and a similar argument as for the classical Fourier transform shows that $f(\lambda) \to 0$ as $Re\lambda \to \pm\infty$, uniformly on $\bar{D}_1$.

Finally we have that the Jacobi transform is injective on L^p ($1 \leq p \leq 2$), cf. [48, Theorem 3.2] for the proof. It uses (2.27) for $f \in L^p \cap L^2$, $g \in \mathcal{D}_{even}$.

Let us now discuss the convolution product of L^p-functions. Definition (7.16) of T_sf can still be used if $f \in L^p$ ($1 \leq p \leq \infty$). An application of Hölder's inequality together with (7.15) then shows that $T_sf \in L^p$ and

$$(7.27) \qquad \|T_sf\|_p \leq \|f\|_p.$$

Next, definition (7.19) of $f*g$ remains valid if $f \in L^p$, $g \in L^q$ such that $1 \le p,q,r \le \infty$ and $p^{-1} + q^{-1} -1 = r^{-1}$. Then, by standard techniques, $f*g \in L^r$ and

$$(7.28) \qquad \| f*g \|_r \le \| f \|_p \| g \|_q .$$

However, because of (7.25) we can do better than in (7.28). Observe that for $f,g \in \mathcal{D}_{even}(\mathbb{R})$, $1 \le p < 2$, $p^{-1}+q^{-1} = 1$ we have

$$(2\pi)^{\frac{1}{2}} \| f*g \|_2 = \| \hat{f} \cdot \hat{g} \|_2 \le \| \hat{g} \|_\infty \| \hat{f} \|_2 \le$$

$$\le \| f \|_2 \| g \|_p \, \sup_{\lambda \in \mathbb{R}} \| \phi_\lambda \|_q = \| f \|_2 \| g \|_p \| \phi_0 \|_q .$$

Thus, for some $A_p > 0$ we have

$$(7.29) \qquad \| f*g \|_2 \le A_p \| f \|_2 \| g \|_p , \quad f \in L^2, \ g \in L^p, \ 1 \le p < 2.$$

This phenomenon was first discovered by Kunze & Stein [90] on $SL(2,\mathbb{R})$. From (6.24) it can be derived (cf. [48, Theor. 5.5]) that

$$(7.30) \qquad \| f*g \|_q \le A_q \| f \|_2 \| g \|_2 , \quad f,g \in L^2, \ 2 < q \le \infty.$$

For more general Δ satisfying (2.35) and eventually (2.36) product formula, generalized translation and convolution were treated by Chébli [20], [21], Trimèche [137]. Chébli obtains a positive convolution kernel, by applying a maximum principle to (7.18). In [21] he also gets a Kunze-Stein phenomenon. Trimèche [137] gets his convolution structure by transplantation from the case $\Delta(t) \equiv 1$, by using the Abel transform. Braaksma & de Snoo [17] and Markett [98] discuss generalized translation with respect to operators $L = d^2/dt^2 + (2\alpha+1)t^{-1} \, d/dt + q(t)$ for certain potentials q. They get estimates for the generalized translation operator by using the Riemann function for the p.d.e. $(L_s-L_t)v(s,t) = 0$.

Flensted-Jensen [42] gives a group theoretic interpretation of the convolution structure for the Jacobi transform if $\alpha \ge \beta \ge -\frac{1}{2}$, $\alpha,\beta \in \frac{1}{2}\mathbb{Z}$. This is done in terms of K-H-invariant functions on $G = O(p,q)$. In very interesting work Badertscher [6,§5] transplants convolution for K-biinvariant functions on $G = O(p,1)$ to convolution for K-H-invariant functions by using his Abel-type transform.

8. ADDITION FORMULA, POSITIVE DEFINITE SPHERICAL FUNCTIONS AND DUAL CONVOLUTION STRUCTURE

In §3.3 we already derived a group theoretic version (3.25) of the addition formula for Jacobi functions. Here an analytic version of the addition formula will be obtained and it will be shown how this formula follows from (3.25) for special α, β. Next follow two applications of the addition formula: the examination of positive definiteness of spherical functions and the occurrence of a positive dual convolution structure. Both applications are connected with the positivity of the expansion coefficients in the addition formula of ϕ_λ for certain λ. The reference for the group theoretic derivation of the addition formula and for the positive definite spherical functions will be [50], while the analytic derivation of the addition formula and the treatment of the dual convolution structure is based on [49].

8.1. The addition formula, analytic form

In order to obtain analytic versions of (3.24), (3.25) for general α, β $(\alpha > \beta > -\frac{1}{2})$ we have to expand

$$(8.1) \qquad |\text{ch } t - r\, e^{i\psi} \text{sh } t|^{i\lambda - \rho}$$

and

$$(8.2) \qquad \phi_\lambda[|\text{ch } s \text{ ch } t - r\, e^{i\psi} \text{sh } s \text{ sh } t|],$$

respectively, as functions of $r\, e^{i\psi}$ on the upper half unit disk. This expansion should be in terms of some complete system of functions which are orthogonal with respect to the measure $dm(r, \psi)$ and which generalize the spherical functions for (K, M) (cf. the group theoretic derivation of (5.28) from (4.4) and of (7.2) from (3.6)). The most appropriate choice of the orthogonal system is given by the functions $\chi_{k,\ell}^{(\alpha, \beta)}$ $(\alpha > \beta > -\frac{1}{2}, k, \ell \in \mathbb{Z}, k \geq \ell \geq 0)$ defined by

$$(8.3) \qquad \chi_{k,\ell}^{(\alpha, \beta)}(r, \psi) :=$$

$$:= R_\ell^{(\alpha - \beta - 1, \beta + k - \ell)}(2r^2 - 1) r^{k - \ell} R_{k - \ell}^{(\beta - \frac{1}{2}, \beta - \frac{1}{2})}(\cos\psi),$$

where $R_n^{(\alpha, \beta)}$ is the renormalized Jacobi polynomial (2.3). They are polynomials in the two variables $r^2, r\cos\psi$, orthogonal with respect to the measure $dm_{\alpha,\beta}$. It can be computed that

(8.4) $\pi_{k,\ell}^{(\alpha,\beta)} := \left(\int_0^1 \int_0^\pi (\chi_{k,\ell}^{(\alpha,\beta)} (r,\psi))^2 dm_{\alpha,\beta} (r,\psi) \right)^{-1} =$

$$= \frac{(2k-2\ell+\beta)(k+\ell+\alpha)(\alpha-\beta)_\ell (2\beta+1)_{k-\ell}(\alpha+1)_k}{(k-\ell+2\beta)(k+\alpha)\ell!(k-\ell)!(\beta+1)_k}.$$

The associated Jacobi functions $\phi_{\lambda,k,\ell}^{(\alpha,\beta)}$ were defined by (4.15). Now we can state the analytic versions of (3.24), (3.25):

Theorem 8.1. Let $\alpha > \beta > -\frac{1}{2}$, then

(8.5) $|\text{ch } t - r\, e^{i\psi} \text{sh } t|^{i\lambda-\rho} =$

$$= \sum_{k=0}^\infty \sum_{\ell=0}^k \phi_{\lambda,k,\ell}^{(\alpha,\beta)}(t)\, \pi_{k,\ell}^{(\alpha,\beta)}\, \chi_{k,\ell}^{(\alpha,\beta)}(r,\psi),$$

(8.6) $\phi_\lambda^{(\alpha,\beta)}[|\text{ch } s \text{ ch } t - r\, e^{i\psi}\text{sh } s \text{ sh } t|] =$

$$= \sum_{k=0}^\infty \sum_{\ell=0}^k \phi_{\lambda,k,\ell}^{(\alpha,\beta)}(s)\, \phi_{\lambda,k,\ell}^{(\alpha,\beta)}(t)\, \pi_{k,\ell}^{(\alpha,\beta)}\, \chi_{k,\ell}^{(\alpha,\beta)}(r,\psi).$$

The double series in (8.6), (8.7) converge absolutely, uniformly for (s,t,r,ψ) in compact subsets of their domain.

For $\alpha = \beta$ or $\beta = -\frac{1}{2}$ the theorem remains valid if one puts $r = 1$ or $\phi = 0, \pi$, respectively in (8,5), (8.6). Then both expansions degenerate to a single series ($\ell=0$ if $\alpha=\beta$, $k-\ell=0$ or 1 if $\beta=-\frac{1}{2}$). The case $\alpha = \beta = 0$ was proved in [145,§15.71] and the case $\alpha = \beta$ by Henrici [73,(80)].

In order to prove Theorem 8.1 we imitate the proof in [82] of the addition formula for Jacobi polynomials. First we show that the expansions formally hold, i.e. that

(8.7) $\int_0^1 \int_0^\pi |\text{ch } t - r\, e^{i\psi}\text{sh } t|^{i\lambda-\rho} \chi_{k,\ell}^{(\alpha,\beta)}(r,\psi)\, dm_{\alpha,\beta}(r,\psi) =$

$$= \phi_{\lambda,k,\ell}^{(\alpha,\beta)}(t),$$

(8.8) $\int_0^1 \int_0^\pi \phi_\lambda^{(\alpha,\beta)}[|\text{ch } s \text{ ch } t - r\, e^{i\psi}\text{sh } s \text{ sh } t|]\cdot$

$$\cdot \chi_{k,\ell}^{(\alpha,\beta)}(r,\psi)\, dm_{\alpha,\beta}(r,\psi) = \phi_{\lambda,k,\ell}^{(\alpha,\beta)}(s)\, \phi_{-\lambda,k,\ell}^{(\alpha,\beta)}(t).$$

For k = ℓ = 0 these formulas coincide with (5.28) and (7.2), respectively. In fact, we will derive (8.7), (8.8) from (5.28), (7.2) by using the following lemma:

Lemma 8.2. Let $f \in C^{\infty}((0,\infty)), \alpha > \beta > -\frac{1}{2}, k, \ell \in \mathbb{Z}, k \geq \ell \geq 0$,

(8.9) $(D_{k,\ell}^{(\alpha,\beta)} f)(x) := \left(\frac{d^2}{dx^2} + \frac{2(\beta+k-\ell)+1}{x}\frac{d}{dx}\right)^{\ell}\left(\frac{d}{d(x^2)}\right)^{k-\ell} f(x).$

Then, for a,b in $\mathbb{R}$, a > |b|:

(8.10) $\int_0^1\int_0^{\pi} f(|a+bre^{i\psi}|) \chi_{k,\ell}^{(\alpha,\beta)}(r,\psi) dm_{\alpha,\beta}(r,\psi) =$

$= \frac{a^{k+\ell}b^{k-\ell}}{2^{2\ell}(\alpha+1)_{k+\ell}} \int_0^1\int_0^{\pi} (D_{k,\ell}^{(\alpha,\beta)} f)(|a+bre^{i\psi}|) \cdot$

$\cdot dm_{\alpha+k+\ell,\beta+k-\ell}(r,\psi).$

Lemma 8.2 is proved by using a Rodrigues type formula for $\chi_{k,\ell}^{(\alpha,\beta)}$ and by integration by parts (cf. [82, Lemma 4.1]). Formula (8.7) follows from (5.28) and (8.10) by using that

(8.11) $D_{k,\ell}^{(\alpha,\beta)}\left(\frac{\Gamma(\alpha+1)x^{i\lambda-\alpha-\beta-1}}{c_{\alpha,\beta}(-\lambda)}\right)$

$= \frac{(-1)^{k+\ell}2^{2k+2\ell}\Gamma(\alpha+k+\ell+1)x^{i\lambda-\alpha-\beta-2k-1}}{c_{\alpha+k+\ell,\beta+k-\ell}(-\lambda)}.$

Formula (8.11) is derived by straightforward differentiation and by substitution of (2.18). Formula (8.8) follows from (7.2) and (8.10) by using that

(8.12) $D_{k,\ell}^{(\alpha,\beta)}\left(\frac{\Gamma(\alpha+1)\Phi_{\lambda}^{(\alpha,\beta)}[x]}{c_{\alpha,\beta}(-\lambda)}\right)$

$= \frac{2^{4k+2\ell}\Gamma(\alpha+k+\ell+1)\Phi_{\lambda}^{(\alpha+k+\ell,\beta+k-\ell)}[x]}{c_{\alpha+k+\ell,\beta+k-\ell}(-\lambda)}$

(by termwise differentiation in (5.71) or by (5.69), (5.73), (5.74)).

The proof of Theorem 8.1 is completed by showing that the right hand sides of (8.5), (8.6) actually converge absolutely and uniformly to the left hand sides. This follows from the following lemma (cf. [49,Theor.3.6]):

Lemma 8.3. Let f be a C^∞-function on the closed upper half unit disk. Then, for each $\kappa > 0$:

$$\hat{f}(k,\ell) := \int_0^1 \int_0^\pi f(re^{i\psi}) \chi_{k,\ell}(r,\psi) \, dm(r,\psi) = O(k^{-\kappa}) \text{ as } k \to \infty,$$

uniformly in ℓ, and

$$f(re^{i\psi}) = \sum_{k=0}^\infty \sum_{\ell=0}^k \hat{f}(k,\ell) \pi_{k,\ell} \chi_{k,\ell}(r,\psi)$$

with absolute and uniform convergence. Moreover, if f depends on an additional parameter s running over a set S such that all partial derivatives of f with respect to $r \cos\psi$, $r \sin\psi$ are uniformly bounded in $(re^{i\psi},s)$ then the absolute convergence of the above series is also uniform on S.

Now we will sketch how (8.5), (8.6) follow from (3.24), (3.25) for special values of α,β. By abuse of notation we consider the spherical function ψ_δ for (K,M) as a function on $K/M = S(\mathbb{F}^n)$. Then (3.22) takes the form

(8.13) $$\phi_{\lambda,\delta}(t) = \int_{S(\mathbb{F})^n} |\text{cht} - y_n \text{sht}|^{i\lambda-\rho} \psi_\delta(y) \, dy.$$

For ψ_δ we have in the three cases $\mathbb{F} = \mathbb{R}, \mathbb{C}, \mathbb{H}$:

$\mathbb{F} = \mathbb{R}$ (see [34,Ch.11]): $\alpha = \frac{1}{2}n-1$, $\beta = -\frac{1}{2}$,

$$\psi_\delta(y) := R_\ell^{(\frac{1}{2}n-3/2, -\frac{1}{2}+k-\ell)} (2y_n^2 - 1) y_n^{k-\ell} \quad (\ell = 0,1,\ldots; k=\ell \text{ or } \ell-1.$$

$\mathbb{F} = \mathbb{C}$ (see [140],[80]): $\alpha = n-1$, $\beta = 0$,

$$\psi_\delta(y) := R_{k \wedge \ell}^{(n-2, |k-\ell|)} (2|y_n|^2 - 1) |y_n|^{k-\ell} \cdot$$
$$\cdot e^{i(k-\ell)(\arg y_n)} (k,\ell = 0,1,\ldots).$$

Hence

$$\text{Re}\psi_\delta(y) = \text{Re}\psi_\delta^\vee(y) =$$
$$= R_\ell^{(n-2, k-\ell)} (2 y_n^2 - 1) |y_n|^{k-\ell} R_{k-\ell}^{(-\frac{1}{2}, -\frac{1}{2})} \left(\frac{\text{Re } y_n}{|y_n|}\right),$$

where $k,\ell \in \mathbb{Z}$, $k \geq \ell \geq 0$.

$\mathbb{F} = \mathbb{H}$ (see [123], [77]) : $\alpha = 2n-1$, $\beta = 1$,

$$\psi_\delta(y) = R_\ell^{(2n-3, k-\ell+1)} (2|y_n|^2 - 1) |y_n|^{k-\ell} R_{k-\ell}^{(\frac{1}{2}, \frac{1}{2})} \left(\frac{\text{Re } y_n}{|y_n|}\right),$$

where $k, \ell \in \mathbb{Z}$, $k \geq \ell \geq 0$.

By inserting these expressions for ψ_δ into (8.14) and by taking the "radial" part of the normalized invariant measure on $S(\mathbb{F}^n)$ we obtain that $\phi_{\lambda,\delta}(t) = \phi_{\lambda,\delta}^\vee(t)$ is given by the left hand side of (8.7) for suitable values of α, β. By applying (8.7) we obtain (8.5), (8.6) from (3.24), (3.25).

A group theoretic derivation of (8.6) in the case $\alpha = \beta = 0, \frac{1}{2}, 1, \ldots$ ($\mathbb{F} = \mathbb{R}$) was given in [139, Chap.10, §3.5]. Durand [29] obtained an addition formula for Jacobi functions of the second kind.

8.2. A criterium for positive definite spherical functions

Remember the definition of positive definite spherical functions on G and their relationship with unitary representations as given in §3.2. For $G = U(1,n;\mathbb{F})$ let $\pi_{\lambda,0}$ ($\lambda \in \mathbb{C}$) be the (unique) irreducible subquotient representation of the representation π_λ (cf.(3.21)) which contains the representation 1 of K. Then precisely those δ from $(K/M)^\wedge$ occur in $\pi_{\lambda,0}$ for which both $\phi_{\lambda,\delta}$ and $\phi_{\bar{\lambda},\delta}$ are nonzero (cf.[84,§3]). Furthermore, the spherical function associated with $\pi_{\lambda,0}$ equals ϕ_λ. The spherical function ϕ_λ is positive definite iff $\pi_{\lambda,0}$ is unitarizable, i.e., if there exists a possible new G-invariant inner product on some G-invariant dense subspace of $H(\pi_{\lambda,0})$. Thus it is important to know which spherical functions ϕ_λ are positive definite. We derive a criterium by using the addition formula (3.25). For f in $C_c(G)$ write

$$f_\delta(s) := \int_K \int_K f(k^{-1} a_s \ell) \psi_\delta(k) dk \, d\ell, \quad \delta \in (K/M)^\wedge.$$

Then it follows from (3.25) that

$$\int_G \phi_\lambda(x^{-1}y) f(x) \overline{f(y)} dx \, dy = \sum_{\delta \in (K/M)^\wedge} d_\delta \cdot$$

$$\cdot \int_0^\infty \phi_{\lambda,\delta}(a_s) \overline{\phi_{\bar{\lambda},\delta}(a_t)} f_\delta(s) \overline{f_\delta(t)} \Delta(s) \Delta(t) ds \, dt.$$

We conclude:

Lemma 8.4. ϕ_λ is positive definite on G iff for all δ in $(K/M)^\wedge$ with $\phi_{\lambda,\delta} \neq 0 \neq \phi_{\bar{\lambda},\delta}$ there is $c_\delta > 0$ such that $\phi_{\bar{\lambda},\delta} = c_\delta \phi_{\lambda,\delta}$.

For real λ we already know that π_λ is unitary and if λ is not real or imaginary then $\overline{\phi_\lambda(t)} = \phi_{\bar\lambda}(t) \neq \phi_\lambda(t)$, so then ϕ_λ cannot be positive definite because of (3.8). So we will use Lemma 8.4 in the case of imaginary λ. In §8.1 we pointed out that $t \mapsto \phi_{\lambda,\delta}(a_t)$ equals the associated Jacobi functions $\phi_{\lambda,k,\ell}$ (cf.(4.15)) for certain k,ℓ. By use of (2.18) it follows from Lemma 8.4 that:

<u>Theorem 8.5.</u> ϕ_λ <u>is a positive definite spherical function on</u> $G = U(1,n;\mathbb{F})$ <u>iff</u>

$$\lambda \in \mathbb{R}, \quad \lambda = \pm i\rho \text{ or } i\lambda \in \begin{cases} (-\rho,\rho) & (\mathbb{F} = \mathbb{R} \text{ or } \mathbb{C}), \\ \\ (-2n+1,2n-1) & (\mathbb{F} = \mathbb{H}). \end{cases}$$

On comparing Theorem 8.5 with [48, Theorem 6.3] we see that, in the cases $\mathbb{F} = \mathbb{R}$, $\mathbb{C}$, a bounded spherical function ϕ satisfies (3.7) for all f in $C_C(G)$ iff ϕ satisfies (3.7) for all f in $C_C(G//K)$ but that this equivalence no longer holds if $\mathbb{F} = \mathbb{H}$.

Theorem 8.5 (for all rank one cases) was earlier proved by Kostant [87], and, for $\mathbb{F} = \mathbb{R}$, by Takahashi [129]. Takahashi [132] deals with the exceptional rank one case.

8.3. Dual convolution structure

Let $G = U(1,n;\mathbb{F})$ and let $f \in L^p(G//K)$ ($1 \leq p < 2$). Then $\hat{f}$ is even, bounded and analytic on $\mathbb{R}$ (cf.§7.2). Hence it follows by a slight variation on the Bochner–Godement and Plancherel–Godement theorem (cf. [39,Ch.1]) that f is positive definite iff $\hat{f}(\lambda) \geq 0$ for $\lambda \geq 0$ and that for such positive definite f we have $\hat{f}$ in $L^1(\mathbb{R}_+;\nu)$ and

$$(8.14) \qquad f(x) = \int_0^\infty \hat{f}(\lambda)\phi_\lambda(x)d\nu(\lambda), \quad x \in G$$

(ν given by (2.26)). Now let $f(x) := \phi_{\lambda_1}(x)\phi_{\lambda_2}(x)$ ($x \in G, \lambda_1, \lambda_2 \in \mathbb{R}$). Then $f \in L^p(G//K)$ for each $p > 1$ (cf.(6.3), (4.8)) and f is positive definite because ϕ_{λ_1} and ϕ_{λ_2} are so. We conclude that

$$(8.15) \quad a(\lambda_1,\lambda_2,\lambda_3) := (\phi_{\lambda_1}\phi_{\lambda_2})^\wedge(\lambda_3) = \int_G \phi_{\lambda_1}(x)\phi_{\lambda_2}(x)\phi_{\lambda_3}(x)dx =$$

$$= \int_0^\infty \phi_{\lambda_1}(t)\phi_{\lambda_2}(t)\phi_{\lambda_3}(t)\Delta(t)dt$$

is well-defined and even, bounded and analytic in λ_3, that
$a(\lambda_1,\lambda_2,\lambda_3) \geq 0$ $(\lambda_1,\lambda_2,\lambda_3 \in \mathbb{R})$ and that

$$(8.16) \qquad \phi_{\lambda_1}(x)\phi_{\lambda_2}(x) = \int_0^\infty \phi_{\lambda_3}(x)a(\lambda_1,\lambda_2,\lambda_3)d\nu(\lambda_3).$$

Now we can define the dual convolution product $f \circ g$ of f,g in
$L^1(\mathbb{R}_+ ;\nu)$ by

$$(8.17) \quad (f \circ g)(\lambda_1) = \int_0^\infty\!\!\int_0^\infty f(\lambda_2)g(\lambda_3)a(\lambda_1,\lambda_2,\lambda_3)d\nu(\lambda_2)d\nu(\lambda_3).$$

Then $f \circ g \in L^1(\mathbb{R}_+ ;\nu)$ and

$$(8.18) \qquad \| f \circ g \|_1 \leq \| f \|_1 \| g \|_1.$$

Let $f \mapsto \overset{\vee}{f}$ denote the inverse spherical Fourier transform

$$(8.19) \qquad \overset{\vee}{f}(x) := \int_0^\infty f(\lambda)\phi_\lambda(x)d\nu(\lambda), \quad f \in L^1(\mathbb{R}_+ ;\nu).$$

Then

$$(8.20) \qquad (f \circ g)^\vee(x) = \overset{\vee}{f}(x)\overset{\vee}{g}(x).$$

In order to develop a dual convolution structure with
nonnegative kernel for the Jacobi transform in the case of
more general α,β we need a substitute without use of representations for the usual positive definiteness proof of
$\phi_\mu\phi_\nu(\mu,\nu \in \mathbb{R})$. We observed that, in the group case, it is sufficient to know that $(\phi_\mu\phi_\nu)^\wedge$ is nonnegative on $\mathbb{R}$. This last
property is equivalent to

$$(8.21) \qquad \int_G\!\!\int_G \phi_\mu(x^{-1}y)\phi_\nu(x^{-1}y)f(x)\overline{f(y)}\,dxdy \geq 0$$

for all f in $\mathcal{D}(G//K)$, so it is not necessary to prove (8.21)
for all f in $C_c(G)$. The left hand side of (8.21) can be rewritten as

$$\int_0^\infty\!\!\int_0^\infty \left(\int_K \phi_\mu(a_{-s}ka_t)\phi_\nu(a_{-s}ka_t)dk\right)f(a_s)\overline{f(a_t)} \cdot$$

$$\cdot \Delta(s)\Delta(t)dsdt.$$

Combination with (3.25) yields

$$\sum_{\delta \in (K/M)^\wedge} d_\delta \left| \int_0^\infty \phi_{\mu,\delta}(a_s)\phi_{\nu,\delta}^{\vee}(a_s)f(a_s)\Delta(s)ds \right|^2,$$

which is nonnegative. ($\overset{\vee}{\delta}$ is contragredient representation to δ.)

We will now imitate this method in cases without group theoretic interpretation. We need the following lemma (cf. [49, Lemmas 4.1, 4.2, 4.3]).

Lemma 8.6. Let $\alpha \geq \beta \geq -\frac{1}{2}$, $1 \leq p < 2$ and $f \in L^p(\mathbb{R}_+ ; \Delta(t)dt)$. Then $\hat{f}(\lambda) \geq 0$ for $\lambda \geq 0$ iff $\int_0^\infty (f*g)(t)\overline{g(t)}\Delta(t)dt \geq 0$ for all g in $\mathcal{D}_{even}(\mathbb{R})$. If f is moreover continuous on $[0,\infty)$ and $\hat{f}(\lambda) \geq 0$ for $\lambda \geq 0$ then $\hat{f} \in L^1(\mathbb{R}_+ ; \nu)$ and (2.25) holds.

If $\lambda_1, \lambda_2 \in \mathbb{R}$ then $\phi_{\lambda_1}\phi_{\lambda_2} \in L^p(\mathbb{R}_+ , \Delta(t)dt)$ for all $p > 1$ (cf.§7.2) and for all g in $\mathcal{D}_{even}(\mathbb{R})$ we have

(8.22)
$$\int_0^\infty ((\phi_{\lambda_1}\phi_{\lambda_2})*g)(t)\overline{g(t)}\Delta(t)dt =$$

$$= \int_0^\infty\int_0^\infty\left(\int_0^1\int_0^\pi \phi_{\lambda_1}[|ch\ s\ ch\ t - r\ e^{i\psi}sh\ s\ sh\ t|]\cdot\right.$$

$$\cdot\phi_{\lambda_2}[|ch\ s\ ch\ t - r\ e^{i\psi}sh\ s\ sh\ t|]dm(r,\psi))\cdot$$

$$\cdot g(s)\overline{g(t)}\Delta(s)\Delta(t)ds\ dt = \sum_{k=0}^\infty \sum_{\ell=0}^k \pi_{k,\ell} \cdot$$

$$\cdot\left|\int_0^\infty \phi_{\lambda_1,k,\ell}(s)\phi_{\lambda_2,k,\ell}(s)g(s)\Delta(s)ds\right|^2 \geq 0,$$

where we used the addition formula (8.6) and the property that $\overline{\phi_{\lambda,k,\ell}(t)} = \phi_{-\lambda,k,\ell}(t)$ if λ is real. Now the properties of the positive dual convolution structure for $L^1(\mathbb{R}_+ ; \nu)$ follow in the same way as in the group case.

The method of applying the addition formula in order to prove the positivity of the dual convolution structure was earlier used in [83] in connection with Jacobi polynomials. The dual convolution kernel (8.15) was explicitly computed by Mizony [104] for $\alpha = \beta = 0$ or $\frac{1}{2}$. It is an open problem to find it for other α,β. Mayer-Lindenberg [100,§3] discusses dual convolution in the spherical rank one case. Nussbaum [109], also working in the rank one case, considers functions ϕ in $C(G//K)$ which satisfy (3.7) for all f in $C_C(G//K)$ (a weaker form of positive definiteness). Such ϕ are inverse Jacobi transforms of certain positive measures on $\mathbb{R}_+ \cup i\mathbb{R}_+$. He shows that radial functions on a ball around 0 in G/K

which are positive definite in this sense have an extension
to a similar function on G/K. Trimèche [137] and Chébli
[21], working with a more general Δ satisfying (2.35) and
(in [21]) (2.36), show that any distribution T satisfying
$<T,f*\tilde{f}> \geq 0$ for all f in $\mathcal{D}_{even}(\mathbb{R})$ (convolution in generalized sense) is the inverse ϕ_λ-transform of appropriate
positive measures on $\mathbb{R}_+$ and $i\mathbb{R}_+$. By use of the Abel transform they reduce this property of T to the case Δ = 1, for
which the result can be found in [54,Ch.2,§6.3,Theor.5].
In [137] the Plancherel formula (2.37) is deduced from this
result.

9. TWO SPECIAL ORTHOGONAL SYSTEMS MAPPED ONTO EACH OTHER BY THE JACOBI TRANSFORM

It is well-known that the functions $x \mapsto e^{-\frac{1}{2}x^2} H_n(x)$ (H_n
Hermite polynomial) form a complete orthogonal system of
eigenfunctions with respect to the Fourier transform and,
similarly, the functions $x \mapsto 2^{\alpha+\frac{1}{2}} e^{-\frac{1}{2}x^2} L_n^\alpha(x^2)$ (L_n^α Laguerre
polynomial) with respect to the Hankel transform. For the
Jacobi transform we cannot expect such a system because
$\phi_\lambda(t)$ is not symmetric in λ,t. Still it would be pleasant
to have two explicit orthogonal systems in $L^2(\mathbb{R}_+;\Delta(t)dt)$
and $L^2(\mathbb{R}_+; |c(\lambda)|^{-2}d\lambda)$ which are mapped onto each other by
the Jacobi transform. In this section I will present such
systems (author's result, unpublished until now). The proofs
will only be sketched.
    First observe that, for Re μ > 0:

$$(9.1) \qquad \int_0^\infty (2cht)^{-\mu-\rho} \phi_\lambda^{(\alpha,\beta)}(t) \Delta_{\alpha,\beta}(t)dt =$$

$$= c_{\alpha,\beta}(-i\mu) \int_{-\infty}^\infty e^{i\lambda s} (2chs)^{-\mu}ds =$$

$$= \frac{2^{\alpha+\beta-\mu}\Gamma(\alpha+1)\Gamma(\frac{1}{2}(\mu-i\lambda))\Gamma(\frac{1}{2}(\mu+i\lambda))}{\Gamma(\frac{1}{2}(\alpha+\beta+1+\mu))\Gamma(\frac{1}{2}(\alpha-\beta+1+\mu))}$$

in view of (5.4), (5.56), (2.18) and [33,1.5(26)]. This
formula is quite useful for evaluating Jacobi transforms of
functions which are given as series in inverse powers of
cht. For instance, by use of (9.1) and (2.15), the integral

$$\int_0^\infty \phi_\mu^{(\gamma,\beta)}(t) \phi_\lambda^{(\alpha,\beta)}(t) \Delta_{\alpha,\beta}(t)dt$$

can be evaluated as a quotient of products of gamma functions. A special case of this evaluation can be found in Sprinkhuizen [126,§8.3], where it is obtained by use of fractional integrals in two variables.

Now fix α,β,δ such that $\alpha \geq \beta \geq -\frac{1}{2}$, $\delta > -1$. For $n = 0,1,2,\ldots$ let

$$(9.2) \qquad r_n(t) := (cht)^{-\alpha-\beta-\delta-2} R_n^{(\alpha,\delta)} (1-2th^2t),$$

where $R_n^{(\alpha,\delta)}$ is a Jacobi polynomial, cf. (2.3). It follows from the orthogonality properties of Jacobi polynomials (cf. [34,§10.8]) that

$$(9.3) \qquad \int_0^\infty r_n(t) r_m(t) \Delta_{\alpha,\beta}(t) dt =$$

$$= \frac{2^{2\alpha+2\beta+1} (\Gamma(\alpha+1))^2 \Gamma(n+\delta+1) n!}{(2n+\alpha+\delta+1) \Gamma(n+\alpha+1) \Gamma(n+\alpha+\delta+1)} \delta_{n,m}$$

and that the system $\{r_n\}$ is complete in $L^2(\mathbb{R}_+;\Delta(t)dt)$. By [34,10.8(13)] and (2.3) $r_n(t)$ can be expanded as

$$(-1)^n \frac{(\delta+1)_n}{(\alpha+1)_n} \sum_{k=0}^n \frac{(-n)_k (n+\alpha+\delta+1)_k}{(\delta+1)_k k!} (ch\, t)^{-\alpha-\beta-\delta-2-2k}$$

On combining this with (9.1) we obtain that

$$(9.4) \qquad \int_0^\infty r_n(t) \phi_\lambda^{(\alpha,\beta)}(t) \Delta_{\alpha,\beta}(t) dt = s_n(t) :=$$

$$= \frac{(-1)^n 2^{2\alpha+2\beta+1} \Gamma(\alpha+1) (\delta+1)_n \Gamma(\frac{1}{2}(\delta+1-i\lambda)) \Gamma(\frac{1}{2}(\delta+1+i\lambda))}{\Gamma(\frac{1}{2}(\alpha+\beta+\delta+2)) \Gamma(\frac{1}{2}(\alpha-\beta+\delta+2)) (\alpha+1)_n} \cdot$$

$$\cdot {}_4F_3\left(\begin{matrix} -n,n+\alpha+\delta+1,\frac{1}{2}(\delta+1-i\lambda),\frac{1}{2}(\delta+1+i\lambda) \\ \delta+1,\frac{1}{2}(\alpha+\beta+\delta+2),\frac{1}{2}(\alpha-\beta+\delta+2) \end{matrix} \Big| 1\right) .$$

In view of (9.3) and (2.27) we must have

$$(9.5) \qquad (2\pi)^{-1} \int_0^\infty s_n(\lambda) s_m(\lambda) |c_{\alpha,\beta}(\lambda)|^{-2} d\lambda = \text{RHS of } (9.3)$$

On the other hand, the right hand side of (9.4) can be expressed in terms of Wilson polynomials

$$(9.6) \qquad p_n(t^2;a,b,c,d) := (a+b)_n (a+c)_n (a+d)_n \cdot$$

$$\cdot {}_4F_3\left(\begin{matrix} -n,n+a+b+c+d-1,a-t,a+t \\ a+b,a+c,a+d \end{matrix} \Big| 1\right),$$

cf. Wilson [146]. For positive a,b,c,d these are orthogonal polynomials in t^2 on $\mathbb{R}_+$ satisfying the orthogonality relations [146,(3.1)]. They are symmetric in a,b,c,d. By (9.4) and (9.6) we have

$$(9.7) \qquad s_n(\lambda) = \frac{(-1)^n 2^{2\alpha+2\beta+1} \Gamma(\alpha+1)(\delta+1)_n |\Gamma(\tfrac{1}{2}(\delta+1-i\lambda))|^2}{\Gamma(\tfrac{1}{2}(\alpha+\beta+\delta+2)) \Gamma(\tfrac{1}{2}(\alpha-\beta+\delta+2))(\alpha+1)_n} \cdot$$

$$\cdot p_n(-\tfrac{1}{4}\lambda^2; \tfrac{1}{2}(\delta+1), \tfrac{1}{2}(\delta+1), \tfrac{1}{2}(\alpha+\beta+1), \tfrac{1}{2}(\alpha-\beta+1))/p_n(0).$$

By substitution of (9.7) into (9.5) we obtain the orthogonality relations [146,(3.1)] for the Wilson polynomials. Thus the Plancherel Theorem 2.4 and the orthogonality relations for the Wilson polynomials imply each other. If, more generally, $\alpha > -1$ and β is arbitrarily real then , by Theorem 2.4, the functions s_n are orthogonal with respect to the measure ν (cf.(2.26)) which may include a discrete part, see [146,(3.3)].

The Hankel transform acting on Laguerre polynomials can be obtained as a limit case of (9.4). Indeed, use of Stirling's formula [33,1.18(2)] gives

$$2^{2(\alpha+\beta+1)} \int_0^\infty {}_1F_1(-n;\alpha+1;t^2) \, {}_0F_1(\alpha+1;-\tfrac{1}{4}\lambda^2 t^2) \cdot$$

$$\cdot t^{2\alpha+1} e^{-\tfrac{1}{2}t^2} dt =$$

$$= \lim_{\delta\to\infty} \delta^{\alpha+1} \int_0^\infty (\operatorname{ch} t)^{-\alpha-\beta-\delta-2} R_n^{(\alpha,\delta)}(1-2\operatorname{th}^2 t) \cdot$$

$$\cdot \phi_{\lambda\delta^{1/2}}^{(\alpha,\beta)}(t) \Delta(t) dt =$$

$$= (-1)^n 2^{2\alpha+2\beta+1} \Gamma(\alpha+1) \lim_{\delta\to\infty} \cdot$$

$$\cdot \frac{\delta^{\alpha+1} (\tfrac{1}{2}(\alpha+\beta+\delta+2))_n \Gamma(\tfrac{1}{2}(\delta+1-i\lambda\delta^{\frac{1}{2}})) \Gamma(\tfrac{1}{2}(\delta+1+i\lambda\delta^{\frac{1}{2}}))}{\Gamma(\tfrac{1}{2}(\alpha+\beta+\delta+2)) \Gamma(\tfrac{1}{2}(\alpha-\beta+\delta+2)) (\tfrac{1}{2}(\alpha-\beta+\delta+2))_n} \cdot$$

$$\cdot {}_4F_3\left(\begin{array}{c} -n, n+\alpha+\delta+1, \tfrac{1}{2}(\alpha+\beta+1+i\lambda\delta^{\frac{1}{2}}), \tfrac{1}{2}(\alpha+\beta+1-i\lambda\delta^{\frac{1}{2}}) \\ \alpha+1, \tfrac{1}{2}(\alpha+\beta+\delta+2), \tfrac{1}{2}(\alpha+\beta+\delta+2) \end{array}\Big| 1\right)$$

$$= 2^{3\alpha+2\beta+2}(-1)^n \Gamma(\alpha+1) e^{-\tfrac{1}{2}\lambda^2} \, {}_1F_1(-n;\alpha+1;\lambda^2).$$

Thus we have obtained the well-known formula (cf.[35,8.9(3)])

$$(9.8) \qquad \int_0^\infty t^{\alpha+\frac{1}{2}} e^{-\tfrac{1}{2}t^2} L_n^\alpha(t^2) J_\alpha(\lambda t)(\lambda t)^{\frac{1}{2}} dt =$$

$$= (-1)^n e^{-\tfrac{1}{2}\lambda^2} \lambda^{\alpha+\frac{1}{2}} L_n^\alpha(\lambda^2).$$

A special case of (9.4) was obtained by Boyer & Ardalan [15] in a group theoretic context. The paper gives the following motivation for the choice of a special orthogonal basis like (8.2). A spherical principal series representation of $0(1,n)$ can be realized both on $H_1 := L^2(0(1,n-1)/0(n-1))$ and on $H_2 := L^2(0(n)/0(n-1))$. A canonical (continuous) basis for H_1 with respect to $0(n-1)$ involves Jacobi functions, while a canonical basis for H_2 with respect to $0(n-1)$, transferred to H_1 by the intertwining operator, involves functions of the form (9.2). Then the overlap functions between these two bases involve functions of the form (9.4).

It is an open problem to extend the results of this section such that Wilson polynomials (9.6) occur without the restriction a = b in the parameters.

10. FURTHER RESULTS

First we list some further references on the harmonic analysis of the Jacobi transform. Hasegawa [62] gives sufficient conditions on f in L^2 in order that $\hat{f} \in L^1$. Schindler [120] derives transplantation theorems in the case $\alpha = \beta$. Stanton & Tomas [127], Vretare [141] and Achour & Trimèche [2], [3] obtain multiplier results by using various techniques (in [2], [3] an analogue of the Littlewood-Paley g-function). Van de Wetering [143] obtains necessary and sufficient conditions for $\hat{f}$ in order that f is variation diminishing. Meaney [101] considers differentiability properties of the inverse Jacobi transform of a L^1-function and applies this to the study of sets of synthesis. Chébli [24] gives a theory of almost periodic functions with respect to differential operators of the form (2.8).

Of course, special cases of the above results allow immediate translation to the context of noncompact symmetric spaces of rank one. The following papers exclusively deal with harmonic analysis on rank one spaces G/K but make essential use of properties of Jacobi functions. Berenstein & Zalcman [12] give criteria for pairs of double cosets KxK, KyK in order that they satisfy the Pompeiu property on a noncompact rank one symmetric space. Kawazoe [79] studies the radial maximal function on G/K. Lohoué & Rychener [97] obtain properties of the resolvent of the Laplacian on G/K by using inversion of the Abel transform.

Samii [119] obtains curious inequalities for the spherical functions on $SO_0(1,n)/SO(n)$ by using an explicit expression for the operator which intertwines the Euclidean and the non-Euclidean Poisson transform for the unit ball in $\mathbb{R}^n$.

Mizony [104] and Grünbaum [59] raise the question whether $\phi_\lambda^{(\alpha,\beta)}(t)$ is eigenfunction of a differential operator in λ, as is the case when $\alpha = \beta = \frac{1}{2}$ (cf.2.11)). Grünbaum (see also [60]) would like to apply such a property to obtain a p.d.o. commuting with the operation of both time and band limiting on a rank one space. However, the answer is probably negative.

Van den Ban [7] considers the integral representation (4.4) as an integral over a contour in $(K/M)_c$, (complexification) and he next deforms the contour. In the rank one case a sume comes out of two integrals which represent $c(\lambda)\phi_\lambda$ and $c(-\lambda)\phi_{-\lambda}$, respectively

Roehner & Valent [116] use Jacobi functions in connection with birth and death processes. They prove the Plancherel theorem for the Jacobi transform by writing it as a Hankel transform followed by a Kontorovich–Lebedev transform. This factorization can be seen from the integral representation

$$\phi_\lambda^{(\alpha,\beta)}(t) = c(\text{sht})^{-\alpha} \int_0^\infty K_{i\lambda}(x) J_\alpha(x\ \text{sh}\ t) x^\beta dx,$$

cf. [34,7.7(31)]. See Faraut [40] and Terras [134] for the occurrence of the Kontorovich–Lebedev transform on rank one spaces.

Berezin & Karpelevič [13] state without proof that the spherical functions for $(G,K) = (SU(n,n+k), S(U(n) \times U(n+k))$ restricted to $A \simeq \mathbb{R}^n$ equal

$$(t_1,\ldots,t_n) \to \frac{c.\det(\phi_{\lambda_i}^{(k,0)}(t_j))_{i,j=1,\ldots,n}}{\Pi_{1 \le i < j \le n}(\text{ch}2t_i - \text{ch}2t_j)}.$$

A rigorous proof was later given by Hoogenboom [74].

Historical remarks about the Mehler–Fock transform in connection with separation of variables problems can be found in Robin [115,Ch.IX]. Sneddon [124] gives applications of this transform.

Centrum voor Wiskunde en Informatica
Postbus 4079
1009 AB Amsterdam, Nederland

REFERENCES

[1] Abel, N.H.: 'Résolution d'un problème de méchanique',
     J. Reine Angew. Math. 1(1826) (in German) = in:
     Oeuvres, Tome I, pp.97-101.
[2] Achour, A. and K. Trimèche: 'La g-fonction de Littlewood-
     Paley associée à l'opérateur de Jacobi', in Séminaire
     d'Analyse harmonique de Tunis, 1980-81, Faculté des
     Sciences, Tunis, 1981, Exposé 28.
[3] Achour, A. and K. Trimèche: 'La g-fonction de Littlewood-
     Paley associée à un opérateur différentiel singulier',
     preprint.
[4] Aomoto, K.: 'Sur les transformations d'horisphère et les
     équations intégrales qui s'y rattachent', J. Fac.Sci.
     Univ. Tokyo Sect. I 14(1967), 1-23.
[5] Askey, R.: 'Orthogonal polynomials and special func-
     tions', Regional Conference Series in Applied Math.
     21, SIAM, Philadelphia, 1975.
[6] Badertscher, E.: 'Harmonic analysis on straight line
     bundles', preprint.
[7] Ban, E.P. van den: 'Asymptotic expansions and integral
     formulas for eigenfunctions on semisimple Lie groups',
     Dissertation, University of Utrecht, 1982.
[8] Bargmann, V.: 'Irreducible unitary representations of
     the Lorentz group', Ann. of Math.(2) 48 (1947), 568-640.
[9] Bellandi Fo, J. and E. Capelas de Oliveira: 'On the
     product of two Jacobi functions of different kinds
     with different arguments', J. Phys. A 15 (1982),
     L447-L449.
[10] Benoist, Y.: 'Analyse harmonique sur les espaces symé-
     triques nilpotents', C.R. Acad.Sci.Paris Ser.I Math.
     296 (1983), 489-492.
[11] Benoist, Y.: 'Espaces symétriques exponentielles',
     Thèse 3me cycle, Université de Paris VII, 1983.
[12] Berenstein, C.A. and L. Zalcman: 'Comment. Math. Helv.
     55 (1980), 593-621.
[13] Berezin, F.A. and F.I. Karpelevič: 'Zonal spherical
     functions and Laplace operators on some symmetric
     spaces', Dokl.Akad.Nauk SSSR 118(1958), 9-12 (in
     Russian).
[14] Berger, M.: 'Les espaces symétriques non compacts,
     Ann.Sci.École Norm.Sup.(4) 74 (1957), 85-177.
[15] Boyer, C.P. & F. Ardalan: 'On the decomposition
     SO(p,1) ⊃ SO(p-1,1) for most degenerate representations',
     J. Math. Phys. 12 (1971), 2070-2075.

[16] Braaksma, B.L.J. and B. Meulenbeld: 'Integral transforms
     with generalized Legendre functions as kernels', Com-
     positio Math. 18(1967), 235-287.
[17] Braaksma, B.L.J. and H.S.V. de Snoo: 'Generalized trans-
     lation operators associated with a singular differen-
     tial operator', in B.D. Sleeman, I.M. Michael (eds.),
     Ordinary and partial differential equations, Lecture
     Notes in Math. 415, Springer, Berlin, 1974, pp.62-77.
[18] Carroll, R.: 'Transmutation, scattering theory and
     special functions', North-Holland, Amsterdam, 1982.
[19] Carroll, R.: 'Some inversion theorems of Fourier type',
     Rev. Roumaine Math. Pures Appl., to appear.
[20] Chébli, H.: 'Sur la positivité des opérateurs de'trans-
     lation généralisée' associés à un opérateur de Sturm-
     Liouville sur [0,∞[', C.R. Acad.Sci. Paris Sér.A-B
     275 (1972), A601-A604.
[21] Chébli, H.: 'Positivité des opérateurs de "translation
     généralisée" associés à un opérateur de Sturm-
     Liouville et quelques applications à l'analyse harmo-
     nique', Thèse, Université Louis Pasteur, Strasbourg,
     1974.
[22] Chébli, H.: 'Sur un théorème de Paley-Wiener associé à
     la décomposition spectrale d'un opérateur de Sturm-
     Liouville sur]0,∞[', J. Funct.Anal. 17 (1974),
     447-461.
[23] Chébli, H.: 'Théorème de Paley-Wiener associé à un
     opérateur différentiel singulier sur (0,∞)', J. Math.
     Pures Appl. (9)58(1979),1-19.
[24] Chébli, H.: 'Sur les fonctions presque-périodiques as-
     sociées à un opérateur différentiel singulier sur
     (0,∞)', preprint.
[25] Dixmier, J.: 'Les C*-algèbres et leurs représentations',
     Gauthier-Villars, Paris, 1969.
[26] Dixmier, J. and P. Malliavin: 'Factorisations de fonc-
     tions et de vecteurs indéfiniment différentiables',
     Bull.Sci.Math.(2) 102 (1978), 305-330.
[27] Duistermaat, J.J.: 'On the similarity between the
     Iwasawa projection and the diagonal part', preprint.
[28] Dunford, N. and J.T. Schwartz: 'Linear operators, Part
     II', Interscience, New York, 1963.
[29] Durand, L.: 'Addition formulas for Jacobi, Gegenbauer,
     Laguerre and hyperbolic Bessel functions of the second
     kind', SIAM J. Math. Anal. 10(1979), 425-437.
[30] Dijk, G. van: 'On generalized Gelfand pairs, a survey
     of results', Proc. Japan Acad. Ser. A Math. Sci., to
     appear.

[31] Ehrenpreis, L. and F.I. Mautner: 'Some properties of
     the Fourier transform on semi-simple Lie groups, I',
     Ann. of Math.(2) 61 (1955), 406-439.
[32] Ehrenpreis, L. and F.I. Mautner: 'Some properties of
     the Fourier transform on semi-simple Lie groups, II',
     Trans.Amer.Math.Soc. 84 (1957), 1-55.
[33] Erdelyi, A., W. Magnus, F. Oberhettinger and
     F.G. Tricomi: 'Higher transcendental functions, Vol.I',
     McGraw-Hill, New York, 1953.
[34] Erdélyi, A., W. Magnus, F. Oberhettinger and
     F.G. Tricomi: 'Higher transcendental functions, Vol.II',
     McGraw-Hill, New York, 1953.
[35] Erdélyi, A., W. Magnus, F. Oberhettinger and
     F.G. Tricomi: 'Tables of integral transforms, Vol.II',
     McGraw-Hill, New York, 1954.
[36] Faraut, J.: 'Opérateurs différentiels symétriques du
     second ordre', in Séminaire de théorie spectrale,
     1974, Institut de Recherche Mathématique Avancée,
     Strasbourg, 1974, Exposé 6.
[37] Faraut, J.: 'Distributions sphériques sur les espaces
     hyperboliques', J. Math. Pures Appl. (9) 58(1979),
     369-444.
[38] Faraut, J.: 'Algèbre de Volterra et transformation de
     Laplace sphérique', in Séminaire d'Analyse harmonique
     de Tunis, 1980-81, Faculté des Sciences, Tunis, 1981,
     Exposé 29.
[39] Faraut, J.: 'Analyse harmonique sur les pairs de
     Guelfand et les espaces hyperboliques', in J.-L. Clerc,
     P. Eymard, J. Faraut, M. Rais, R. Takahashi, Analyse
     harmonique, C.I.M.P.A., Nice, 1982, Ch.IV.
[40] Faraut, J.: 'Un théorème de Paley-Wiener pour la trans-
     formation de Fourier sur un espace riemannien symé-
     trique de rang un', J. Funct. Anal. 49(1982),
     230-268.
[41] Flensted-Jensen, M.: 'Paley-Wiener type theorems for a
     differential operator connected with symmetric spaces',
     Ark. Mat. 10(1972), 143-162.
[42] Flensted-Jensen, M.: 'Spherical functions on rank one
     symmetric spaces and generalizations', Proc.Sympos.
     Pure Math. 26(1973), 339-342.
[43] Flensted-Jensen, M.: 'A proof of the Plancherel formula
     for the universal covering group of SL(2,R) using
     spectral theory and spherical functions', in Séminaire
     de Théorie Spectrale, 1972-73, Institut de Recherche
     Mathématique Avancée, Strasbourg, 1973, Exposé 4.

[44] Flensted-Jensen, M.: 'Spherical functions on a simply
     connected semisimple Lie group. II. The Paley-Wiener
     theorem for the rank one case', Math. Ann. 228 (1977),
     65-92.
[45] Flensted-Jensen, M.: 'Spherical functions on a real
     semisimple Lie group. A method of reduction to the
     complex case', J. Funct. Anal. 30 (1978), 106-146.
[46] Flensted-Jensen, M.: 'Discrete series for semisimple
     symmetric spaces', Ann. of Math. (2) 111 (1980),
     253-311.
[47] Flensted-Jensen, M.: 'Harmonic analysis on semisimple
     symmetric spaces-A method of duality', in R. Herb e.a.
     (eds.), Proceedings Maryland 1982-83, vol. III,
     Lecture Notes in Math., Springer, to appear.
[48] Flensted-Jensen, M. and T.H. Koornwinder: 'The convolu-
     tion structure for Jacobi function expansions', Ark.
     Mat. 10(1973), 245-262.
[49] Flensted-Jensen, M. and T.H. Koornwinder: 'Jacobi func-
     tions: the addition formula and the positivity of the
     dual convolution structure, Ark. Mat. 17 (1979),
     139-151.
[50] Flensted-Jensen, M. and T.H. Koornwinder: 'Positive
     definite spherical functions on a non-compact, rank one
     symmetric space', in P. Eymard, J. Faraut, G. Schiffman,
     R. Takahashi (eds.), Analyse harmonique sur les
     groupes de Lie, II, Lecture Notes in Math. 739,
     Springer, Berlin, 1979, pp.249-282.
[51] Flensted-Jensen, M. and D.L. Ragozin: 'Spherical func-
     tions are Fourier transforms of L_1-functions', Ann.
     Sci. École Norm. Sup. (4) 6 (1973), 457-458.
[52] Fock, V.A.: 'On the representation of an arbitrary
     function by an integral involving Legendre's function
     with a complex index', C.R. (Doklady) Acad. Sci.
     URSS(N.S.) 39(1943), 253-256.
[53] Gangolli, R.: 'On the Plancherel formula and the Paley-
     Wiener theorem for spherical functions on semisimple
     Lie groups', Ann. of Math. (2) 93 (1971), 150-165.
[54] Gelfand, I.M. and N.Ja. Vilenkin: 'Generalized func-
     tions, Vol. 4, Applications of harmonic analysis',
     Moscow, 1961 (in Russian) = Academic Press, New York,
     1964.
[55] Gindikin, S.G. and F.I. Karpelevič: 'Plancherel
     measure for Riemann symmetric spaces of nonpositive
     curvature, Dokl. Akad. Nauk SSSR 145(1962), 252-255
     (in Russian)= Soviet Math. Dokl. 3(1962), 962-965.

[56] Gindikin, S.G. and F.I. Karpelevič: 'One problem of
     integral geometry', in Pamyati N.G. Chebotareva,
     Izdatelstvo Kazanskov Universiteta, 1964 (in Russian)=
     Selecta Math. Soviet. 1(1981), 169-184.
[57] Godement, R.: 'Introduction, aux travaux de A. Selberg',
     in Séminaire Bourbaki, Paris, 1957, Exposé 144.
[58] Götze, F.: 'Verallgemeinerung einer Integral transfor-
     mation von Mehler-Fock durch den von Kuipers und
     Meulenbeld eingeführten Kern $P_k^{m,n}(z)$', Nederl.Akad.
     Wetensch. Proc. Ser.A 68 = Indag Math. 27 (1965),
     396-404.
[59] Grünbaum, F.A.: 'The limited angle problem in tomo-
     graphy and some related mathematical problems', in
     Proceedings Internat.Colloq. Luminy (France), May 1982,
     North-Holland, Amsterdam, to appear.
[60] Grünbaum, F.A.: 'Band and time limiting, recursion rela-
     tions and some nonlinear evolution equations', in this
     volume.
[61] Harish-Chandra: 'Spherical functions on a semi-simple
     Lie group, I,II', Amer. J. Math. 80 (1958), 241-310,
     553-613.
[62] Hasegawa, Y.: 'On the integrability of Fourier-Jacobi
     transforms', Ark. Mat. 16(1978), 127-139.
[63] Heine, E.: 'Handbuch der Kugelfunctionen, Zweiter Band',
     Berlin, 1881.
[64] Helgason, S.: 'Differential geometry and symmetric
     spaces', Academic Press, New York, 1962.
[65] Helgason, S.: 'An analogue of the Paley-Wiener theorem
     for the Fourier transform on certain symmetric spaces',
     Math. Ann. 165(1966), 297-308.
[66] Helgason, S.: 'A duality for symmetric spaces, with
     applications to group representations', Adv. in Math.
     5(1970), 1-154.
[67] Helgason, S.: 'Analysis on Lie groups and homogeneous
     spaces', Regional Conference Series in Math. 14,
     Amer. Math. Soc., Providence, R.I., 1972.
[68] Helgason, S.: 'Eigenspaces of the Laplacian; integral
     representations and irreducibility', J. Funct. Anal.
     17 (1974), 328-353.
[69] Helgason, S.: 'A duality for symmetric spaces with
     applications to group representations, II. Differen-
     tial equations and eigenspace representations', Adv.
     in Math. 22(1976), 187-219.
[70] Helgason, S.: 'Differential geometry, Lie groups and
     symmetric spaces', Academic Press, New York, 1978.

[71] Helgason, S.: 'Topics in Harmonic analysis on homoge-
     neous spaces', Birkhäuser, Boston, 1981.
[72] Helgason, S.: 'Groups and geometric analysis, I',
     Academic Press, New York, to appear.
[73] Henrici, P.: 'Addition theorems for Legendre and
     Gegenbauer functions', J. Rational Mech. Anal. 4(1955),
     983-1018.
[74] Hoogenboom, B.: 'Spherical functions and differential
     operators on complex Grassmann manifolds', Ark. Mat.
     20 (1982), 69-85.
[75] Hoogenboom, B.: 'Intertwining functions on compact Lie
     groups', Dissertation, University of Leiden, 1983.
[76] Johnson, K.D.: 'Composition series and intertwining
     operators for the spherical principal series, II',
     Trans. Amer. Math. Soc. 215 (1976), 269-283.
[77] Johnson, K.D. and N.R. Wallach: 'Composition series and
     intertwining operators for the spherical principal
     series', Trans.Amer.Math.Soc. 229(1977), 137-174.
[78] Kashiwara, M., A. Kowata, K. Minemura, K. Okamoto,
     T. Oshima and M. Tanaka: 'Eigenfunctions of invariant
     differential operators on a symmetric space', Ann. of
     Math.(2) 107(1978), 1-39.
[79] Kawazoe, T.: 'Maximal functions on non-compact rank one
     symmetric spaces. Radial maximal functions and atoms',
     preprint.
[80] Koornwinder, T.H.: 'The addition formula for Jacobi
     polynomials. II. The Laplace type integral represen-
     tation and the product formula', Report TW 133/72, Math.
     Centrum, Amsterdam, 1972.
[81] Koornwinder, T.H.: 'A new proof of a Paley-Wiener type
     theorem for the Jacobi transform', Ark. Mat. 13 (1975),
     145-159.
[82] Koornwinder, T.H.: 'Jacobi polynomials, III. An analy-
     tic proof of the addition formula', SIAM J. Math. Anal.
     6 (1975), 533-543.
[83] Koornwinder, T.H.: 'Positivity proofs for linearization
     and connection coefficients of orthogonal polynomials
     satisfying an addition formula', J. London Math. Soc.
     (2) 18(1978), 101-114.
[84] Koornwinder, T.H.: 'The representation theory of
     SL(2,ℝ), a global approach', Report ZW 145/80, Math.
     Centrum, Amsterdam, 1980.
[85] Koornwinder, T.H.: 'The representation theory of
     SL(2,ℝ), a non-infinitesimal approach', Enseign. Math.
     (2) 28(1982), 53-90.

[86] Koornwinder, T.H. (ed.): 'The structure of real semi-simple Lie groups', MC Syllabus 49, Math. Centrum, Amsterdam, 1982.

[87] Kostant, B.: 'On the existence and irreducibility of certain series of representations', in I.M. Gelfand (ed.), Lie groups and their representations, Halsted Press, New York, 1975, pp. 231-329.

[88] Kosters, M.T.: 'Spherical distributions on an exceptional hyperbolic space of type F_4', Report ZW 161/81, Math. Centrum, Amsterdam, 1981.

[89] Kosters, M.T.: 'Spherical distributions on rank one symmetric spaces', Dissertation, University of Leiden, 1983.

[90] Kunze, R.A. and E.M. Stein: 'Uniformly bounded representations and harmonic analysis of the 2×2 real unimodular group', Amer. J. Math. 82 (1960), 1-62.

[91] Langer, R.E.: 'On the asymptotic solutions of ordinary differential equations with reference to the Stokes phenomenon about a singular point', Trans. Amer. Math. Soc. 37(1935), 397-416.

[92] Lebedev, N.N.: 'Parseval's formula for the Mehler-Fock transform', Dokl. Akad. Nauk SSSR 68 (1949), 445-448 (in Russian).

[93] Lebedev, N.N.: 'Some integral representations for products of sphere functions', Dokl. Akad. Nauk SSSR 73 (1950), 449-451 (in Russian).

[94] Lebedev, N.N.: 'Special functions and their applications', Moscow, revised ed., 1963 (in Russian) = Dover, New York, 1972.

[95] Lewis, J.B.: 'Eigenfunctions on symmetric spaces with distribution-valued boundary forms', J. Funct. Anal. 29 (1978), 287-307.

[96] Lions, J.L.: 'Equations différentielles-opérationnels et problèmes aux limites', Springer, Berlin, 1961.

[97] Lohoué, N. and Th. Rychener: 'Die resolvente von Δ auf symmetrischen Räumen von nichtkompakten Typ', Comment. Math. Helv. 57(1982), 445-468.

[98] Markett, C.: 'Norm estimates for generalized translation operators associated with a singular differential operator', preprint.

[99] Matsushita, O.: 'The Plancherel formula for the universal covering group of $SL(2,\mathbb{R})$', Sc. Papers College Gen. Ed. Univ. Tokyo 29(1979), 105-123.

[100] Mayer-Lindenberg, F.: 'Zur Dualitätstheorie symmetrischer Paare', J. Reine Angew. Math. 321 (1981), 36-52.

[101] Meaney, C.: 'Spherical functions and spectral syn-
      thesis', preprint.
[102] Mehler, F.G.: 'Ueber die Vertheilung der statischen
      Elektricität in einem von zwei Kugelkalotten begrenz-
      ten Körper', J. Reine Angew. Math. 68 (1868), 134-150.
[103] Mehler, F.G.: 'Über eine mit den Kugel- und Cylinder-
      functionen verwandte Function und ihre Anwendung in
      der Theorie der Elektricitätsvertheilung', Math. Ann.
      18 (1881), 161-194.
[104] Mizony, M.: 'Algèbres et noyaux de convolution sur le
      dual sphérique d'un groupe de Lie semi-simple, non-
      compact et de rang 1', Publ. Dép. Math. (Lyon) 13
      (1976), 1-14.
[105] Mizony, M.: 'Une transformation de Laplace-Jacobi',
      SIAM J. Math. Anal. 14 (1983), 987-1003.
[106] Mizony, M.: 'Analyse harmonique hyperbolique: represen-
      tations et contractions des groupes $SO_0(1,n)$', pre-
      print.
[107] Molčanov, V.F.: 'The Plancherel fromula for the pseudo-
      Riemannian space $SL(3,\mathbb{R})$ /$GL(2,\mathbb{R})$ ', Sibirsk. Mat. Ž.
      23 (1982) no. 5, 142-151 (in Russian) = Siberian Math.
      J. 23 (1983), 703-711.
[108] Nostrand, R.G. van: 'The orthogonality of the hyper-
      boloid functions', J. Math. Phys. 33 (1954), 276-282.
[109] Nussbaum, A.E.: 'Extension of positive definite func-
      tions and representation of functions in terms of
      spherical functions in symmetric spaces of noncom-
      pact type of rank 1', Math. Ann. 215 (1975), 97-116.
[110] Nussbaum, A.E.: 'Paley-Wiener theorem associated with
      a certain singular Sturm-Liouville operator', pre-
      print.
[111] Olevskiĭ, M.N.: 'On a generalization of Bessel func-
      tions', C.R. (Doklady) Acad.Sci. URSS(N.S.) 40 (1943),
      5-10.
[112] Olevskiĭ, M.N.: 'On the representation of an arbitrary
      function in the form of an integral with a kernel con-
      taining a hypergeometric function', Dokl. Akad. Nauk
      S.S.S.R. 69 (1949), 11-14 (in Russian).
[113] Olver, F.W.J.: 'Asymptotics and special functions',
      Academic Press, New York, 1974.
[114] Pukanszky, L.: 'The Plancherel formula for the univer-
      sal covering group of $SL(R,2)$', Math. Ann. 156 (1964).
[115] Robin, L.: 'Fonctions sphériques de Legendre et fonc-
      tions sphéroidales, tome III', Gauthier-Villars,
      Paris, 1959.

[116] Roehner, B. and G. Valent: 'Solving the birth and death processes with quadratic asymptotically symmetric transition rates', SIAM J. Appl. Math. 42 (1982), 1020-1046.

[117] Rosenberg, J.: 'A quick proof of Harish-Chandra's Plancherel theorem for spherical functions on a semisimple Lie group', Proc. Amer. Math. Soc. 63 (1977), 143-149.

[118] Rudin, W.: 'Functional analysis', McGraw-Hill, New York, 1973.

[119] Samii, H.: 'Les transformations de Poisson dans la boule hyperbolique', Thèse 3me cycle, Université de Nancy I, 1982.

[120] Schindler, S.: 'Some transplantation theorems for the generalized Mehler transforms and related asymptotic expansions', Trans. Amer. Math. Soc. 155 (1971), 257-291.

[121] Schmid, W.: 'Representations of semi-simple Lie groups', in M.F. Atiyah (ed.), Representation theory of Lie groups, Cambridge University Press, Cambridge, 1979, pp. 185-235.

[122] Sekiguchi, J.: 'Eigenspaces of the Laplace-Beltrami operator on a hyperboloid', Nagoya Math. J. 79 (1980), 151-185.

[123] Smith, R.T.: 'The spherical representations of groups transitive on S^n', Indiana Univ. Math. J. 24 (1974), 307-325.

[124] Sneddon, I.N.: 'The use of integral transforms', McGraw-Hill, New York, 1972.

[125] Sprinkhuizen-Kuyper, I.G.: 'A fractional integral operator corresponding to negative powers of a certain second order differential operator', J. Math. Anal. Appl. 72 (1979), 674-702.

[126] Sprinkhuizen-Kuyper, I.G.: 'A fractional integral operator corresponding to negative powers of a second order partial differential operator', Report TW 191/79, Math. Centrum, Amsterdam, 1979.

[127] Stanton, R.J. and P.A. Tomas: 'Expansions for spherical functions on noncompact symmetric spaces', Acta Math. 140 (1978), 251-271.

[128] Stein, E.M. and S. Wainger: 'Analytic properties of expansions, and some variants of Parseval-Plancherel formulas', Ark. Mat. 5 (1963), 553-567.

[129] Takahashi, R.: 'Sur les réprésentations unitaires
      des groupes de Lorentz généralisés', Bull. Soc. Math.
      France 91 (1963), 289-433.
[130] Takahashi, R.: 'Fonctions sphériques dans les groupes
      Sp(n,1)', in J. Faraut (ed.), Théorie du potentiel
      et analyse harmonique, Lecture Notes in Math. 404,
      Springer, Berlin, 1974, pp. 218-238.
[131] Takahashi, R.: 'Spherical functions in $Spin_0(1,d)/$
      Spin(d-1) for d = 2,4 and 8', in J. Carmona,
      M. Vergne (eds.), Non-commutative harmonic analysis,
      Lecture Notes in Math. 587, Springer, Berlin, 1977,
      pp. 226-240.
[132] Takahashi, R.: 'Quelques résultats sur l'analyse har-
      monique dans l'espace symétrique non compact de rang 1
      du type exceptionnel', in P. Eymard, J. Faraut,
      G. Schiffman, R. Takahashi (eds.), Analyse harmonique
      sur les groupes de Lie, II, Lecture Notes in Math.
      739, Springer, Berlin, 1979, pp. 511-567.
[133] Takahashi, R.: 'SL(2,$\mathbb{R}$)', in J.-L. Clerc, P. Eymard,
      J. Faraut, M. Rais, R. Takahashi, Analyse harmonique,
      C.I.M.P.A., Nice, 1982, Ch.III.
[134] Terras, A.: 'Noneuclidean harmonic analysis', SIAM
      Rev. 24 (1982), 159-193.
[135] Thomas, E.G.F.: 'The theorem of Bochner-Schwartz-
      Godement for generalised Gelfand pairs', preprint.
[136] Titchmarsh, E.C.: 'Eigenfunction expansions associated
      with second-order differential equations, Part I',
      Oxford University Press, London, 2nd ed., 1962.
[137] Trimèche, K.: 'Transformation intégrale de Weyl et
      théorème de Paley-Wiener associés à un opérateur
      différentiel singulier sur (0,∞)', J. Math. Pures
      Appl.(9) 60 (1981), 51-98.
[138] Vilenkin, N.Ja.: 'Special functions connected with
      class 1 representations of groups of motion in spaces
      of constant curvature', Trudy Moskov. Mat. Obšč.
      12 (1963), 185-257 (in Russian) = Trans. Moscow
      Math. Soc. 12 (1963), 209-290.
[139] Vilenkin, N.Ja.: 'Special functions and the theory of
      group representations', Moscow, 1965 (in Russian) =
      Amer. Math. Soc. Transl. of Math. Monographs, Vol. 22,
      Amer. Math. Soc., Providence, R.I., 1968.
[140] Vilenkin, N.Ja. and R.L. Šapiro: 'Irreducible repre-
      sentations of the group SU(n) of class I relative to
      SU(n-1)', Izv. Vysš. Učebn. Zaved. Matematika (1967),
      no. 7 (62), 9-20 (in Russian) = Amer. Math. Soc.
      Transl.(2) 113(1979), 187-200.

[141] Vretare, L.: 'On L_p Fourier multipliers on certain symmetric spaces', Math. Scand. 37 (1975), 111-121.

[142] Wallach, N.R.: 'Harmonic analysis on homogeneous spaces', Dekker, New York, 1973.

[143] Wetering, R.L. van de: 'Variation diminishing Fourier-Jacobi transforms', SIAM J. Math. Anal. 6 (1975), 774-783.

[144] Weyl, H.: 'Über gewöhnliche lineare Differential-gleichungen mit singulären Stellen und ihre Eigen-funktionen (2.note)', Göttinger Nachrichten (1910), 442-467 = Gesammelte Abhandlungen I, 222-247.

[145] Whittaker, E.T. and G.N. Watson: 'Modern analysis', Cambridge University Press, Cambridge, 4th ed., 1927.

[146] Wilson, J.A.: 'Some hypergeometric orthogonal polyno-mials', SIAM J. Math. Anal. 11 (1980), 690-701.

Dennis Stanton

ORTHOGONAL POLYNOMIALS AND CHEVALLEY GROUPS

1. INTRODUCTION

This paper is a survey of recent work on orthogonal polynomials and Chevalley groups. The orthogonal polynomials
we emphasize are those given by basic hypergeometric series,
or q‑series. The group theoretic significance of these
polynomials is that they are the spherical functions for
Chevalley groups over the finite field $GF(q)$.

Delsarte's thesis [12] made it clear that there is a
great interplay between orthogonal polynomials and certain
combinatorial questions. One question he asked was: Given
a finite metric space X , find some necessary conditions
so that there is a subset $Y \subset X$ whose pairwise distances
have certain properties. With some assumptions about X
and Y , the conditions can be stated with orthogonal polynomials. Many examples of spaces X can be constructed
from Chevalley groups over a finite field. The related polynomials are given by basic hypergeometric series.

In effect, the physicist's sphere S^2 has been replaced
by a combinatorialist's space X . The classical theory of
spherical harmonics on S^2 is replaced by a theory of harmonics on X . The associated group actions on S^2 and X
are very similar. There is some irony here. The fields $\mathbb{R}$
and $\mathbb{C}$ have been replaced by a finite field $GF(q)$. There

R. A. Askey et al. (eds.), Special Functions: Group Theoretical Aspects and Applications, 87–128.
© *1984 by D. Reidel Publishing Company.*

are interesting special functions for classical groups over
a finite field.

The group theoretic view that this paper stresses is
spherical functions. It is well‑known that spherical
functions on symmetric spaces are orthogonal polynomials
[31], [47]. The most classical example is ultraspherical
polynomials. In this case the group $G = SO(n)$ acts on
the sphere S^{n-1}. Then the functions on the sphere de‑
compose: $L^2(S^{n-1}) = \displaystyle\bigoplus_{\ell=0}^{\infty} \text{Harm}(\ell)\big|_{S^{n-1}}$ where each

$\text{Harm}(\ell)\big|_{S^{n-1}}$ is an irreducible, inequivalent representation

of $SO(n)$. Because $S^{n-1} \cong SO(n-1)\backslash SO(n)$, each
$\text{Harm}(\ell)\big|_{S^{n-1}}$ contains a function f_ℓ (unique up to scalar)

which is right $SO(n-1)$‑invariant. Letting $SO(n-1)$ be
be the rotations about the north pole of S^{n-1}, $f_\ell(p)$ only
depends upon the angle θ that p makes with the north

pole. In fact $f_\ell(p) = P_\ell^{(\frac{n-3}{2}, \frac{n-3}{2})}(\cos\theta)$, [31], the
ultraspherical polynomial.

This is but a special case of a more general point of
view of special functions and group theory. We took
$G = SO(n)$, $H = SO(n-1)$ and considered functions on the
double cosets HgH. One can take other decompositions and
other invariance properties to define other special functions
on groups. There are many applications. Yet in this paper
we shall wee that spherical functions naturally lead to
orthogonal polynomials on finite groups.

Some other interpretations of special functions (and orthogonal polynomials) and group theory should be mentioned. One successful approach is to explicitly parametrize elements of a Lie group, and then compute matrix elements of an ir- reducible representation of that group [46]. The matrix elements can be orthogonal polynomials. There has been no work in this direction for finite Chevalley groups.

Another highly successful and productive method is by separation of variables [36]. Here a Lie algebra of symme- tries of a partial differential equation is computed. The Lie algebra is realized by differential operators. Diagonal- izing certain operators yields special functions (sometimes orthogonal polynomials). Then facts about these functions can be found by exponentiating - applying an element of the symmetry group. In fact, it is possible to start with the special functions and build a Lie algebra based upon the differential - recurrence relations they satisfy [35]. This method has been applied [37] to the functions which are spherical functions for Chevalley groups. Only the formal properties of the polynomials (not the group theory) has been used. But there has been no attempt at using the Chevalley basis for the Lie algebra over GF(q) . This is another possible avenue of inquiry.

Finally, the Clebsh - Gordon coefficients [3], [33] can be renormalized to be orthogonal polynomials. There is a connection between them and finite groups [22], [33]. A start on this problem for Chevalley groups was given by Dunkl [23].

The rest of the paper is organized in the following
way. In Section 2 a general theory for finite groups and
orthogonal polynomials is given. For Weyl groups and
Chevalley groups this theory is applied in sections 3 and
4. The intertwining functions obtained are listed in
Tables 1 and 2. Section 5 gives the analogues of harmonic
polynomials via posets.

2. FINITE GROUPS AND ORTHOGONAL POLYNOMIALS

In this section we present the general machinery for finite
groups to give orthogonal polynomials. Some relevant
sources are [5], [6], [12], [28], [49].

First we recall some facts about orthogonal polynomials
[44]. The classical orthogonal polynomials are sequences
of real polynomials $\{p_n(x)\}_{n=0}^{\infty}$, degree $(p_n(x)) = n$,
such that

$$\int_a^b p_n(x)p_m(x)\, w(x)\, dx = 0 \quad , \quad n \neq m . \tag{2.1}$$

The weight function $w(x)$ is non-negative on $[a,b]$. For
example, the Legendre polynomials $P_n(x)$ satisfy

$$\int_{-1}^1 P_n(x)P_m(x)\, dx = 0 , n \neq m . \tag{2.2}$$

However, in (2.1) the weight function $w(x)\, dx$ can be replaced by $d\alpha(x)$ [44]. For finite groups we are concerned
with $\alpha(x)$ constant, except for a finite number of jumps.
Then the orthogonality relation becomes

$$\sum_{i=0}^N p_n(x_i)\, p_m(x_i)\, w(x_i) = 0 , n \neq m , \tag{2.3}$$

for some positive weights $w(x_i)$. It is easy to see that

(2.3) implies that $\{p_0, p_1, \ldots, p_N\}$ is an independent set
of functions on the set $\{x_0, x_1, \ldots, x_N\}$. Thus for
$m > N$ p_m must vanish on $\{x_0, \ldots, x_N\}$. So, in the discrete orthogonality relation (2.3), it is customary to speak
of the corresponding finite sequence of orthogonal polynomials
$\{p_n(x)\}_{n=0}^{N}$.

For certain sets of discrete orthogonal polynomials,
we shall give a finite group theoretic interpretation of
(2.3). Recall the group theoretic proof of (2.2). Let
$SO(3)$ be the group of rotations, which acts upon the unit
sphere $S^2 \subset \mathbb{R}^3$. Then

$$L^2(S^2) = \bigoplus_{\ell=0}^{\infty} \text{Harm}(\ell)\big|_{S^2} , \qquad (2.4)$$

and each $\text{Harm}(\ell)$ is an irreducible representation of
$SO(3)$. Each $\text{Harm}(\ell)$ has a unique (up to scalar multiple)
function $f_\ell \in L^2(S^2)$ such that $f_\ell(p)$ is constant along
the circle $p \cdot (0,0,1) = \cos\theta$. This function $f_\ell(p)$ is
the principle spherical harmonic, $f_\ell(p) = P_\ell(\cos\theta)$.
As functions on S^2, $f_m \perp f_n$, because they are in distinct
irreducible representations of $SO(3)$. The measure on
S^2 is $\frac{1}{4\pi} \sin\theta \, d\varphi \, d\theta$. We obtain (2.2) with $x = \cos\theta$.

Thus (2.2) is an orthogonality for functions on the
double coset space $SO(2)\backslash SO(3)/SO(2)$. For finite groups
G , counting measure on the double cosets $H\backslash G/H$ will
supply the measure for (2.3).

Other special functions have orthogonality relations.
The distinguishing property of orthogonal polynomials is the

three - term recurrence relation

$$xp_n(x) = a_n p_{n+1}(x) + b_n p_n(x) + c_n p_{n-1}(x) \ , \ 0 \leq n \leq N-1$$

(2.5)

The constants a_n, b_n , and c_n have certain positivity restrictions that depend upon the normalization of the polynomials. Favard's Theorem [44] is a converse. Finite groups
will automatically give an interpretation of (2.5). This is
not true for Lie groups. There differential equations (via
the Lie algebra) are given for the polynomials. Not all interesting sets of orthogonal polynomials have nice differential equations. Finite groups are better attuned to the
polynomial nature of these special functions.

Now we come to the general hypotheses for finite groups
G . Let G be a finite group of automorphism of a finite
metric space X . Assume that the metric d on $X \times X$ is
integer valued. Note that X naturally forms the vertices
of a graph whose edges are (x_1, x_2) with $d(x_1, x_2) = 1$.
We make the hypotheses (*) about G and X

(*) $\begin{cases} (1) & \text{the metric } d \text{ on } X \times X \text{ coincides with the graph} \\ & \text{distance on } X \times X , \\ (2) & G \text{ is two - point homogeneous on } X ; \text{ i.e., given} \\ & (x_1, x_2) \in X \times X , \ (y_1, y_2) \in X \times X \text{ with } d(x_1, x_2) = \\ & d(y_1, y_2) , \text{ then there is an element } g \in G \text{ such} \\ & \text{that } (x_1 g, \ x_2 g) = (y_1, y_2) . \end{cases}$

The hypothesis (*)(2) is analogous to the action of
$G = SO(3)$ on $X = S^2$ (see [48]). In particular, if
$x_1 = x_2$ and $y_1 = y_2$, (*)(2) implies that G is transitive on X . Fix a base point $x^{(0)} \in X$ and define the

stablizer subgroup $H = \{g \in G: x^{(0)}g = x^{(0)}\}$, so that
$X \cong H\backslash G$. Just as the action of $G = SO(3)$ on $X = S^2$
yields the induced representation of G on $L^2(S^2)$, in (*)
the action of G on $L^2(X)$ (all complex-valued functions
on X) is precisely the permutation representation of G
induced from the subgroup H . Under (*), the following
theorem describes this representation.

Theorem 2.6 Let G and X satisfy (*) , where the maximum
distance in $X \times X$ is N . Then

$$L^2(X) = V_{\lambda_0} \oplus \cdots \oplus V_{\lambda_N} ,$$

where $\{V_{\lambda_i}\}_{i=0}^N$ are distinct irreducible representations
of G .

   Since the representation $L^2(X)$ is multiplicity free,
each V_{λ_i} has a unique function $f_{\lambda_i} \in V_{\lambda_i}$ such that

$f_{\lambda_i}(x^{(0)}) = 1$ and $f_{\lambda_i}(xh) = f_{\lambda_i}(x)$ for all $x \in X$, $h \in H$.
The functions f_{λ_i} are the spherical functions of $L^2(X)$.

They are constant on the H-orbits on $X \cong H\backslash G$, so each
f_{λ_i} can be identified with a function on the double cosets
$H\backslash G/H$.

   The spherical functions satisfy the orthogonal relation
(2.3). There is a 1-1 correspondence between the double
cosets $H\backslash G/H$ and the G-orbits on $X \times X$. Since there are
$N+1$ distances, the H-orbits on X are

$$\Omega_k = \{x \in X : d(x^{(0)},x) = k\} , \ 0 \le k \le N .$$

If $x \in \Omega_k$, we can identify $f_{\lambda_i}(\Omega_k) = f_{\lambda_i}(x)$. The orthogonality $f_{\lambda_i} \perp f_{\lambda_j}$ in $L^2(X)$ is

$$\frac{1}{|X|} \sum_{k=0}^{N} f_{\lambda_i}(\Omega_k)\overline{f_{\lambda_j}(\Omega_k)}|\Omega_k| = 0 , \ 0 \le i \neq j \le N . \quad (2.7)$$

It is well-known [24; p. 103] that the L^2-norm can be given by the dimensions of the irreducible constituents, so

$$\frac{1}{|X|} \sum_{k=0}^{N} f_{\lambda_i}(\Omega_k)f_{\lambda_j}(\Omega_k)|\Omega_k| = \delta_{ij}(\dim V_{\lambda_j})^{-1} . \quad (2.8)$$

Equations (2.7) and (2.8) resemble (2.3) with $w(x_k)$ replaced by $|\Omega_k|$. The functions f_{λ_i} are real valued. We do not yet have polynomials.

Equation (2.8) implies that the matrix

$$M_{ik} = f_{\lambda_i}(\Omega_k)[\dim V_{\lambda_i} \cdot |\Omega_k| / |X|]^{1/2}$$

is orthogonal by rows. The orthogonality by columns implies

$$\frac{1}{|X|} \sum_{i=0}^{N} f_{\lambda_i}(\Omega_k)f_{\lambda_i}(\Omega_j)(\dim V_{\lambda_i}) = \delta_{kj}|\Omega_k|^{-1} . \quad (2.9)$$

Thus, both (2.8) and (2.9) are orthogonality relations of the form (2.3). We need to find the polynomial nature of the functions $f_{\lambda_i}(\Omega_k)$ by looking at a proof of Theorem 2.6.

The commuting algebra $\mathfrak{U}$ of $L^2(X)$ is the algebra of $|X|x|X|$ complex matrices which commute with all of the permutation matrices g ,

$$\mathfrak{U} = \{A | A : L^2(X) \to L^2(X) , \ gA = Ag \ \forall \ g \in G\} \quad (2.10)$$

It is well - known [11], [49] that if $\mathfrak{A}$ is commutative, then $L^2(X)$ is multiplicity free. Since the action of G on X satisfies (*)(2), it is easy to check that $\mathfrak{A}$ is spanned by the matrices $\{D_0, D_1, \ldots, D_N\}$ where

$$D_i(x,y) = \begin{cases} 1 & d(x,y) = i \\ 0 & d(x,y) \neq i \end{cases}, \quad (x,y) \in X \times X .$$

The following three - term recurrence relation is easy to show:

$$D_1 D_i = c_{i+1} D_{i+1} + a_i D_i + b_{i-1} D_{i-1} , \quad 0 \le i \le N \qquad (2.11)$$

where $c_{i+1} = |\{z : d(x,z) = 1, d(y,z) = i\}|$, for fixed (x,y)

$\qquad\qquad d(x,y) = i+1$,

$\qquad\qquad a_i = |\{z : d(x,z) = 1, \ d(y,z) = i\}|$, for fixed (x,y),

$\qquad\qquad d(x,y) = i$,

$\qquad\qquad b_{i-1} = |\{z : d(x,z) = 1, \ d(y,z) = i\}|$, for fixed (x,y),

$\qquad\qquad d(x,y) = i-1$.

Equation (2.11) implies that D_i is a polynomial of degree i in D_1, $D_i = p_i(D_1)$, because $c_i \neq 0$, $0 < i \le N$. Therefore $\mathfrak{A}$ is commutative. The three term recurrence (2.5) is inter- preted as special structure constants in the commuting algebra.

We need a connection between the algebra $\mathfrak{A}$ and the spherical functions to transfer the polynomial property to the f_{λ_i}'s . The irreducible representations V_{λ_i} in Theorem 2.6 can be described by the algebra $\mathfrak{A}$. Any $A \in \mathfrak{A}$ is con- stant on V_{λ_i} , so we put

$$V_{\lambda_i} = \{f \in L^2(X) : D_1 f = \lambda_i f\} . \qquad (2.12)$$

Because $\mathfrak{A}$ is $(N+1)$ - dimensional and generated by D_1 , the λ_i's of (2.12) are distinct. Each λ_i solves (2.11) for $i = N$:

$$\lambda_i p_N(\lambda_i) = a_N p_N(\lambda_i) + b_{N-1} p_{N-1}(\lambda_i) \; .$$

In (2.12) choose the spherical function $f_{\lambda_i} \in V_{\lambda_i}$. If we evaluate the eigenvalue equation $D_1 f_{\lambda_i} = \lambda_i f_{\lambda_i}$ at $x \in \Omega_k$, we find

$$\lambda_i f_{\lambda_i}(\Omega_k) = \gamma_k f_{\lambda_i}(\Omega_{k+1}) + \alpha_k f_{\lambda_i}(\Omega_k) + \beta_k f_{\lambda_i}(\Omega_{k-1}) \; , \quad (2.13)$$

where

$$\gamma_k = |\{z : d(x,z) = 1, \; d(z,x^{(0)}) = k+1\}| \; ,$$

$$\alpha_k = |\{z : d(x,z) = 1, \; d(z,x^{(0)}) = k\}| \; , \text{ and}$$

$$\beta_k = |\{z : d(x,z) = 1, \; d(z,x^{(0)}) = k-1\}| \; .$$

Again (2.13) implies that $f_{\lambda_i}(\Omega_k)$ is a polynomial of degree k in λ_i . By a suitable change of normalization (2.13) and (2.11) are identical. In (2.11) if we consider all triangles (x,y,z) we find

$$|X| |\Omega_{i+1}| c_{i+1} = |X| |\Omega_i| \gamma_i \; ,$$

$$|X| |\Omega_i| a_i = |X| |\Omega_i| \alpha_i \; , \text{ and}$$

$$|X| |\Omega_{i-1}| b_{i-1} = |X| |\Omega_i| \beta_i \; .$$

Thus,

$$|\Omega_k| f_{\lambda_i}(\Omega_k) = p_k(\lambda_i) \; . \qquad\qquad (2.14)$$

Theorem 2.15 In Theorem 2.6 let f_{λ_i} be the spherical function.for V_{λ_1} . Let $\Omega_k = \{x \in X : d(x^{(0)},x) = k\}$. Then $|\Omega_k| f_{\lambda_i}(\Omega_k) = p_k(\lambda_i)$, for some orthogonal polynomial $p_k(x)$

of degree k .

The dual orthogonality relation (2.9) is the orthogonality relation (2.3) for the polynomials $p_k(x)$. Equation
(2.11) (or (2.13)) is the three - term recurrence relation for
the polynomials $p_k(x)$. For the spherical functions $f_{\lambda_i}(\Omega_k)$,
it can be interpreted as a second order difference equation.
We shall see, that for all of the examples in this paper,
$f_{\lambda_i}(\Omega_k) = q_i(\mu_k)$ for some polynomial $q_i(x)$ of degree i .
Then (2.8) is the orthogonality relation for the polynomials
$q_i(x)$.

A few more comments are in order to describe the con-
nection between $\mathfrak{U}$ and the f_{λ_i}'s . The f_{λ_i}'s can be
considered as functions on group G such that $f_{\lambda_i}(h_1 g h_2) =$
$f_{\lambda_i}(g)$. They form an algebra $\mathfrak{B}$ under convolution on the
group. $\mathfrak{B}$ is isomorphic to $\mathfrak{U}$. Another basis for $\mathfrak{B}$ is
given by the characteristic functions of the double cosets
Hg_kH corresponding to Ω_k . Roughly speaking, under the
isomorphism Hg_kH maps to D_k and f_{λ_i} maps to an idem-
potent J_i . Because f_{λ_i} is a linear combination of
Hg_kH , an expression for J_i in terms of D_k also gives
the values of f_λ [12], [6,p.132].

In summary, some of the striking similarities (see
[27; Chap. 8]) between the spherical functions for $L^2(X)$
and $L^2(S^{n-1})$ are:

(1) In $L^2(S^{n-1})$ any invariant differential operator
is a polynomial in the spherical Laplacian. In $L^2(X)$ any

operator commuting with the action of G is a polynomial in $D_1 - |\Omega_1| I$.

(2) $\Delta_{sph} f = \lambda f$ is a second order differential equation for the ultraspherical polynomials. $(D_1 - |\Omega_1| I) f = \lambda f$ is a second order difference equation for discrete orthogonal polynomials.

(3) The spherical Laplacian Δ_{sph} can be obtained from a mean value operator T_θ . The operator T_θ averages a function f at an angle θ from a point p . Then, up to a scalar, $\Delta_{sph} = \lim_{\theta \to 0} \dfrac{T_\theta - I}{\theta^2}$. This corresponds to the discrete difference quotient $D_1 - |\Omega_1| I$, since one is the minimal distance.

(4) For $f \in \text{Harm}(\ell)|_{S^{n-1}}$, $T_\theta f = p_\ell^{(\frac{n-3}{2}, \frac{n-3}{2})} (\cos \theta)$.

For $f \in V_{\lambda_\ell}$, $D_k f = p_k(\lambda_\ell) f$.

(5) The eigenvalues of Δ_{sph} are non-positive. The eigenvalues of $D_1 - |\Omega_1| I$ are non-positive.

3. ORTHOGONAL POLYNOMIALS FOR WEYL GROUPS

In this section we apply the general theory of section 2. We take $G = W$, the Weyl group associated with a Chevalley group, and $X = W_J \backslash W$, where W_J is a maximal parabolic subgroup of W . As the next section shows, this allows a natural q-analogue by taking the corresponding Chevalley group over $GF(q)$.

Before describing the groups and the associated metric spaces, we give some basic facts about the classical discrete orthogonal polynomials [2], [3], [26] that will occur. To agree with (2.8) and (2.9), we write the orthogonality relation as

$$\frac{1}{|X|} \sum_{k=0}^{N} p_i(X_k)p_j(X_k)w(x_k) = \delta_{ij}\, h_j^{-1}. \tag{3.1}$$

The <u>Hahn</u> <u>polynomials</u> $Q_j(x;\; \alpha,\beta,N)$

The Hahn polynomials $Q_j(x;\alpha,\beta,N)$ are polynomials in x of degree j . They satisfy (3.1) with $x_k = k$,

$$|X| = \binom{-\alpha-\beta-2}{N},$$

$$w(x_k) = \binom{-\alpha-1}{k}\binom{-\beta-1}{N-k}, \text{ and} \tag{3.2}$$

$$h_j = \frac{(-N)_j(\alpha+1)_j(\alpha+\beta+1)_j}{j!(N+\alpha+\beta+2)_j(\beta+1)_j}\; \frac{2j+\alpha+\beta+1}{\alpha+\beta+1}\; (-1)^j. \tag{3.3}$$

Our notation is the usual one: for shifted factorials: $(a)_k = \Gamma(a+k)/\Gamma(a)$, and for the binomial coefficients $\binom{a}{k} = (-a)_k(-1)^k/k$! Note that the weight function $w(x_k)$ is positive if $-1<\alpha,\beta$ or $\alpha,\beta<-N$. For finite groups α and β will be negative integers so that (3.2) has combinatorial meaning. An explicit formula is

$$Q_j(x;\alpha,\beta,N) = {}_3F_2\left(\begin{array}{c}-j, j+\alpha+\beta+1, -x\\ \alpha+1, -N\end{array}\middle|\, 1\right), \; 0\le j\le N. \tag{3.4}$$

The <u>dual</u> <u>Hahn</u> (<u>Eberlein</u>) <u>polynomials</u> $E_j(\lambda(x);\alpha,\beta;N)$
Put $\lambda(x) = x(x+\alpha+\beta+1)$ and note that

$$(-x)_m(x+\alpha+\beta+1)_m = \prod_{\ell=0}^{m-1} (-\lambda(x)+\ell(\ell+\alpha+\beta+1)) \tag{3.5}$$

is a polynomial in $\lambda(x)$ of degree m. We see from (3.4)
that for $j = 0,1,\ldots,N$ $Q_x(j;\alpha,\beta,N)$ is a polynomial in
$\lambda(x)$ of degree j. Define $E_j(\lambda(x);\alpha,\beta,N) = Q_x(j,\alpha,\beta,N)$,
$0 \leq j \leq N$. The orthogonality relation (3.1) is given by
(3.2) and (3.3) with $w(x_k)$ and h_j interchanged. The
three-term recurrence for the Eberlein polynomials is the
same as the second-order difference equation for Hahn polynomials. It is

$$-\lambda(x)E_k(\lambda(x)) = k(N+\beta+1-k)E_{k-1}(\lambda(x)) +$$

$$(N-k)(\alpha+1+k)E_{k+1}(\lambda(x)) - \{k(N+\beta+1-k)$$

$$+ (N-k)(\alpha+1+k)\}E_k(\lambda(x)) , \quad 0 \leq k \leq N . \qquad (3.6)$$

The symmetric Krawtchouk polynomials $K_j(x;1/2,N)$

These are again polynomials in x of degree j. They
satisfy (3.1) with $x_k = k$, $|X| = 2^N$,

$$w(x_k) = \binom{N}{k} \quad , \text{ and} \qquad (3.7)$$

$$h_j = \binom{N}{j} . \qquad (3.8)$$

An explicit formula for these polynomials is

$$K_j(x;1/2,N) = {}_2F_1\left(\begin{matrix} -j, -x \\ -N \end{matrix} \middle| 2\right) , 0 \leq j \leq N . \qquad (3.9)$$

Since (3.9) implies that $K_j(x;1/2,N)$ is symmetric in x
and j for $x,j = 0,1,\ldots,N$, these polynomials are self-
dual. This explains $w(x_k) = h_k$. The second-order difference equation (or three-term recurrence) is

$$-2x K_k(x) = k K_{k-1}(x) + (N-k) K_{k+1}(x) - N K_k(x) \qquad (3.10)$$

There are Krawtchouk polynomials $K_j(x;p,N)$, whose
weight function is $w(x_k) = \binom{N}{k}p^k(1-p)^{-k}$. For $p = t/(t+1)$
they are spherical functions for finite groups that are not

Weyl groups [18].

Now we come to the group theoretic interpretation of these polynomials form Weyl groups.

A Chevalley group of type A_{v-1} has the symmetric group $G = S_v = W$ as its Weyl group [7], [8]. A maximal parabolic subgroup W_J of W for $J = \{1,2,\ldots,v-1\} - \{n\}$ is $W_n = S_n \times S_{v-n}$. The coset space $W_n \backslash W \cong X$, the set of all n - subsets of a v - set. Note that W_n stabilizes the set $\{1,2,\ldots,n\} = x(0)$, and by symmetry we can assume that $n \leq v-n$. The G - invariant metric on X is $d(x,y) = n - |x \cap y|$. The hypotheses (*) are satisfied by G and X.

The orbits of W_n on X are

$$\Omega_k = \{x \in X: |x \cap x^{(0)}| = n-k\} \tag{3.11}$$

and clearly $|X| = \binom{v}{n}$ and

$$|\Omega_k| = \binom{n}{n-k}\binom{v-n}{k} . \tag{3.12}$$

From (3.2), we could expect that the spherical functions are the Hahn polynomials $Q_j(k;n-v-1,-n-1;n)$. We check that the eigenvalue equation (2.13) agrees with (3.6). A computation of the coefficients $\alpha_k, \beta_k, \gamma_k$ in (2.13) yields

$$\lambda\, f(\Omega_k) = (n-k)(v-n-k)f(\Omega_{k+1}) + k^2\, f(\Omega_{k-1})$$
$$+ \{n(v-n) - (n-k)(v-n-k) - k^2\} f(\Omega_k) , \quad 0 \leq k \leq n . \tag{3.13}$$

From (3.6), we see that $f_{\lambda_i}(\Omega_k)$ is an Eberlein polynomial $E_k(\lambda(i); n-v-1, -n-1, n)$ in $\lambda(i)$, where the eigenvalues of D_1 are $\lambda_i = n(v-n) - \lambda(i) = n(v-n) - i(i-v-1)$,

$0 \leq i \leq n$. Equivalently, $f_{\lambda i}(\Omega_k)$ is a Hahn polynomial $Q_i(k; n-v-1, -n-1; n)$.

It is clear from the representation theory of the symmetric group S_v [30] that in the decomposition $L^2(X) = \overset{n}{\underset{k=0}{\oplus}} V_{\lambda_k}$, the modules V_{λ_k} correspond to the irreducible representations whose associated partition has at most two rows. In fact, $V_{\lambda_k} \cong [v-k,k] = \mu$. From (3.3) we see that $\dim V_{\lambda_k} = \binom{v}{k} - \binom{v}{k-1}$, which agrees with the hook formula. In this direction one other formula can be mentioned. Let χ^μ be the character of V_{λ_k} . The spherical function f_{λ_k} for V_{λ_k} can be found from χ^μ by

$$f_{\lambda_k}(g) = \frac{1}{|W_n|} \underset{h \in W_n}{\Sigma} \chi^\mu(gh) \quad . \tag{3.14}$$

The substitution $g_x = (1, n+1) \ldots (x, n+x)$, $0 \leq x \leq n$ implies

$$_3F_2 \left(\begin{array}{c} -k,\ k-v-1,\ -x \\ n-v,\ -n \end{array} \middle| 1 \right) = \frac{1}{n!(v-n)!} \underset{h \in S_n \times S_{v-n}}{\Sigma} \chi^\mu(g_x h) . \tag{3.15}$$

One can ask if there are any relationships between these Hahn polynomials and the ultraspherical polynomials. For example, Bessel functions can be obtained as a limit of ultraspherical polynomials. Geometrically, this corresponds to a very large sphere being nearly flat. For Hahn polynomials

$$\lim_{N \to \infty} Q_k(Nx; \alpha, \beta, N) = c\ P_k^{(\alpha,\beta)}(1-2x) \quad , \tag{3.16}$$

where $P_k^{(\alpha,\beta)}(x)$ are Jacobi polynomials, $\alpha, \beta > -1$. Our

Hahn polynomials have $\alpha = n - v - 1 < -n$ and $\beta = -n - 1 < -n$, so if $n \to \infty$, then $\alpha, \beta \to -\infty$. However, a transformation [39, Eq. 2.9] for Hahn polynomials implies

$$Q_k(x;n-v-1,-n-1,n) = \frac{x!}{(n-v)_x} Q_{n-k}(x;0,v-2n;n),$$

$$x = 0, 1, \ldots, n. \tag{3.17}$$

Combined with (2.8), (3.17) gives the orthogonality relation for $Q_{n-k}(x; 0, v - 2n, n)$. If $k = n - \ell$, $v - 2n = s$, and $n \to \infty$, then $Q_{n-k}(nx; 0, v - 2n, n) \to c\, P_\ell^{(0,s)}(1 - 2x)$. It is not clear if this has a geometric interpretation. In what sense does the S_{2n} - module $V_{\lambda_{n-\ell}} \subset L^2(X_n)$ correspond to $\mathrm{Harm}(\ell) \subset L^2(S^2)$ for n large? The space X_n is the set of all n - subsets of $\{1, 2, 3, \ldots, 2n\}$. As $n \to \infty$, the second order difference equation approaches the second order differential equation for Legendre polynomials.

A more general set of Hahn polynomials can be found by considering intertwining functions [21] rather than spherical functions. Given a subgroup K of G, the intertwining functions are those functions in V_{λ_k} which are invariant under K. They are functions on the double cosets $H \backslash G / K$. For the symmetric group the choice of $K = W_m$, $n \leq m \leq v - n$, gives exactly one such function (up to a scalar multiple) in each V_{λ_k}. If the double cosets are parametrized properly, these functions are $Q_k(x;n-v-1, -m-1, n)$

A Chevalley group of type B_N or C_N has the hyper-octahedral group $G = S_N \cdot 2^N = W$ as its Weyl group [7], [8]. The maximal parabolic subgroup W_J of W corresponding to $J = \{1, 2, \ldots, N\} - \{k\}$ is $W_k = S_k \times (S_{N-k} \cdot 2^{N-k})$.

We concentrate on the case $k = N$, so that $W_N \backslash W \cong X$, the set of all N-tuples of 0's and 1's. (Note that W_N stablizes $(0, 0, ..., 0)$.) The G-invariant metric on X is the Hamming metric, $d(x,y) = \#$ of places where x and y differ. The set X is the set of vertices of an N-dimensional cube.

The orbits of W_N on X are $\Omega_k = \{x \in X: x \text{ has } k \text{ 1's}\}$, and clearly $|X| = 2^N$ and $|\Omega_k| = \binom{N}{k}$. The eigenvalue equation (2.13) is

$$\lambda f(\Omega_k) = (N-k)f(\Omega_{k+1}) + k f(\Omega_{k-1}) . \qquad (3.18)$$

So (3.10) implies that $\lambda_i = N - 2i$, and

$$f_{\lambda_i}(\Omega_k) = K_k(i; 1/2, N) . \qquad (3.19)$$

These polynomials are self-dual, and $\dim V_{\lambda_k} = \binom{W}{k}$.

A Chevalley group of type D_N has a subgroup W of $S_N \cdot 2^N$ as its Weyl group [7], [8]. Here W consists of all elements of $S_N \cdot 2^N$ with an even number of sign changes. We shall write $W = S_N \cdot 2^{N-1}$. The maximal parabolic subgroup W_J for $J = \{1, 2, ..., N\} - \{k\}$ is $W_k = S_k \times (S_{N-k} \cdot 2^{N-k-1})$, $1 \le k \le N-2$, and $W_{N-1} \cong W_N \cong S_N$. For $W_N \backslash W \cong X$, X consists of all N-tuples of 0's and 1's with an even $\#$ of 1's. For $W_{N-1} \backslash W \cong Y$, Y consists of all N-tuples of 0's and 1's with an odd $\#$ of 1's. Take $W_N \backslash W \cong X$, so that $W_N = S_N$ fixes $(0, 0, ..., 0)$. The metric is half of the Hamming metric.

Again the orbits on W_N on X are

$\Omega_k = \{x \in X: x \text{ has } 2k \text{ 1's}\}$, clearly $|X| = 2^{N-1}$ and
$|\Omega_k| = \binom{N}{2k}$. This time the eigenvalue equation (2.13) is

$$\mu_i f(\Omega_k) = \frac{1}{2}(N-2k)(N-2k-1) f(\Omega_{k+1}) +$$

$$k(2k-1) f(\Omega_{k-1}) + 2k(N-2k) f(\Omega_k) . \tag{3.20}$$

For N even, the solution is given by Hahn polynomials

$$f_{\mu_i}(\Omega_k) = Q_i(k;(-N-1)/2, N/2) , \mu_i = 2\lambda(i) +$$

$$N(N-1)/2 = 2i(i-N) + N(N-1)/2 , 0 \le i \le N/2 .$$

For N odd, the solution is again Hahn polynomials

$$f_{\mu_i}(\Omega_k) = Q_i(k; -N/2-1, -N/2, (N-1)/2) \tag{3.22}$$

for the same value of μ_i . In any case we have

$$f_{\mu_i}(\Omega_k) = {}_3F_2\left(\begin{matrix} -k,-i, i-N \\ (1-N)/2, -N/2 \end{matrix} \middle| 1\right), \quad 0 \le i, k \le [N/2] . \tag{3.23}$$

The group W is closely related to the hyperoctahedral group, so it is reasonable to assume that the Hahn polynomials in (3.23) are related to the symmetric Krawtchouk polynomials. The operators D_1 and D_2 for the hyperoctahedral group satisfy $D_1^2 = 2D_2 + nD_0$. The operator D_2 satisfies $D_2 f = \mu_j f$, $f \in V_{\mu_j}$ for type D_N . Because $\lambda_j^2 - N = 2\mu_j$ $= \lambda_{N-j}^2 - N$, we see that $V_{\lambda_j}\big|_X = V_{\lambda_{N-j}}\big|_X = W_{\mu_j}$. Thus $K_j(2k;1/2,N)$ and $K_{N-j}(2k;1/2,N)$ are also spherical functions for V_{μ_j} . The result is

$$_2F_1\left(\begin{matrix} -j, -2k \\ -N \end{matrix} \middle| 2\right) = {}_3F_2\left(\begin{matrix} -j, j-N, -k \\ (1-N)/2, -N/2 \end{matrix} \middle| 1\right),$$

$$0 \le j , k \le [N/2] . \tag{3.24}$$

This transformation can be derived from [4, Eq. 4.6(3)] .

Only the exceptional Weyl groups of types E, F, and G
remain. Since no infinite families of polynomials result,
we do not give the results. There is one other infinite
family of finite Coxeter groups - the dihedral groups $D_N = G$.
Here we let $C_2 \backslash G \cong X$, a regular N-gon , which the
usual graphical distance between nodes. The polynomials that
result are Chebychev polynomials [6, p. 17], [42].

For the hyperoctahedral group G , the double cosets
$S_k \times (S_{N-k} \cdot 2^{N-k}) \backslash G / S_\ell \times (S_{N-\ell} \cdot 2^{N-\ell})$ gives polynomials in
two variables [18, Th. 4.2] for the spherical functions if
$1 \leq k$, $\ell < n$. The algebra $\mathfrak{U}$ is commutative, and has two
generators and relations. The results of this section are
summarized in Table 1.

4. ORTHOGONAL POLYNOMIALS FOR CHEVALLEY GROUPS

We apply section 2 to Chevalley groups over GF(q) . These
groups are "made to order" for the program of section 2.
This is because the program worked for Weyl groups and there
is a 1-1 correspondence between the double cosets $W_J \backslash W / W_K$
and $G_J \backslash G / G_K$, where $G(G_J)$ is a Chevalley group
(parabolic subgroup) corresponding to the Weyl group W
(parabolic subgroup W_J [7], [8]). Also, the corresponding
algebras of spherical functions are isomorphic [11]. Thus,
whatever happened in section 3 applies to Chevalley groups
in section 4. This is the process of "taking the q-analogue".

The weight functions involved will contain the q - binomial coefficients $\begin{bmatrix} n \\ k \end{bmatrix}_q$. Recall [1] that

$$\begin{bmatrix} n \\ k \end{bmatrix}_q = \prod_{\ell=0}^{k-1} \frac{(1-q^{n-\ell})}{(1-q^{\ell+1})} = \frac{(q^n;q^{-1})_k}{(q;q)_k} \tag{4.1}$$

where

$$(a;q)_k = \prod_{\ell=0}^{k-1} (1 - aq^\ell) , \tag{4.2}$$

and that $\begin{bmatrix} n \\ k \end{bmatrix}_q$ is the number of k - dimensional subspaces of an n - dimensional vector space over $GF(q)$.

We list here the basic facts for q - Hahn and q - Krawtchouk polynomials. The notation does not entirely agree with previous papers [20], [23], [38], [39]. In essence, we have exponentiated all of the parameters.

<u>The q - Hahn polynomials</u> $Q_j(q^{-x};q^\alpha,q^\beta,N;q)$

The q - Hahn polynomials $Q_j(q^{-x};q^\alpha,q^\beta,N;q)$ are polynomials in q^{-x} of degree j . They satisfy (3.1) with $x_k = q^{-k}$, $|X| = \begin{bmatrix} -\alpha-\beta-2 \\ N \end{bmatrix}_q$,

$$w(x_k) = \begin{bmatrix} -\alpha-1 \\ k \end{bmatrix}_q \begin{bmatrix} -\beta-1 \\ N-k \end{bmatrix}_q q^{k(-\beta-1-N+k)} , \text{ and } \tag{4.3}$$

$$h_j = \frac{(q^{-N};q)_j(q^{\alpha+1};q)_j(q^{\alpha+\beta+1};q)_j}{(q;q)_j(q^{\beta+1};q)_j(q^{N+\alpha+\beta+2};q)_j} \frac{(1-q^{2j+\alpha+\beta+1})}{(1-q^{\alpha+\beta+1})} (-1)^j$$

$$q^{j(N-\alpha) - \binom{j+1}{2}} \tag{4.4}$$

Again α and β will be negative integers satisfying $\alpha, \beta < -N$ So that (4.3) has combinatorial meaning. An explicit formula is

$$Q_j(q^{-x};q^{\alpha},q^{\beta},N;q) = {}_3\phi_2\left(\begin{array}{c} q^{-j},q^{j+\alpha+\beta+1},q^{-x} \\ q^{\alpha+1}, \quad q^{-N} \end{array}\middle| q;q\right),$$

$$0 \leq j \leq N .$$

$$(4.5)$$

The dual q-Hahn (q-Eberlein) polynomials
$E_j(\lambda(x);q^{\alpha},q^{\beta},N;q)$

As in section 3, we see that $Q_x(q^{-j};q^{\alpha},q^{\beta},N;q)$ is a
polynomial in $\lambda(x) = (1-q^{-x})(1-q^{x+\alpha+\beta+1})$ of degree j .
So we define $E_j(\lambda(x);q^{\alpha},q^{\beta},N;q) = Q_x(q^{-j},q^{\alpha},q^{\beta},N;q)$,
$j = 0,1, \ldots, N$. The orthogonality relation (3.1) has $w(x_k)$
and h_j interchanged. The three-term recurrence relation is

$$-q^{-\alpha-\beta-1}\lambda(x)E_k(\lambda(x)) = (1-q^k)(1-q^{k-1-N-\beta})E_{k-1}(\lambda(x))$$

$$+q^{-N-\beta}(q^N-q^k)(q^{-\alpha-1}-q^k) - \{(1-q^k)(1-q^{k-1-N-\beta})$$

$$+q^{-N-\beta}(q^N-q^k)(q^{-\alpha-1}-q^k)\}E_k(\lambda(x)) .$$

$$(4.6)$$

The q-Krawtchouk polynomials $K_n(q^{-x}; q^c, N;q)$

These again are polynomials in q^{-x} of degree j . They
satisfy (3.1) with $x_k = q^{-k}$, $|X| = (-q^c;q)_N$,

$$w(x_k) = \begin{bmatrix} N \\ k \end{bmatrix}_q q^{\binom{k}{2}+ck} , \quad \text{and}$$

$$(4.7)$$

$$h_j = \frac{(q^{-N};q)_j(-q^{-N-c};q)_j}{(q;q)_j(-q^{1-c};q)_j} \frac{(1+q^{2j-N-c})}{(1+q^{-N-c})} (-1)^j q^{j(2N+c-j)} .$$

$$(4.8)$$

An explicit formula is

$$K_j(q^{-x};q^c,N;q) = {}_3\phi_2\left(\begin{array}{c} q^{-j},-q^{j-N-c},q^{-x} \\ 0,\ q^{-N} \end{array}\middle| q;q\right), 0 \le j \le N .$$

$$(4.9)$$

We do not give the dual polynomials. The second - order difference equation is

$$\lambda(j)K_j(q^{-x}) = (1-q^x)K_j(q^{1-x}) - q^c(q^N-q^x)K_j(q^{-1-x})$$

$$-\{(1-q^x)-q^c(q^N-q^x)\}K_j(q^{-x}) , \ 0 \le x \le N .$$ $$(4.10)$$

where $\lambda(j) = (1-q^{-j})(q^{N+c}+q^j)$.

The affine q - Krawtchouk polynomials $K_j^{Aff}(q^{-x};q^{-a},N;q)$

These are polynomials in q^{-x} of degree j . They satisfy (3.1) with $x_k = q^{-k}$, $|X| = q^{aN}$,

$$w(x_k) = [\begin{smallmatrix} N \\ k \end{smallmatrix}]_q(q^a;q^{-1})_k(-1)^k q^{\binom{k}{2}} , \quad \text{and} \quad (4.11)$$

$$h_j = w(x_j) .$$ $$(4.12)$$

An explicit formula is

$$K_j^{Aff}(q^{-x};q^{-a},N;q) = {}_3\phi_2\left(\begin{array}{c} q^{-j},q^{-x},0 \\ q^{-a},q^{-N} \end{array}\middle| q;q\right), 0 \le j \le N . (4.13)$$

It is clear from (4.13) that these polynomials are symmetric in x and j for $0 \le x, j \le N$, thus they are self - dual. The three - term recurrence is

$$\lambda(x)K_k(q^{-x}) = (q^N - q^k)(q^a - q^k)K_{k+1}(q^{-x}) +$$

$$q^{k-1}(q^k-1)K_{k-1}(q^{-x}) - \{(q^N-q^k)(q^a-q^k)+q^{k-1}(q^k-1)\}K_k(q^{-x})$$

$$(4.14)$$

where $\lambda(x) = -q^{a+N}(1-q^{-x})$.

It is clear that as $q \to 1$, the q-Hahn and q-Krawtchouk polynomials become the Hahn and Krawtchouk polynomials. If q^c is replaced by c in (4.7), as $q \to 1$, $w(x_k) \to \binom{N}{x}c^x$. So we really have a q-analogue of the general Krawtchouk polynomials, whose weight function is $w(x_k) = \binom{N}{k}p^k(1-p)^{N-k}$ (take $c = p/(1-p)$.). All of the spherical functions will have the q^c form, thus are q-analogues of the symmetric Krawtchouk polynomials. The affine q-Krawtchouk polynomials do not have a limit as $q \to 1$. Note that $w(x_k) \to 0$ as $q \to 1$ for $k = 1, 2, \ldots, N$. If we replace q^a by a , then $w(x_k) \to \binom{N}{k}(a-1)^k$. However, all of the spherical functions will have the q^a form. Delsarte calls these polynomials q-Krawtchouk [14], [16], [17].

Both sets of q-Krawtchouk polynomials are limiting cases of the q-Hahn polynomials. Replace q^α and q^β by a and b in the q-Hahn polynomials. If $b \to \infty$ and $a \to 0$ so that $abq = -q^{-N-c}$, we obtain the q-Krawtchouk polynomials. If $\alpha = -a-1$ and $q^\beta = b \to 0$, we obtain the affine q-Krawtchouk polynomials. There are also explicit formulas for both sets as ${}_2\phi_1$'s . We have used the ${}_3\phi_2$'s to emphasize the symmetry and the relations to the q-Hahn polynomials.

Now we come to the group theoretic interpretation of these polynomials from Chevalley groups over $GF(q)$.

For a Chevalley group of type A_{v-1} , let $G = GL_v(q)$ and $G_J = G_n$ (for $J = \{1, 2, \ldots, v-1\} - \{n\}$) be a maximal parabolic subgroup. Then [7] G_n is the stablizer

of a fixed n-dimensional subspace $x^{(0)} \subseteq V$ $(\dim V = v)$,
and $G_n \backslash G \cong X$, the set of all n-dimensional subspaces of
V. Again we assume $n \leq v-n$. The G-invariant metric on
X is $d(x,y) = n - \dim(x \cap y)$. The hypotheses $(*)$ of
section are satisfied by G and X.

The orbits of G_n on X are $\Omega_k = \{x \in X: \dim(x \cap x^{(0)}) = n-k\}$. Clearly $|X| = [\begin{smallmatrix} v \\ n \end{smallmatrix}]_q$ and

$$|\Omega_k| = [\begin{smallmatrix} n \\ n-k \end{smallmatrix}]_q [\begin{smallmatrix} v-n \\ k \end{smallmatrix}]_q \, q^{k^2} . \qquad (4.15)$$

The eigenvalue equation (2.13) is

$$(q-1)^2 \lambda \, f(\Omega_k) = (q^n - q^k)(q^{v-n} - q^k)q \, f(\Omega_{k+1}) +$$
$$(q^k - 1)(q^k - 1) \, f(\Omega_{k-1}) + \{(q^n - 1)(q^v - 1)q -$$
$$(q^n - q^k)(q^{v-n} - q^k)q - (q^k - 1)(q^k - 1)\} \, f(\Omega_k) . \qquad (4.16)$$

The solutions are q-Hahn polynomials [15], [20]

$$f_{\lambda_i}(\Omega_k) = Q_i(\, q^{-k}; q^{n-v-1}, q^{-n-1}, n; q) \qquad (4.17)$$

with $(q-1)^2 \lambda_i = (q^n - 1)(q^{v-n} - 1)q - q^{1+v}(1 - q^{-i})(1 - q^{i-v-1})$.
Note also that (4.4) implies that $\dim V_{\lambda_i} = [\begin{smallmatrix} v \\ i \end{smallmatrix}]_q - [\begin{smallmatrix} v \\ i-1 \end{smallmatrix}]_q$.

For the hyperoctahedral group the situation is more
interesting. There are five Chevalley groups whose Weyl group
is $S_N \cdot 2^N$. These correspond to types B_N, C_N, $^2D_{N+1}$,
and $^2A_{2N}$. These are groups which preserve a non-singular
bilinear form B with certain symmetry properties. Because
the form B is over a finite field, there are several
standard forms [8, Ch.1]. If B is symmetric, there are two:
$G = SO_{2N+1}(q)$ (type B_N) or $G = SO^-_{2N+2}(q)$ (type $^2D_{N+1}$).

If B is skew - symmetric and two does not divide q , G = $Sp_{2N}(q)$ (type C_N) . If B is hermitian, the G = $SU_{2N+1}(q^2)$ (type $^2A_{2N}$) or G = $SU_{2N}(q^2)$ (type $^2A_{2N-1}$) . In each case the subscript on the group gives the dimension of the latent vector space V on which G acts.

In each case, V has subspaces W such that $B(w_1,w_2) = 0$ for all $w_1, w_2 \in W$. Such a subspace is called isotropic. The dimension of a maximal isotropic subspace is N . For $J = \{1, 2, \ldots, N\} - \{k\}$, the maximal parabolic subgroup $G_J = G_k$ is the subgroup of G fixing a k - dimensional isotropic subspace $x^{(0)}$. In fact [7], [38], $G_k \backslash G \cong X_k$, all k - dimensional isotropic subspaces of V . For k = N , X_N is the q - analogue of an N - dimensional cube.

The q - Krawtchouk polynomials that result from these five cases are not identical. The parameter c shifts slightly (see Table 2). Nevertheless it is possible to derive the three - term recurrence uniformly (see [38]). This requires some groundwork. Instead we concentrate on type C_N . This is the simplest case because the form is skew - symmetric, $B(v,w) = -B(w,v)$, so all vectors v are isotropic, $B(v,v) = 0$.

It is not even immediately clear what the value of $|X_N|$ is. It is a q - analogue of 2^N . Counting all pairs $(\alpha,\beta) \in X_k \times X_{k+1}$ with $\alpha \subset \beta$ we find

$$|X_k| \cdot (q^{2(N-k)} - 1) = |X_{k+1}| (q^{k+1} - 1) \tag{4.18}$$

which implies that

$$|X_N| = (1+q)(1+q^2) \cdots (1+q^N) = (-q;q)_N . \qquad (4.19)$$

The G_N-orbits on X_N are again $\Omega_k = \{x \in X_N :$ $\dim(x \cap x^{(0)}) = N-k\}$. It is not hard to see that $|\Omega_k| = [\begin{smallmatrix} N \\ k \end{smallmatrix}]_q \, q^{\binom{k+1}{2}}$, the weights for the q - Krawtchouk polynomials $K_j(q^{-k};q^1,N;q)$. To verify this the eigenvalue equation (2.13) is

$$(q-1)\lambda f(\Omega_k) = (q^N-q^k)q \, f(\Omega_{k+1}) + (q^k-1) \, f(\Omega_{k-1}) +$$

$$\{(q^N-1)q - (q^N-q^k)q - (q^k-1)\} \, f(\Omega_k) \quad 0 \le k \le N . \quad (4.20)$$

From (4.10) the solutions for $\lambda_j(q-1) = q(q^N-1) - (1-q^{-j})(q^{N+1}+q^j)$ are

$$f_{\lambda_j}(\Omega_k) = K_j(q^{-k};q^1,N;q) . \qquad (4.21)$$

From (4.8) we have

$$\dim V_{\lambda_j} = \frac{(q^{-N};q)_j \, (-q^{-N-1};q)_j}{(q;q)_j \, (-1;q)_j} \, \frac{(1+q^{2j-N-1})}{(1+q^{-N-1})} \, (-1)^j q^{j(2N+1-j)} .$$

The other four cases follow in the same way. For the unitary groups, q is replaced by q^2 because the vector space is over the field $GF(q^2)$. For type B_N the polynomials agree with type C_N.

Next we come to type D_N, $G = SO_N^+(q)$, whose Weyl group W is $S_N \cdot 2^{N-1}$. In section 3 we saw that W had two maximal parabolic subgroups (W_{N-1} and W_N) so that $W_{N-1}\backslash W \cong X$, $W_N\backslash W \cong Y$, and $X \cup Y =$ all N-tuples of 0's and 1's. $G = SO_{2N}^+(q)$ furnished the q-analogue: G has two maximal parabolic subgroups [38] G_{N-1} and G_N such that $G_{N-1} \backslash G \cong X$, $G_N \backslash G \cong Y$, and $X \cup Y = X_N =$ all maximal isotropic subspaces of V. Moreover, if $x_1, x_2 \in X$, $y_1, y_2 \in Y$, then $\dim(x_1 \cap x_2) = N - 2k$, $\dim(y_1 \cap y_2) = N - 2j$, and $\dim(x_1 \cap y_1) = N - 2\ell - 1$ for some integers k, j, and ℓ.

Let G act on X with invariant metric $d(x_1, x_2) = \frac{1}{2}(N - \dim(x_1 \cap x_2))$. The orbits of G_{N-1} are $\Omega_k = \{x \in X_N : \dim(x \cap x^{(0)}) = N - 2k\}$. The values are $|X| = (1+q) \cdots (1+q^{N-1})$ and

$$|\Omega_k| = \left[{N \atop 2k} \right]_q q^{\binom{2k}{2}}. \tag{4.22}$$

As in section 3, the three-term recurrence implies that the spherical functions are

$$f_{\mu_j}(\Omega_k) = \begin{cases} Q_j(q^{-2k}; q^{-N-1}, q^{-N-1}, N/2; q^2) & N \text{ even} \\ Q_j(q^{-2k}; q^{-N-2}, q^{-N}, (N-1)/2; q^2) & N \text{ odd}, \end{cases} \tag{4.23}$$

or

$$f_{\mu_j}(\Omega_k) = {}_3\phi_2 \left({q^{-2j}, \ q^{2j-2N}, q^{-2k} \atop q^{1-N}, \ q^{-N}} \, \middle| \, q^2; q^2 \right). \tag{4.24}$$

where $(q-1)(q^2-1)\mu_j = q(q^N-1)(q^{N-1}-1) - q^{2N}(1-q^{-2j})(1-q^{2j-2N})$.

We can amalgamate X and Y into one orbit by letting $O_{2N}^+(q)$ act rather than $SO_{2N}^+(q)$. For this action $|X \cup Y| = 2(1+q) \cdots (1+q^{N-1})$, $|\Omega_k| = \left[{N \atop k} \right] q^{\binom{k}{2}}$, and the spherical

functions are the q - Krawtchouk polynomials $K_j(q^{-k}, q^0, N; q)$
with $(q-1)\lambda_j = q^{N-j} - q^j$. This time we verify that

$$D_1^2 = (q+1)D_2 + \frac{q^N - 1}{q-1} D_0$$

and $\lambda_j^2 = (q+1)\mu_j + \frac{q^{N-1}}{q-1}$. The q-analogue of (3.24) is

$$_3\phi_2 \left(\begin{array}{c} q^{-j}, -q^{j-N}, q^{-2k} \\ 0, q^{-N} \end{array} \middle| q; q \right) =$$

$$_3\phi_2 \left(\begin{array}{c} q^{-2j}, q^{2j-2N}, q^{-2k} \\ q^{1-N}, q^{-N} \end{array} \middle| q^2; q^2 \right), \quad 0 \leq k, j \leq [N/2] . \quad (4.25)$$

Equation (4.25) can be derived from the generating function
for q - Krawtchouk polynomials.

There are three examples for affine q - Krawtchouk
polynomials that do not come from the double coset correspondence. Yet they are still related to Chevalley groups. In
type A_{v-1} , fix a basis $\{e_1, \ldots, e_v\}$ for V . Relative
to this basis, if G_n fixes $\{e_1, \ldots, e_n\}$, G_n can be
written in block form $\left(\begin{array}{cc} A & 0 \\ S & B \end{array} \right)$, where $A \in GL_n(q)$, $B \in$
$GL_{v-n}(q)$ and S is any $(v-n) \times n$ matrix over $GF(q)$. The
subgroup G_n has the Levi decomposition [8, p. 118]
$G_n = U_n L_n$, where $U_n = \left(\begin{array}{cc} 0 & 0 \\ S & 0 \end{array} \right)$ and $L_n = \left(\begin{array}{cc} A & 0 \\ 0 & B \end{array} \right)$. The
subgroup $L_n \cong GL_n(q) \times GL_{v-n}(q)$ acts naturally on U_n via
$S \cdot (A,B) = BSA^{-1}$, so that G_n is the semi - direct product.

Put $X =$ the set of $(v-n) \times n$ matrices over $GF(q)$, so that G_n acts transitively on X. Clearly L_n fixes the zero matrix $x^{(0)}$. The G-invariant metric on X is $d(x,y) = \text{rank}(x-y)$, and G and X satisfy $(*)$. Clearly the orbits of H on X are $\Omega_k = \{x \in X : \text{rank}(x) = k\}$, $0 \le k \le n$, $|X| = q^{n(v-n)}$, and $|\Omega_k| =$

$$[{}^n_k]_q (q^{v-n};q^{-1})_k q^{\binom{k}{2}} (-1)^k .$$

The eigenvalue equation (2.13) for the spherical functions is

$$\lambda(q-1)f(\Omega_k) = (q^n - q^k)(q^{v-n} - q^k)f(\Omega_{k+1}) +$$
$$q^{k-1}(q^k-1)f(\Omega_{k-1}) + \{(q^n-1)(q^{v-n}-1) -$$
$$(q^n-q^k)(q^{v-n}-q^k) - q^{k-1}(q^k-1)\} f(\Omega_k) . \qquad (4.26)$$

The solutions from (4.14) are [14]

$$f_{\lambda_j}(\Omega_k) = K_j^{Aff}(q^{-k}; q^{n-v}, n; q) \qquad (4.27)$$

for $(q-1)\lambda_j = (q^n-1)(q^{v-n}-1) - q^v(1-q^{-j})$. Because these polynomials are self-dual, $\dim V_{\lambda_j} = |\Omega_j|$. Whenever X itself is an abelian normal subgroup of G, the spherical functions are symmetric and $\dim V_{\lambda_j} = |\Omega_j|$.

We will briefly mention the other two examples. These again come the Levi decomposition of G_N for types D_N and ${}^2A_{2N-1}$. These two examples allow a nice parametrization of the subgroup U_N. If X is the set of $N \times N$ skew-symmetric matrices over $GF(q^2)$ (2 does not divide q), then $A \in GL_N(q)$ acts on X by $X \mapsto A^T XA$. The rank of a skew - symmetric matrix is even, so the G - invariant metric for (*) is $d(x,y) = \frac{1}{2}$ rank $(x-y)$. Again $\Omega_k = \{x \in X :$ rank$(x) = 2k\}$, $|X| = q^{N(N-1)/2}$, and [17] $|\Omega_k| =$ $\begin{bmatrix} N \\ k \end{bmatrix}_{q^2} (q^{N-1};q^{-2})_k \, q^{k(k-1)}(-1)^k$. The spherical functions are

$$f_{\lambda_j}(\Omega_k) = \begin{cases} K_j^{Aff}(q^{-2k},q^{1-N},N/2;q^2) & \text{N even} \\ K_j^{Aff}(q^{-2k},q^{-N},(N-1)/2;q^2) & \text{N odd.} \end{cases} \quad (4.28)$$

As in (4.25), this represents one formula as a $_3\phi_2$.

For X the set of N x N hermitian matrices over $GF(q^2)$, then $A \in GL_N(q^2)$ acts on X by $X \to \overline{A}^T X A$, where bar (—) is the automorphism of $GF(q^2)$ over $GF(q)$ of order 2 . Again, $d(x,y) = $ rank$(x-y)$, $\Omega_k = \{x \in X: $ rank$(x) = k\}$, $|X| = q^{N^2}$, and [41]

$$|\Omega_k| = \begin{bmatrix} N \\ k \end{bmatrix}_{-q} (-(-q)^N; -q^{-1})_k (-q)^{\binom{k}{2}} .$$

(Note that the binomial coefficient has base $-q$.) The spherical functions are

$$f_{\lambda_j}(\Omega_k) = K_j^{Aff}((-q)^{-k}; -(-q)^{-N}, N; -q) . \quad (4.29)$$

A reasonable question to ask is: what happens to the symmetric matrices (type C_N)? In this case the action $X \mapsto A^T X A$ is not indexed by the rank. There is an A such that $A^T X A = $ diag$(a_1,\dots,a_r, 0,\dots,0)$, where $r = $ rank (X) . However, the value of $a_1 \dots a_r \in GF(q)$ (squares in $GF(q)$ does not change under A . The double cosets are doubly indexed and (*) does not hold. The spherical functions are

linear combinations of affine q - Krawtchouk polynomials [29].
By combining these double cosets, it is possible to form an
algebra $\mathfrak{A}$, based upon a graph, whose eigenvalues $p_k(\lambda)$
are precisely (4.28) [25]. Yet the automorphism group of
the graph does not satisfy (*).

For types B_N and $^2A_{2N}$ it does not appear that the
Levi decomposition gives a natural additive group of matrices.

In section 3 we mentioned another infinite family of
Coexeter groups; the dihedral groups, which acted on a regular n - gon X .   It is possible to define a generalized
n - gon which should have all of the properties of a q - analogue of X . This gives an algebra of matrices $\mathfrak{A}$. The
polynomials obtained are q - analogues of Chebychev polynomials [42]. Even though the group G and the hypothesized
structures X do not exist, the orthogonality relations are
true. If $n \to \infty$, the measure obtained is given in [42]. It
has an absolutely continuous part and two jumps. Cartier
found similar measures by find spherical functions on trees
[9].

The results are summarized in Table 2. The intertwining
functions for $G_k \backslash G / G_\ell$ and isotropic subspaces are polynomials in two variables if $1 \leq k$, $\ell \leq N$. They are given
in [38, Th. 6.23].

5. POSETS

The standard decomposition of $L^2(S^{n-1})$ consists of spherical harmonics

$$L^2(S^{n-1}) = \bigoplus_{\ell=0}^{\infty} \text{Harm}(\ell)\big|_{S^{n-1}} , \qquad (5.1)$$

where $\text{Harm}(\ell)$ is the set of all homogeneous polynomials in $x_1, \ldots, x_n$ of degree ℓ whose Laplacian is zero. In sections 3 and 4 we gave finite analogues of $\text{Harm}(\ell)\big|_{S^{n-1}}$, V_{λ_ℓ}. The eigenvalue equation $D_1 f = \lambda_\ell f$ characterized $f \in V_{\lambda_\ell}$.

In this section we briefly outline an exact analogue of (5.1). The details can be found in [43].

We need analogues of $\text{Hom}(\ell)$ and the Laplacian Δ. These are furnished by embedding X as the top level of a ranked partially ordered set (poset) P (see [43]). For each rank ℓ of P, there is a space of complex-valued functions on P called $\text{Hom}(\ell)$. The analogue of Δ is a lowering map $d: \text{Hom}(\ell) \to \text{Hom}(\ell-1)$ which is defined from the covering relations of P. If P and $\text{Aut}(P)$ have appropriate properties, then the subspace of functions $\text{Harm}(\ell) = \text{Hom}(\ell) \cap \ker d$ can be restricted to X. The result is $\text{Harm}(\ell)\big|_X = V_{\lambda_\ell}$.

The posets P for each example are summarized in Table 3. For types $A - D$, the chain complexes of P are the buildings associated with the Chevalley group over $GF(q)$.

6. REMARKS

There is much interest in classifying spaces X for which
(*) holds. A list of the known distance regular graphs is
given in [10]. It is widely believed that all such graphs
of sufficient width are related to the ones in this paper.

We have not given any combinatorial applications. Many
are given in [5], [6], [12]. See also Seidel's paper in
this volume. This is not surprising because the spherical
functions are eigenvalues of graphs. The set of N tuples
of 0's and 1's is the binary Hamming scheme and is important in coding theory. Maximal isotropic subspaces of type
D_N are called self-dual codes and have many applications.

The spherical functions f_k satisfy the product formula

$$f_k(g_1)f_k(g_2) = \frac{1}{|H|} \sum_{h \in H} f_k(g_1 h g_2) \ . \qquad (6.1)$$

For finite groups this becomes the linearization result for
the dual polynomials.

$$p_x(\lambda(k)) \ p_y(\lambda(k)) = \sum_z a_{xyz} \ p_z(\lambda(k)) \ . \qquad (6.2)$$

The structure constants a_{xyz} of the algebra $\mathfrak{A}$ are easily
computed by counting the number of triangles in X with
given sides [12]. The Krein conditions [28] imply positivity
of the structure constants b_{xyz}

$$p_k(\lambda(x)) \ p_k(\lambda(y)) = \sum_z b_{xyz} \ p_k (\lambda(z)) \ . \qquad (6.3)$$

Convolution sturctures can be given for the polynomials [26].

The product formula (6.1) is the first term of an addition theorem for the spherical function f_k . The G - modules are restricted back to stablizer subgroup H and decomposed. When this is carried out for types A_{v-1} , D_N , C_N , and $^2A_{2N-1}$, the irreducible representations of H that occur in the affine q - Krawtchouk polynomials are necessary [20], [40].

If X is an abelian normal group of G , the decomposition of $L^2(X)$ can be found from certain induced characters of X . This allows explicit formulas for the spherical functions from the characters of X . Fourier transforms can also be used.

For the spaces X and posets P there is ample opportunity to mimic classical analysis, e.g. Radon transforms, potential theory, although no one has done so.

It seems mysterious that the spherical functions are also polynomials. The duals are guaranteed to be polynomials. A theorem of Leonard [34] states that the most general orthogonal polynomials whose duals are polynomials are the $_4\phi_3$ polynomials of Askey and Wilson [3]. There remain many interesting special cases of these polynomials which have no group theoretic interpretations.

Another interesting question is to interpret the spherical functions by permutation statistics. This might be possible in view of the well - known correspondence between $B\backslash G$ and S_v .

School of Mathematics, University of Minnesota, Minneapolis, MN 55455
Partially supported by NSF grant MCS 83 - 00872

Table 1   Intertwining functions for Weyl groups

Type	Group G	Subgroup H	Subgroup K	Intertwining Functions	References
A_{v-1}	S_v	$S_n \times S_{v-n}$ $(n \leq v-n)$	$S_m \times S_{v-m}$ $(n \leq m \leq v-n)$	$Q_j(x;n-v-1,-m-1,n)$ Hahn polynomials	[18], [20], [21]
B_N or C_N	$S_N \cdot 2^N$	S_N	$S_m \times (S_{N-m} \cdot 2^{N-m})$	$K_\ell(x;\ 1/2,m)$ Krawtchouk polynomials	[18], [21], [32], [45]
D_N	$S_N \cdot 2^{N-1}$	S_N	S_N	$K_j(2x;1/2,N) =$ $Q_j(x,(-N-1)/2,(-N-1)/2,N/2)$	[38]
$I_2(N)$	D_N	C_2	C_2	$T_j(x)$ Chebychev polynomials	[6], [42]

Table 2   Intertwining functions for Chevalley groups

Type	Group G	Subgroup H	Subgroup K	Intertwining Functions	Reference
A_{v-1}	$GL_v(q)$	$G_n(2n \leq v)$	$G_m(n \leq m \leq v-n)$	$Q_j(q^{-k}, q^{n-v-1}, q^{-m-1}, n; q)$	[15], [20], [23]
B_N	$SO_{2N+1}(q)$	G_N	$G_m(1 \leq m \leq N)$	$K_j(q^{-k}; q^1, m; q)$	[38]
C_N	$Sp_{2N}(q)$	G_N	$G_m(1 \leq m \leq N)$	$K_j(q^{-k}; q^1, m; q)$	[38], [40]
$^2D_{N+1}$	$\overline{SO}_{2N+2}(q)$	G_N	$G_m(1 \leq m \leq N)$	$K_j(q^{-k}; q^2, m; q)$	[38]
$^2A_{2N-1}$	$SU_{2N}(q^2)$	G_N	$G_m(1 \leq m \leq N)$	$K_j(q^{-2k}; q, m; q^2)$	[38], [40]
$^2A_{2N}$	$SU_{2N+1}(q^2)$	G_N	$G_m(1 \leq m \leq N)$	$K_j(q^{-2k}; q^3, m; q^2)$	[38]
D_N	$SO_{2N}(q)$	G_N	G_N	$K_j(q^{-2k}; -1, N; q) = Q_j(q^{-2k}, q^{-N-1}, q^{-N-1}, N/2, q^2)$	[38], [40]
A_{v-1}	G_n	$GL_n(q) \times GL_{v-n}(q)$	same	$K_j^{Aff}(q^{-k}, q^{n-v}, n; q)$	[14], [41]
D_N	G_N	$GL_N(q)$	same	$K_j^{Aff}(q^{-2k}, q^{1-N}, N/2; q^2)$	[17], [41]

Table 2 (cont.)   Intertwining functions for Chevalley groups

Type	Group G	Subgroup H	Subgroup K	Intertwining Functions	Reference
$^2A_{2N-1}$	G_N	$GL_N(q^2)$	same	$K_j^{Aff}((-q)^{-k},(-q)^N,N;-q)$	[41]
				q – Chebychev	[42]

these, in general do not exist

Table 3

Posets

Type	Group G	Poset P	Inclusion	Reference
A_{v-1}	S_v	Boolean algebra	Set inclusion	[13], [19]
B_N, C_N	$S_N \cdot 2^N$	N-tuples of 0's,1's,2's	$\alpha \subset \beta \iff \alpha_i \neq 2 \Rightarrow \alpha_i = \beta_i$	[18], [38]
A_{v-1}	$GL_v(q)$	all subspaces of V	subspace inclusion	[13], [15], [20]
$B_N, C_N, D_N, {}^2D_{N+1}, {}^2A_{2N-1}, {}^2A_{2N}$	$Sp_{2N}(q)$, etc.	all isotropic subspaces of V	subspace inclusion	[38]
A_{v-1}	G_n	$\{(\alpha,T): \alpha \subset V, T: \text{lin. } \alpha \to W\}$	$(\alpha,T) \subset (\beta,S) \iff \alpha \subset \beta$ and $S\|_\alpha = T$	[13], [41]
$D_N, {}^2A_{2N-1}$	G_N		similar to the above, with a restriction on t	[41]

REFERENCES

[1] Andrews, G.: 1974, Applications of basic hypergeometric functions, SIAM Rev. 16, 441-484.

[2] Askey, R. and J. Wilson: Some basic hypergeometric orthogonal polynomials that generalize Jacobi polynomials, A.M.S. Memoir, to appear.

[3] Askey, R. and J. Wilson: A set of orthogonal polynomials that generalize the Racah coefficients or 6-j symbols, SIAM J. Math. Anal. 10, 1008-1016.

[4] Bailey, W.: 1935, Generalized Hypergeometric Series, Cambridge University Press, Cambridge.

[5] Bannai, E. and T. Ito: Algebraic Combinatorics, Part I, Association Schemes, Benjamin Lecture Notes Series, to appear.

[6] Biggs, N.: 1974, Algebraic Graph Theory, Cambridge Tracts in Mathematics, No. 67, Cambridge Univ. Press, Cambridge.

[7] Bourbaki, N.: 1968, Groupes et algebres de Lie, IV, V, VI, Hermann, Paris.

[8] Carter, R.: 1972, Simple Groups of Lie Type, Wiley-Interscience, London.

[9] Cartier, P.: 1973, 'Harmonic analysis on trees', in C.C. Moore, (ed.), Harmonic Analysis on Homogeneous Spaces, Proc. Symp. Pure Math., No. 26, Amer. Math. Soc., Providence, pp. 419-424.

[10] Cohen, A. and A. Neumaier: The known distance regular graphs, preprint.

[11] Curtis, C., N. Iwahori and R. Kilmoyer: Hecke algebras and characters of parabolic type of finite groups with (B,N) pairs.

[12] Delsarte, P.: 1973, An algebraic approach to the association schemes of coding theory, Philips Res. Repts. Suppl. 10.

[13] Delsarte, P.: 1976, Association schemes and t-designs in regular semilattices, J. Comb. Th. A 20, 230-243.

[14] Delsarte, P.: 1978, Bilinear forms over a finite field, with applications to coding theory, J. Comb. Th. A 25, 226-241.

[15] Delsarte, P.: 1978, Hahn polynomials, discrete harmonics, and t-designs, SIAM J. Appl. Math 34, 157-166.

[16] Delsarte, P.: 1976, Properties and application of the recurrence $F(i+1, k+1, n+1) = q^{k+1} F(i, k+1, n) - q^k F(i, k, n)$, SIAM J. Appl. Math. 31, 262 - 270.

[17] Delsarte, P. and J.M. Goethals, 1975, Alternating bilinear forms over $GF(q)$, J. Comb. Th. A 19, 26 - 50.

[18] Dunkl, C.: 1976, A Krawtchouk polynomial addition theorem and wreath products of symmetric groups, Indiana Univ. Math. J. 25, 335 - 358.

[19] Dunkl, C.: 1978, An addition theorem for Hahn polynomials: The spherical functions, SIAM J. Math. Anal. 9, 627 - 637.

[20] Dunkl, C.: 1977, An addition theorem for some q - Hahn polynomials, Monatsh. Math. 85, 5 - 37.

[21] Dunkl, C.: Spherical function on compact groups and applications to special functions, Symposia Mathematica 22, 145 - 161.

[22] Dunkl, C.: 1981, A difference equation and Hahn polynomials in two variables, Pac. J. Math. 92, 57 - 71.

[23] Dunkl, C.: 1980, Orthogonal polynomials in two variables of q - Hahn and q - Jacobi type, SIAM J. Alg. Disc. Math. 1, 137 - 151.

[24] Dunkl, C. and D. Ramirez, 1971, Topics in Harmonic Analysis, Appleton - Century - Crofts, New York.

[25] Egawa, Y.: Association schemes of quadratic forms, to appear.

[26] Gasper, G.: 1975, 'Positivity and special function', in R. Askey (ed.), Theory and Application of Special Functions, Academic Press, New York, 375 - 433.

[27] Helgason, S.: 1962, Differential Geometry and Symmetric Spaces, Academic Press, New York.

[28] Higman, D.: 1975, Coherent configurations; Part I: Ordinary representation theory, Geom. Ded. 4, 1 - 32.

[29] Hodges, J.: 1956, Exponential sums for symmetric matrices in a finite field, Arch. Math. 7, 116 - 121.

[30] James, G. and A. Kerber, 1981, 'The Representation Theory of the Symmetric Group', in G.C. Rota (ed.), Encylopedia of Mathematics and its Applications, Vol 16, Addison - Wesley, Reading, Mass.

[31] Koornwinder, T.: 1975, 'Two variable analogues of the classical orthogonal polynomials', in R. Askey (ed.), Theory and Application of Special Function, Academic Press, New York, 435 - 495.

[32] Koornwinder, T.: 1982, Krawtchouk polynomials, a unification of two different group theoretic interpretations, SIAM J. Math. Anal. 13, 1011 - 1023.

[33] Koornwinder, T.: 1981, Clebsch - Gordon coefficients
      for SU(2) and Hahn polynomials, Math. Centrum Rep.
      160.

[34] Leonard, D.: 1982, Orthogonal polynomials, duality,
      and association schemes, SIAM J. Math. Anal. 13,
      656 - 663.

[35] Miller, W.: 1968, Lie theory and special functions,
      Academic Press, New York.

[36] Miller, W.: 1977, 'Symmetry and Separation of Vari-
      ables', in G.C. Rota (ed.), Encyclopedia of Mathema-
      tics and its Applications, Vol. 4, Addison - Wesley,
      Reading, Mass.

[37] Miller, W.: 1970, Lie Theory and q - difference equa-
      tions, SIAM J. Math. Anal. 1, 171 - 188.

[38] Stanton, D.: 1980, Some q - Krawtchouk polynomials on
      Chevalley groups, Amer. J. Math. 102, 625 - 662.

[39] Stanton, D.: 1980, Product formulas for q - Hahn poly-
      nomials, SIAM J. Math. Anal. 11, 100 - 107.

[40] Stanton, D.: 1981, Three addition theorems for some
      q - Krawtchouk polynomials, Geom. Ded. 10, 403 - 425.

[41] Stanton, D.: 1981, A partially ordered set and q -
      Krawtchouk polynomials, J. Comb. Th. A 30, 276 - 284.

[42] Stanton, D.: 1983, Generalized n - gons and Chebychev
      polynomials, J. Comb. Th. A 34, 15 - 27.

[43] Stanton, D.: Harmonics on posets, preprint.

[44] Szegö, G.: 1975, Orthogonal Polynomials, Amer. Math.
      Soc. Colloquium Pub., Vol. 23, Providence.

[45] Vere - Jones, D.: 1971, Finite bivariate distributions
      and semi - groups of non - negative matrices, Quart. J.
      Math., Oxford (2), 22, 247 - 270.

[46] Vilenkin, N.: 1968, Special Functions and the Theory
      of Group Representations, Translations of Amer. Math.
      Soc. Vol. 22.

[47] Vretare, L.: 1976, Elementary spherical functions on
      symmetric spaces, Math. Scand. 39, 343 - 358.

[48] Wang, H.: 1952, Two - point homogeneous spaces, Ann.
      of Math. 55, 177 - 191.

[49] Wielandt, H.: 1964, Finite Permutation Groups,
      Academic Press, New York.

L.C. Biedenharn,[1] R.A. Gustafson,[2] M.A. Lohe,[1] J.D. Louck,[3] and S.C. Milne[2]

SPECIAL FUNCTIONS AND GROUP THEORY IN THEORETICAL PHYSICS[*†]

1. INTRODUCTION

The subject of special functions, even where restricted to the narrower subject of special functions in theoretical physics, is far too broad for one to attempt any meaningfully complete survey. Our aim is accordingly rather more limited: we will attempt to review the important main themes in the subject and then show how recent advances, some to be presented at this conference, have generalized, broadened and extended these themes.

The name "special functions" is possibly something of a misnomer; Askey in his 1975 lectures [4] cites Turan's suggestion that perhaps "useful functions" may better capture the real meaning. 'Certainly this is how special functions entered theoretical physics, as those very useful functions that arose from separation of variables in the partial differential equations defined by physical problems. Special functions were accordingly both numerous and, as a discipline, somewhat chaotic.

This situation--at least for theoretical physics--changed markedly in the 1960's. Although Klein, Lie and Cartan had indicated the symmetry origins of some special functions, it was the 1955 lectures of Eugene Wigner [5] that, at least for physicists, brought order out of chaos. Wigner pointed out that large classes of special functions arise as matrix elements of the representations of the symmetry groups so fundamental to theoretical physics. Thus, for example, the Legendre and Jacobi functions stem from representations of the quantal rotation group, SU(2); the Bessel functions from representations of the Euclidean motion group, E(2); and the Laguerre and Hermite functions from the Heisenberg group.

Not only did the special functions themselves receive order, but so did their relations: addition theorems arise uniformly from the group multiplication rule; differential

R. A. Askey et al. (eds.), Special Functions: Group Theoretical Aspects and Applications, 129-162.
© 1984 by D. Reidel Publishing Company.

equations arise as limits of the multiplication by infinitesimal elements (generators); and integral and completeness
relations appear as the orthonormality relations for matrix
elements (as typified by the Peter-Weyl theorem). Moreover,
one could understand intuitively such results as Mehler's
formula (asymptotically relating the Legendre functions to
Bessel functions) as the contraction of groups: $SO(3) \to E(2)$,
that is, the rotation group $SO(3)$ acts on the sphere S^2 which
asymptotically (large radius) is replaced by a tangent plane
on which the contracted group $E(2)$ acts.

This point of view for special functions was developed
further in the monograph by Talman [6] (based on the Wigner
lectures) and the more extensive monograph by Vilenkin [7]
(based partly on Gel'fand's work). A very general setting for
the spherical functions (Helgason [8]) considers a group G and
a two-point homogeneous space acted on by the group. The
spherical functions are orthogonal polynomials of $_2F_1$ type
(Askey [4]) with the variable being the distance function.

One should not be too categoric about this 'group representation viewpoint', important as it is, for there are special
functions that do not fall into this context. Askey [4] gave
different settings for some special functions and noted, for
example, that statistical mechanics and critical phenomena
involve special functions arising from certain positivity
constraints and inequalities. Another example is furnished by
the Appell functions (F_2 and F_3) that occur in physics in the
theory of Coulomb excitation (Biedenharn and Brussaard [9])
(energy transfer by the electromagnetic field of accelerated
charged particles during scattering) and do not appear in the
context of matrix elements of group representations.

A widening of the group representation context appears in
the book by Miller [10] which is based on the recognition that
the special functions defined by separation of variables in
second-order linear partial differential equations can be
characterized as eigenfunctions defined by commuting elements
in the enveloping algebra of the Lie algebra of the symmetry
group of the equation. This structure encompasses naturally
the factorization method of Schrödinger (see Ref. [11]), the
approaches of Truesdell [12] and Weisner [13] and leads to a
uniform approach to hypergeometric special functions as well
as nonhypergeometric functions (Mathieu, Lamé, ...). One may
discern a trend toward algebraic, as opposed to purely group-

theoretic, methods in the approach to spherical functions in
terms of <u>convolution algebras</u> (Helgason [8]).

Let us turn now from these well-established basic ideas to
more recent, less widely known, approaches. The main theme is
that we now go beyond group representations and Lie algebras,
and beyond even the enveloping algebra of a Lie algebra. We
may gain an intuitive idea of this by considering the prototype
for generalizations: Abelian groups defining the elementary
transcendentals. The characters (representations) obey the
product rule given by

$$\chi^j(a)\chi^j(b) = \chi^j(ab). \tag{1.1}$$

This multiplication is generalized for compact groups to the
matrix product law for representations given by

$$D^j(a) \, D^j(b) = D^j(ab), \tag{1.2}$$

which characterized the approach ascribed to Wigner and
Vilenkin above. Moreover, the characters of an Abelian group
obey a <u>second</u> multiplication rule expressed by

$$\chi^j(a) \, \chi^{j'}(a) = \chi^{j+j'}(a), \tag{1.3}$$

where we have indexed the distinct characters by the symbol j.
One recognizes Eq. (1.3) as the <u>product law of the dual group</u>.

Can one generalize relation (1.3) to non-Abelian groups?
No really satisfactory answer is known, although there are some
fundamental results (for example, Tannaka duality). The Van
Kampen-Pontriagin theorem shows that the dual space to the
group space of a compact group is discrete; thus we are seek-
ing some kind of product structure on discrete spaces. This
structure <u>cannot be that of</u> a <u>group</u>, as the specific examples
to be discussed below will show, but the structure <u>does</u> become
a group in a well-defined asymptotic limit. The sequence of
discrete Abelian groups Z_n as n becomes large can be taken to
limit to the Abelian rotation group $R_2(\theta)$ of the plane. Some
suitably generalized, but discrete, analog of this for non-
Abelian groups would, <u>if it exists!</u>, be of great help to
theoretical physics.

Let us, however, proceed straightforwardly, despite the
lack of clear guidance. A result for the quantal rotation

group, very similar in structure to Eq. (1.3), is known. This
is the Wigner product law, where in analogy to Eq. (1.3),
different representations for the same group element are to be
multiplied. If we express this result in matrix element form
we have the following relation:

$$D^{j_1}_{m_1 m_1'}(R)\ D^{j_2}_{m_2 m_2'}(R) = \sum_{j_3 m_3 m_3'} C^{j_1 j_2 j_3}_{m_1 m_2 m_3}\ D^{j_3}_{m_3 m_3'}(R)\ C^{j_1 j_2 j_3}_{m_1' m_2' m_3'}. \quad (1.4)$$

Here the $D^{j}_{mm'}(R)$ are the matrix elements of the representation
D^j for the group element R; the (real, orthonormal) coefficients $C^{:::}_{:::}$ are the Wigner-Clebsch-Gordan coefficients (see,
for example, Ref. [14]).

We can achieve a form analogous to Eq.(1.3) if, using
orthonormality, we put the Wigner-Clebsch-Gordan coefficients
on the left-hand side, thus obtaining:

$$\sum_{\substack{m_1 m_1' \\ m_2 m_2'}} C^{j_1 j_2 j_3}_{m_1 m_2 m_3} C^{j_1 j_2 j_3}_{m_1' m_2' m_3'}\ D^{j_1}_{m_1 m_1'}(R) D^{j_2}_{m_2 m_2'}(R) = D^{j_3}_{m_3 m_3'}(R). \quad (1.5)$$

Now let us regard the two coefficients $C^{j_1 j_2 j_3}_{m_1 m_2 m_3}$ each as effec-
ting a product; that is, we regard the left-hand side above
as the formal product:

$$\left[D^{j_1} \begin{smallmatrix} ⓦ \\ ⓦ \end{smallmatrix} D^{j_2} \right]^{j_3} = D^{j_3}. \quad (1.6)$$

(Here each of the symbols ⓦ denotes a $C^{:::}_{:::}$ coefficient, with
a sum over the relevant indices.) In this form, we can discern
that Eq. (1.6) is a generalization of Eq. (1.3), and consti-
tutes a valid identity that extends the Abelian group model to
the non-Abelian group SU(2). [Since SU(2) is compact the dual
space (indexed by j,m,m') is discrete. Thus instead of a
simple monomial product as in Eq. (1.3) we now have a complicated sum of products effected on the dual space.]

Our reason for writing the Wigner product law in this
curious way is to motivate two valid identities--each a formal
analog of Eq. (1.6)--that involve algebraic elements outside
the enveloping algebra of SU(2) whose matrix elements are
special functions of generalized hypergeometric type.

Let us define the Hilbert space H to consist of a direct
sum of irrep spaces of SU(2), each irrep space occurring once

and only once. Thus the irrep j has, as a basis, the $(2j+1)$ orthonormal vectors $|jm\rangle$, $m = j$, $j-1$, ..., $-j$ for each $j = 0, \frac{1}{2}$, 1,... . The space H is the direct sum: $H = \Sigma \oplus |jm\rangle$. We define the (J,M,Δ) <u>Wigner operator</u> to be the operator in H given by

$$
\left\langle \begin{array}{c} J+\Delta \\ 2J \quad 0 \\ J+M \end{array} \right\rangle = \sum_{jm} |j+\Delta, m+M\rangle \ C^{j \ J \ j+\Delta}_{m \ M \ m+M} \ \langle jm|. \tag{1.7}
$$

By taking matrix elements in H, one sees that this operator has the WCG coefficients $C^{...}_{...}$ as matrix elements.

The product of two Wigner operators is

$$
\left\langle \begin{array}{c} J'+\Delta' \\ 2J' \quad 0 \\ J'+M' \end{array} \right\rangle \left\langle \begin{array}{c} J+\Delta \\ 2J \quad 0 \\ J+M \end{array} \right\rangle = \sum \underline{\underline{W}}^{JJ'J''}_{\Delta\Delta'\Delta''} \ C^{JJ'J''}_{MM'M''} \left\langle \begin{array}{c} J''+\Delta'' \\ 2J'' \quad 0 \\ J''+M'' \end{array} \right\rangle, \tag{1.8}
$$

where $\underline{\underline{W}}^{...}_{...}$ denotes an SU(2) <u>invariant operator</u> whose matrix elements are the Racah, or 6-j coefficients.

We can invert this last equation to define the (symbolic) <u>Wigner operator product law</u>

$$
\left[\left\langle \begin{array}{c} \bullet \\ 2J' \quad 0 \\ \bullet \end{array} \right\rangle \begin{array}{c} \circledR \\ \\ \circledW \end{array} \left\langle \begin{array}{c} \bullet \\ 2J \quad 0 \\ \bullet \end{array} \right\rangle \right]^{J''} = \left\langle \begin{array}{c} \bullet \\ 2J'' \quad 0 \\ \bullet \end{array} \right\rangle, \tag{1.9}
$$

where $\circledW$ denotes the Wigner product (using the 3-j coefficient) and $\circledR$ denotes the Racah product (using, in effect, the 6-j coefficients).

There is a similar product law involving two Racah operators (and two Racah products $\circledR$).

Thus we have three product laws in SU(2) that are formal extensions of the Abelian product law [(Eq. (1.3)] for the dual group. Each of these three product laws is an identity in special functions involving, respectively, the $_2F_1$, $_3F_2$ and $_4F_3$ special functions. (This is discussed at length in Ref. [15]).

Moreover, there is a remarkable series of asymptotic relations: $6\text{-}j \sim 3\text{-}j \sim D^j$ that carry these three special functions (and these three product laws) into each other.

Let us note explicitly that the Wigner operators constitute an algebraic structure that goes beyond the concept of
the enveloping algebra of the Lie group SU(2), although the
enveloping algebra, and the Lie algebra, are contained as
subalgebras.

It is clear, we believe, even from this brief sketch that
the extension of these ideas to groups other than SU(2) holds
forth the promise of yielding unsuspected families of
(possibly new) special functions and relations. We will
illustrate this by discussing briefly how this hope is being
realized in recent work on SU(3) and SU(n) groups.

2.   THE PROBLEM OF DEFINING WIGNER-CLEBSCH-GORDAN
COEFFICIENTS FOR SU(3)

The "Wigner coefficient problem" for SU(3) may be described
as follows: The unitary irreducible representations (irreps)
of $G = U(3)$ are given by

$$D^\lambda = \{D^\lambda(g) \mid g \in G\} , \qquad (2.1)$$

where $D^\lambda(g)$ is a unitary matrix of dimension $\text{Dim}\lambda$ in which
$\lambda \in P$ is any element of the set of partitions P defined by

$$P = \{[\lambda_1, \lambda_2, \lambda_3] \mid \lambda_i \in Z, \ \lambda_1 \geq \lambda_2 \geq \lambda_3\}. \qquad (2.2)$$

The irreducible representations given by $[\lambda_1+d, \ \lambda_2+d, \ d]$
each $d \in Z$ = set of all integers, are all equivalent to the
irrep $[\lambda_1, \ \lambda_2, \ 0]$ for the group $G = SU(3)$. We will denote
irreps (always taken to be unitary) of G by $\mu, \ \nu, \ \lambda, \ \ldots$.
Then the Kronecker product of D^μ and D^ν, which we denote by
$D^\mu \otimes D^\nu$ is completely reducible by a unitary transformation U,
that is,

$$U (D^\mu \otimes D^\nu)U^\dagger = \sum_\lambda \oplus g(\mu\nu\lambda)D^\lambda , \qquad (2.3)$$

where the right-hand side of this expression denotes a (direct)
sum of block diagonal matrices in which a given D^λ is repeated
a number of times equal to $g(\mu\nu\lambda)$. We will denote the abstract
Clebsch-Gordan series corresponding to the explicit reduction
(2.3) of the Kronecker product by

$$\mu \times \nu = \sum_{\lambda} g(\mu\nu\lambda)\lambda .$$ (2.4)

A set of Wigner coefficients for $G = SU(3)$ comprises the elements of the matrix U. With an appropriate indexing of rows and columns of U, which has dimension $Dim\mu \times Dim\nu$, we have

$$C^{\mu\nu\lambda;\tau}_{mn\ell} = \underbrace{U_{\lambda\ell\tau;mn}}_{\text{rows;columns}} , \quad (\text{suppressing } \mu \text{ and } \nu)$$ (2.5)

in which $m = 1, \ldots, Dim\mu$; $n = 1, \ldots, Dim\nu$; $\ell = 1, \ldots, Dim\lambda$; and τ is an index which enumerates the number of occurrences of λ in $\mu\nu\lambda$, that is, $\tau = 1, \ldots, g_{\mu\nu\lambda} \equiv g(\mu\nu\lambda)$.

The preceding is, no doubt, familiar. At first glance there would appear to be little if any connection with special functions, and the problem might seem, perhaps, essentially to be solved through the explicitly known representation functions. In a certain numerical sense, where one regards μ and ν as specified, this is true. But if one takes a broader view of these coefficients and their relationship to the intertwining number function of the group and to a comprehensive theory of tensor operators, one realizes that there is a deep and elegant structure underlying this problem. This is the aspect of the problem that we would like to motivate in this survey; it is best brought out through the concept of a tensor operator with respect to a group G, as we next describe.

(a) Assume we are given a separable Hilbert space H such that

$$H = \sum_{\lambda} \oplus H_{\lambda} ,$$ (2.6)

where H_{λ} is a vector space that is invariant and irreducible under the action of $G = SU(3)$ given by $U_g : H_{\lambda} \to H_{\lambda}$, each $g \in G$, where U_g is a unitary operator on H defined by

$$U_g \left| {\lambda \atop \ell} \right\rangle = \sum_{\ell'} D^{\lambda}_{\ell'\ell}(g) \left| {\lambda \atop \ell'} \right\rangle ,$$ (2.7)

where $\left| {\lambda \atop \ell} \right\rangle$, $\ell = 1, \ldots, Dim\lambda$ is an orthonormal basis of H_{λ}.

(b) Define a unit tensor operator in H by

$$T^{\mu}_{m}(\tau) : H_{\nu} \to H_{\lambda} , \quad \lambda \in \mu \times \nu ,$$ (2.8)

$$T^{\mu}_{m}(\tau) \left| {\nu \atop n} \right\rangle = \sum_{\ell} C^{\mu\nu\lambda;\tau}_{mn\ell} \left| {\lambda \atop \ell} \right\rangle ,$$ (2.9)

where $C_{\cdot\cdot\cdot}^{\cdot\cdot\cdot}$ is defined in (2.5).

(c) The tensor operator (2.8) satisfies the transformation property:

$$U_g \, T_m^\mu(\tau) U_g^{-1} = \Sigma \, D_{m'm}^\mu(g) T_m^\mu(\tau),$$

each $\tau = 1, \ldots, g_{\mu\nu\lambda}$.
                                                                        (2.10)

(d) Moreover, we have a set of bounded linear maps on H:

$$\left\{ T_m^\mu(\tau) \; \middle| \; \begin{array}{l} m = 1, \ldots, \text{Dim}\mu \\ \tau = 1, \ldots, g_{\mu\nu\lambda} \end{array} \right\},$$

which is a basis for all linear maps: $H_\nu \to H_\lambda$.

Remark: The use of the multiplicity index τ is ad hoc-- it makes the concept of tensor operator depend not only on the irrep label μ of the operator, but also on the pair $\{H_\nu, H_\lambda\}$ of subspaces in H. There should exist an "intrinsic" labelling of an operator of "type" μ that depends only on the operator itself; in general, a tensor operator map between the pair of subspaces $\{H_\nu, H_\lambda\}$ has no meaning for a second pair $\{H_{\nu'}, H_{\lambda'}\}$.

The intrinsic label problem was resolved early in the 1960's in Ref. [16]. Let us first discuss the label itself and then its significance. Each tensor operator is labelled not only by a set of labels that specify its transformation property under the action of the group [this is the label m in (2.10)], but also by the addition labels

$$\begin{pmatrix} \alpha \\ \mu \end{pmatrix} = \begin{pmatrix} & \alpha_1 & \\ \alpha_2 & & \alpha_3 \\ \mu_1 & \mu_2 & \mu_3 \end{pmatrix}$$
                                                                        (2.11)

that specifies the shift action of the operator in the Hilbert space H. In the pattern (2.11), which we call an operator pattern, the labels α_i may assume all integral values consistent with

$$\mu_1 \geq \alpha_2 \geq \mu_2 \geq \alpha_3 \geq \mu_3, \quad \alpha_2 \geq \alpha_1 \geq \alpha_3,$$
                                                                        (2.12)

where the partition $[\mu_1\mu_2\mu_3]$ is specified (these relations are referred to as the betweenness conditions). With each pattern of the form (2.11), there is also associated a shift pattern $\Delta = [\Delta_1, \Delta_2, \Delta_3]$ defined by

$$\Delta_1 = \alpha_1, \quad \Delta_2 = \alpha_2+\alpha_3-\alpha_1, \quad \Delta_3 = \mu_1+\mu_2+\mu_3-\alpha_2-\alpha_3. \tag{2.13}$$

It is the operator pattern (2.11) that replaces the arbitrary multiplicity index τ in (2.10), and we now denote a tensor operator by the explicit notation

$$T\begin{pmatrix} \alpha \\ \mu \\ m \end{pmatrix}. \tag{2.14}$$

This operator still undergoes the transformation (2.10)--this is the meaning of the label m --under the group action. In addition, the significance of the shift pattern Δ of (2.13) for the operator pattern (2.11) is defined in the following manner: The action of the tensor operator in H is given by

$$T\begin{pmatrix} \alpha \\ \mu \\ m \end{pmatrix} : \begin{cases} H_\lambda \to H_{\lambda+\Delta}, & \text{if } \lambda+\Delta \in \mu \times \lambda \\ \text{or} \\ H_\lambda \to 0, & \text{if } \lambda+\Delta \notin \mu \times \nu \end{cases}. \tag{2.15}$$

Specifically,

$$T\begin{pmatrix} \alpha \\ \mu \\ m \end{pmatrix}\Big| \begin{matrix} \lambda \\ n \end{matrix}\Big\rangle \equiv \sum_{n'} C\left[\begin{pmatrix} \lambda+\Delta \\ n' \end{pmatrix}\begin{pmatrix} \alpha \\ \mu \\ m \end{pmatrix}\begin{pmatrix} \lambda \\ n \end{pmatrix}\right]\Big| \begin{matrix} \lambda+\Delta \\ n' \end{matrix}\Big\rangle, \tag{2.16}$$

where $C^{...}_{...}$ is a numerical Wigner coefficient.

How many operator patterns $\begin{pmatrix} \alpha \\ \mu \end{pmatrix}$ are there having given irrep labels $[\mu_1\mu_2\mu_3]$ and given shift pattern $[\Delta_1\Delta_2\Delta_3]$?

The answer is provided by the formula:

$$M = \begin{cases} (\mu_2-\mu_3+1)-(\delta_1+\delta_2+\delta_3) & \text{for } \Delta_1+\Delta_2+\Delta_3 = \sum_i \mu_i, \\ & \mu_1 \geq \Delta_i \geq \mu_3; \quad i \\ 0 & \text{otherwise,} \end{cases} \tag{2.17}$$

where $\delta_i \equiv \max(0, \mu_2 - \Delta_i)$.

One must prove, of course, that operator patterns--the set of $\begin{pmatrix} \alpha \\ \mu \end{pmatrix}$ patterns--really work. For this we require properties of the intertwining number function $I_{\mu,\Delta}$ with values defined on the set P of partitions by

$$I_{\mu,\Delta}(\lambda) = g_{\mu,\lambda,\lambda+\Delta}. \tag{2.18}$$

Here we regard μ and Δ as <u>specified</u> and seek the set of values
$\{I_{\mu,\Delta}(\lambda)\,|\,\lambda\in P\}$. The numbers in this set are precisely 0,1,
...;M; the next step is to exhibit in an appropriate graph
the "level" subsets of P defined by

$$P_k = \{\lambda\in P\,|\,I_{\mu,\Delta}(\lambda)=k\}, \quad k=0,1,\ldots,M. \tag{2.19}$$

The coordinate system most appropriate for graphing the
level subsets (2.19) of the intertwining function $I_{\mu,\Delta}$ is the
barycentric system shown in Fig. 1 (also called the
Möbius plane):

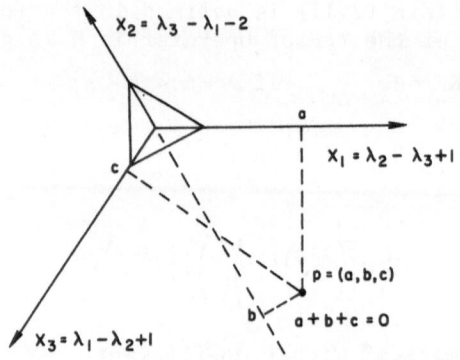

Figure 1. The 3-simplex (equilateral triangle) in R^2 defines
the three coordinate axes used for graphing the intertwining
function.

The translation properties $I_{\mu+a,\Delta+a} = I_{\mu,\Delta}$ and $I_{\mu,\Delta}(\lambda+a)$
$= I_{\mu,\Delta}(\lambda)$, a=(a,a,a), are important properties used in graphing
the intertwining function shown in Fig. 2.

The role of the intertwining function in completing the
definition of a canonical tensor operator can be inferred from
the special case of $I_{[630],[333]}$ shown in Fig. 3.

Let L_1, L_2, L_3 denote the set of lattice points belonging
to the bent solid lines denoted I = 1, I = 2, I = 3 in Fig. 3.
Also let T_0 denote the lattice points in the equilateral triangle shown in the shaded area, that is, the set of points

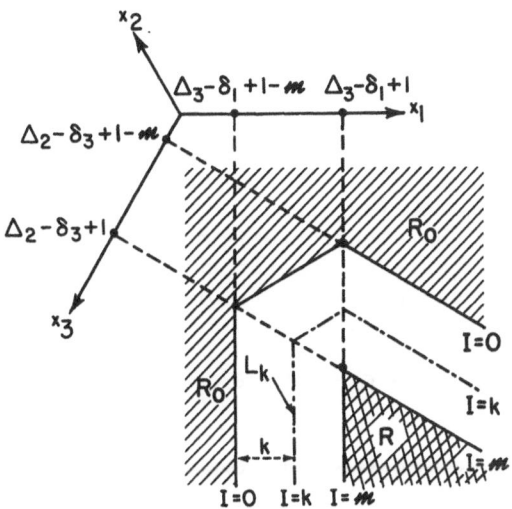

Figure 2. Graph of the intertwining function. The intertwining function $I_{\mu,\Delta}$ has values as follows on the lattice points in the Möbius plane: value M in the cross-hatched region R, including the bent solid boundary line; value $k(k=0,1,\ldots,M)$ on the bent line L_k; and value 0 in the shaded region R_0.

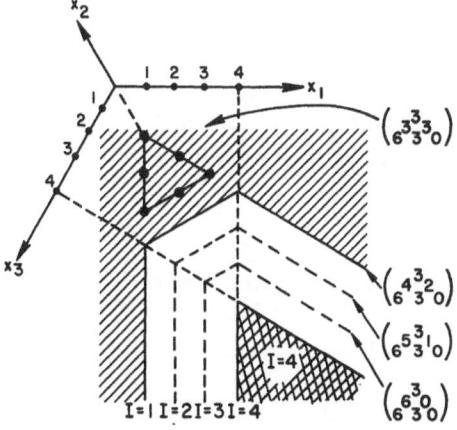

Figure 3. Graph of the intertwining function $I_{[630],[333]}$.

$(1, -2, 1)$ ●

　　　　　　　　　　　● $(2, -3, 1)$

$(1, -3, 2)$ ●　　　　　　　●$(3, -4, 1)$.

　　　　　　　　　● $(2, -4, 2)$

$(1, -4, 3)$ ●

We then require that the Wigner operators (unit tensor operators) in the set having irrep labels [630] and shift pattern $\Delta = [333]$, namely, the four tensor operators labelled by the patterns in Fig.3, have characteristic null spaces described as follows (<u>characteristic</u> denotes that the entire irrep space is in the null space and not just certain vectors in the space):

$$T\begin{pmatrix} & 6 & & 3 & & 0 & \\ 6 & & 3 & & 0 & \\ & & & \bullet & & \end{pmatrix} \quad : \quad H_\lambda \to 0 \text{ for all } x \in L_1 \cup L_2 \cup L_3 \cup T_0 \equiv N_1,$$

$$T\begin{pmatrix} & 5 & & 3 & & 1 & \\ 6 & & 3 & & 0 & \\ & & & \bullet & & \end{pmatrix} \quad : \quad H_\lambda \to 0 \text{ for all } x \in L_1 \cup L_2 \cup T_0 \equiv N_2,$$

$$T\begin{pmatrix} & 4 & & 3 & & 2 & \\ 6 & & 3 & & 0 & \\ & & & \bullet & & \end{pmatrix} \quad : \quad H_\lambda \to 0 \text{ for all } x \in L_1 \cup T_0 \equiv N_3,$$

$$T\begin{pmatrix} & 3 & & 3 & & 3 & \\ 6 & & 3 & & 0 & \\ & & & \bullet & & \end{pmatrix} \quad : \quad H_\lambda \to 0 \text{ for all } x \in T_0 \equiv N_4. \tag{2.20}$$

Recall that $(x_1, x_2, x_3) = (\lambda_2 - \lambda_3 + 1, \lambda_3 - \lambda_1 - 2, \lambda_1 - \lambda_2 + 1)$. Clearly the characteristic null spaces of these operators are nested (strictly decreasing):

$$N_1 \supset N_2 \supset N_3 \supset N_4 \quad . \tag{2.21}$$

The results described above for a special case generalize: The tensor operators in the multiplicity set

$$\begin{pmatrix} \Gamma_1 \\ \mu_1\mu_2\mu_3 \end{pmatrix}, \begin{pmatrix} \Gamma_2 \\ \mu_1\mu_2\mu_3 \end{pmatrix}, \dots, \begin{pmatrix} \Gamma_M \\ \mu_1\mu_2\mu_3 \end{pmatrix} \tag{2.22}$$

corresponding to distinct operator patterns having the same shift pattern $\Delta = [\Delta_1, \Delta_2, \Delta_3]$ are distinguished by the nested characteristic null spaces

$$N_1 \supset N_2 \supset \dots \supset N_M \quad , \tag{2.23}$$

each of which can be identified explicitly from the graph of the intertwining function $I_{\mu, \Delta}$.

An explicit procedure, based on intrinsic properties of the tensor operators themselves, can also be given for making the assignment of operator patterns to characteristic null spaces, but this result will not be given here, since our main purpose is to indicate the origin of the functions discussed in the sections to follow.

A canonical set of tensor operators (hence, Wigner coefficients) is then a set that is labelled by operator patterns and associated with the characteristic null space as described above. Thus, not only do the operator patterns enumerate all irreducible tensor operators, but they do so in a way that gives exactly a basis for all transformations between any two irrep spaces; equivalently, this solves the Wigner coefficient problem for SU(3) in a structurally meaningful way and eliminates the vague and arbitrary "multiplicity index."

This review, although very incomplete, suffices for our purpose of indicating the origin of the special functions that are the topic of the subsequent discussion: They arise in the explicit construction of the canonical Wigner coefficients discussed in this section.

Remarks: (a) An elegant global construction for the set of all tensor operators in SU(3) has recently been given by Flath [17]. The key to this construction is an embedding of SU(3) in SO(8). A very particular realization of the complex so_8 Lie algebra g is constructed based on polynomial differential operators in six variables. A unitary subset of g belongs to a real form of the D_4 Lie algebra and (it is shown) generates a single unitary irrep V of the group SO(6,2). The algebra of all operators carrying $V \rightarrow V$ is then found to be a simple algebra, A, a quotient algebra of the enveloping algebra of so_8.

(b) Can one hope to develop similar results for SU(n) (n > 3), or even for all Lie groups? The elegance of Flath's SU(3) result argues otherwise, for the construction admits an S_3 automorphism that accords nicely with the S_3 outer automorphism of the Lie group SO(8). For SU(4) the structure must admit the S_4 group, but there is no simple Lie group with such an automorphism! Something unusual must happen in going from SU(3) to SU(4), and it is not at all clear that the Lie subalgebra analogous to $g \in A$ is even finitely generated.

3. THE FORM OF THE G_q^t FUNCTIONS OCCURRING IN THE DIRECT
    CALCULATION OF THE CANONICAL WIGNER COEFFICIENTS

The implementation of the characteristic null space properties
of the canonical tensor operators leads to the vanishing of
certain of the so-called isoscalar factors in the Wigner
coefficients. Using these vanishing isoscalar factors, one
can then give a complete algorithm for the calculation of all
SU(3) Wigner coefficients (see Refs. [18–21]). This procedure
is, however, very difficult to implement to obtain complete
algebraic expressions for the coefficients. Thus far we have
been able to obtain such closed expressions only for a certain
SU(3) invariant part of the coefficient--a part that can be
identified as essentially the norm of the Wigner operator.
It is to these SU(3) invariants that we now direct our
attention, giving the very complicated form in which the
functions G_q^t first emerge. Henceforth, we use the notation
$[\mu_1,\mu_2,\mu_3]$ = $[p,q,0]$.

Preliminary to giving the functions G_q^t, it is convenient
to introduce a number of notations and definitions that will
be used throughout the remainder of Sections 3 and 4:

q is an integer that may assume values $0,1,2,\ldots$;
p is an integer that may assume for each q the values
     q, $q+1$, $\ldots$;
Δ is the 3-tuple of integers $(\Delta_1,\Delta_2,\Delta_3)$ with domain
     given by $0 \le \Delta_i \le p+q$, $\Delta_1 + \Delta_2 + \Delta_3 = p+q$;   (3.1)

n is the 3-tuple of integers (n_1,n_2,n_3) with domain
     given by

$$n_i \in \{0,1,\ldots,q\} \cap \{q-\Delta_i,q-\Delta_i+1,\ldots,p-\Delta_i\}, \text{ and}$$
$$n_1+n_2+n_3 = q;$$ (3.2)

$x = (x_1,x_2,x_3)$ is an arbitrary point in the Möbius plane;

$(z)_a = z(z+1)\ldots(z+a-1)$ for each $a = 0,1,2,\ldots$ and
     indeterminate z is Pochhammer's notation for a
     rising factorial;

$[z]_a = z(z-1)\ldots(z-a+1)$ is a falling factorial;
$\binom{z}{a} = [z]_a/a!$;

$$M = q+1-\delta_1-\delta_2-\delta_3, \quad \delta_i = \max (0, q-\Delta_i), \text{ each } i = 1,2,3.$$

We now introduce the following <u>polynomial</u> in the variables x_1, x_2, x_3:

$$D_n^2(\Delta;x) = \prod_{i=1}^{3} n_i!(\Delta_i-q+n_i)! \cdot (-1)^{p+q+\Delta_2+1}$$

$$\cdot \frac{\prod_{\substack{ijk \\ \text{cyclic}}} (x_i+q-\Delta_k-n_j)_{p-\Delta_i-n_i+1} \, (x_i-n_j)_{q-n_i+1}}{\prod_{\substack{ijk \\ \text{cyclic}}} (x_i + \Delta_j - \Delta_k)(x_i - n_j + n_k)} , \qquad (3.3)$$

in which $\prod_{\substack{ijk \\ \text{cyclic}}} A_{ijk}$ means the product of terms $A_{123}A_{231}A_{312}$.

Concerning this definition of $D_n^2(\Delta;x)$, two points need to be noted: The factors in the denominator do divide the numerator, so that the expression <u>is</u> a polynomial in x_1, x_2, x_3 (the expression is somewhat simpler if we do not carry out the explicit cancellations); for $x_1 \geq 1$ and $x_3 \geq 1$, hence, $x_2 \leq -2$, we have $D_n^2(\Delta;x) \geq 0$, which accounts both for the notation using a square and the presence of the phase factor $(-1)^{p+q+\Delta_2+1}$ [for $(x_1,x_2,x_3) = (\lambda_2 - \lambda_3 + 1, \lambda_3 - \lambda_1 - 2, \lambda_1 - \lambda_2 + 1)$, this is the region in which the irrep labels $(\lambda_1,\lambda_2,\lambda_3)$ obey $\lambda_1 \geq \lambda_2 \geq \lambda_3 \geq 0$].

We next introduce the polynomials $F_n^s(\delta;x)$ in x_1, x_2, x_3 obtained from the following expansion:

$$(y+1+\tfrac{1}{3}(x_2-x_3)+\delta_1)_{n_1-\delta_1} (y+1+\tfrac{1}{3}(x_3-x_1)+\delta_2)_{n_2-\delta_2}$$

$$\times (y+1+\tfrac{1}{3}(x_1-x_2)+\delta_3)_{n_3-\delta_3} = \sum_{s=1}^{M} y^{M-s} \, F_n^s(\delta;x). \qquad (3.4)$$

Using the polynomials $F_n^s(\delta;x)$ and the $D_n^2(\Delta;x)$, we then define

$$V_n^s(\Delta;x) = F_n^s(\delta;x)/D_n(\Delta;x) \text{ for } x_1 \geq 1, \ x_3 \geq 1, \ x_2 \leq -2, \quad (3.5)$$

each $s = 1,2,\ldots,M$, and for each $n = (n_1,n_2,n_3)$ in its domain of definition. The notation (V^r,V^s) for $r,s = 1,2,\ldots, M$ then denotes the "inner product" defined by

$$(V^r,V^s) = \sum_n F_n^r(\delta;x)F_n^s(\delta;x)/D_n^2(\Delta;x). \qquad (3.6)$$

Concerning this definition of (V^r, V^s), we note that it is
defined even when the conditions on x in Eq.(3.5) are violated,
since $D_n^2(\Delta;x)$ is a polynomial--in this case, it is Eq.(3.6)
itself that is used for the definition of (V^r, V^s). The summation over n in Eq.(3.6) is to be carried out over all n in
the domain of definition given in Eq.(3.2). It should also
be noted that we have suppressed in the notation (V^r, V^s) the
fact that this quantity depends on p, q, Δ, and x.

We require still one more definition before introducing
the G_q^t functions of interest, namely, the polynomial factors
defined by

$$L_s(\Delta;x) = [x_1 x_2 x_3/(x_1+\Delta_2-\Delta_3)(x_2+\Delta_3-\Delta_1)(x_3+\Delta_1-\Delta_2)]$$

$$\cdot \prod_{\substack{ijk \\ cyclic}} (x_i + 1)_{\Delta_j-s+1} (-x_i + 1)_{\Delta_k-s+1} , \tag{3.7}$$

each $s = 1, 2, \ldots, M$. Again the denominator factor always
divides the numerator.

Finally, we define the functions $G_q^t(\Delta;x)$ by

$$G_q^t(\Delta;x) = (\#) \prod_{s=1}^{t} L_s(\Delta;x) \cdot \det \begin{bmatrix} (V^1,V^1)(V^1,V^2) \cdots (V^1,V^t) \\ (V^2,V^1)(V^2,V^2) \cdots (V^2,V^t) \\ \vdots \\ (V^t,V^1)(V^t,V^2) \cdots (V^t,V^t) \end{bmatrix}, \tag{3.8}$$

each $t = 1, 2, \ldots M$, in which $(\#)$ denotes a quantity that is
independent of x_1, x_2, x_3, that is, depends only on p, q, Δ_i, t
(we will not require the explicit form of $\#$ for this discussion).

The result given by Eq.(3.8) is very complicated, and
there would appear to be little hope for much simplification,
let alone a proof that G_q^t is, in fact, a polynomial in the
x_1, x_2, x_3 with remarkable symmetries and patterns of zeros.
Fortunately, the simplest case of the general result, namely,

$$G_q^1(\Delta;x) = (\#) L_1(\Delta;x) (V^1,V^1), \tag{3.9}$$

with

$$(V^1, V^1) = \sum_n \frac{1}{D_n^2(\Delta;x)} \tag{3.10}$$

was studied first (see Refs. [18,19]) and it proved possible to obtain the explicit expression for G_q^1 and find its many nice properties. It was the encouragement coming from these results--and the explicit calculations of Alcaras, et al. [22] --that led us to believe that the complicated expression (3.8) is also a polynomial of a similar sort.

The polynomials $G_q^t(\Delta;x)$ of Eq.(3.8) may be considered as special functions of SU(3). The special functions defined in this way have interesting symmetries, which are most easily grasped by slightly changing the notation: We write $G_q^t(\Delta;x) = G_q^t(A)$, where A denotes the matrix array[1] given by

$$A = \begin{pmatrix} \Delta_1-t+1 & \Delta_2-t+1+x_1 & \Delta_3-t+1-x_1 \\ \Delta_2-t+1 & \Delta_3-t+1+x_2 & \Delta_1-t+1-x_2 \\ \Delta_3-t+1 & \Delta_1-t+1+x_3 & \Delta_2-t+1-x_3 \end{pmatrix}. \tag{3.11}$$

As discussed at the beginning of this section the functions $G_q^t(A)$ characterize the set of SU(3) tensor operators with irrep label [p,q,0] and shift pattern $(\Delta_1, \Delta_2, \Delta_3)$. The properties of the function $G_q^t(A)$ include the following:

(a) The function G_q^t is a polynomial in the variables of the matrix array.

(b) The polynomial is non-negative in the region of the Möbius plane corresponding to $\lambda_1 \geq \lambda_2 \geq \lambda_3$ (lexical region).

(c) The polynomial is invariant under all determinantal symmetries of the array A (row interchange, column interchange, and transposition).

(d) The polynomial G_q^t(each $t = 1,2,\ldots,q$) has value zero on a set of lattice points in the lexical region of the Möbius plane, this set of points being exactly that of the weight space diagram for irrep [q-t, 0, -t+1] of U(3), where the multiplicity of a weight space point is the order of the corresponding zero. (The "extremal" functions G_q^1 and G_q^q in the family $\{G_q^t | t = 1,2,\ldots,q\}$ have zeros, respectively, at the lattice points of the triangle and inverted triangle corresponding to the weight space diagrams [q-1, 0, 0] and [0, 0, -q+1].) (These remarkable properties were first noted by Lohe.)

(e) The polynomials $\{G_q^t\}$ satisfy reduction formulas [see Eqs.(4.6) and (4.7)] such that the subset of polynomials for $t = 1,2,\ldots,M$ split all multiplicities for tensor operators $[p,q,0]$ with shift pattern $(\Delta_1,\Delta_2,\Delta_3)$; that is, each of the tensor operators of this type, M in number, is uniquely associated with these polynomials (actually to a ratio of successive polynomials).

There is a wealth of special function information contained in the G_q^t functions. Let us illustrate this remark.

Consider first the quantal rotation group SU(2). The map from the Wigner operators, Eq.(1.7), to the invariant denominator function vanishes on the characteristic null space (the null space consisting of entire representations). For SU(2), the space of all irreps, each exactly once, is represented by the positive integers, that is, $2j+1 = p_{12}-p_{22} \equiv x_{12} > 0$. Because of the underlying S_2 symmetry, this becomes the integer points on the real line. The characteristic null spaces are found to be of the form of a "factorial power:" $[x_{12}]_n = x_{12}!/(x_{12}-n)!$. For SU(2), such considerations are so elementary and obvious as to have escaped attention, but they are essential for generalizations. Even for SU(2) they are important: the tensor operators in SU(2) may be constructed entirely from null space considerations (see Ref. [15]).

Let us return now to the tensor operators in SU(3). The space of all irreps, each exactly once, is now represented by lattice points in the Möbius plane defined by $x_1 = p_{23}-p_{33}$, $x_2 = p_{33}-p_{13}$, $x_3 = p_{13}-p_{23}$, which automatically satisfy the constraint $x_1 + x_2 + x_3 = 0$.

Let us consider tensor operators--transforming as the SU(3) irrep $[p,q,0]$--that effect the shift $(\Delta_1\Delta_2\Delta_3)$. Then for a special class of these operators (the "stretched" operators, $t=1$) the map tensor operator $\to$ invariant defines the function $G_q^1(\Delta)$. For $q=1$, that is for the operators $[p,1,0]$, we find the explicit result

$$G_1^1(\Delta;x) = -\Delta_1\Delta_2(\Delta_1 + x_3)(\Delta_2 - x_3)$$
$$-\Delta_2\Delta_3(\Delta_2 + x_1)(\Delta_3 - x_1) - \Delta_3\Delta_1(\Delta_3 + x_2)(\Delta_1 - x_2) \qquad (3.12)$$
$$- \Delta_1\Delta_2\Delta_3(\Delta_1 + \Delta_2 + \Delta_3).$$

We notice that this polynomial defines exactly 6 zeros,

only one of which $(x_1 = \Delta_3, x_3 = \Delta_2)$ lies in the lexical region.
The overall S_3 symmetry yields six 'triangles' of zeros, each
'triangle' having one point.

The G_q^1 functions are defined by two recursion relations,
which have the form of contiguous relations in that the parameters and arguments are shifted by integers. If we ignore
these shifts in the recursion formulas, then we find that the
solution is just $[G_1^1(\Delta;x)]^q$, that is the qth ordinary power of
the first solution.

This suggests the actual solution is to be found by some
sort of symbolic or <u>umbral</u> power, and this is indeed correct.
Let us consider sequences of polynomials of <u>binomial</u> <u>type</u> (see
Ref. [19]), that is, polynomials $p_n(x)$ obeying the rule

$$p_n(x+y) = \sum_k \binom{n}{k} p_k(x)\, p_{n-k}(y), \tag{3.13}$$

and extend this rule to multinomials.

Let us regard the terms in Eq.(3.12) as the terms in a
quadrinomial and expand out the qth power, collecting together
the powers of $\Delta_1, \Delta_2, \Delta_3$, but leaving the powers in (Δ_1+x_3),
etc. as well as $(\Delta_1+\Delta_2+\Delta_3)$ undisturbed:

$$[G_1^1(\Delta;x)]^q = (-1)^q q!\, \sum_{(k)} \Delta_1^{q-k_1} \Delta_2^{q-k_2} \Delta_3^{q-k_3} (\Delta_1+\Delta_2+\Delta_3)^{k_4}$$
$$\cdot\, (\Delta_2+x_1)^{k_1} (\Delta_3-x_1)^{k_1} (\Delta_3+x_2)^{k_2} (\Delta_1-x_2)^{k_2} \tag{3.14}$$
$$\cdot\, (\Delta_1+x_3)^{k_3} (\Delta_2-x_3)^{k_3} \big/ (k_1)!\,(k_2)!\,(k_3)!\,(k_4)!,$$

where the sum is over all non-negative integers $(k) = (k_1 k_2 k_3 k_4)$
which add to q, that is, $k_1+k_2+k_3+k_4 = q$. In this result, we
now make the replacements

$$\Delta_i^{q-k_i}/(q-k_i)! \to \binom{\Delta_i}{q-k_i},$$

$$(\Delta_2+x_1)^{k_1}/k_1! \to \binom{\Delta_2+x_1}{k_1}, \text{ etc.,}$$

with the single <u>exception</u>: We replace $(\Delta_1+\Delta_2+\Delta_3)^{k_4}/(k_4)!$ by

$$\binom{\Delta_1+\Delta_2+\Delta_3-k_1-k_2-k_3}{k_4}$$

The result is the following polynomial in x_1, x_2, x_3, which we denote by $G_q^1(\Delta;x)$:

$$
G_q^1(\Delta;x) = (-1)^q q! \sum_{(k)} \binom{\Delta_1 + \Delta_2 + \Delta_3 - k_1 - k_2 - k_3}{k_4}
$$

$$
\cdot (k_1)!(q-k_1)! \binom{\Delta_1}{q-k_1}\binom{\Delta_2 + x_1}{k_1}\binom{\Delta_3 - x_1}{k_1}
$$

$$
\cdot (k_2)!(q-k_2)! \binom{\Delta_2}{q-k_2}\binom{\Delta_3 + x_2}{k_2}\binom{\Delta_1 - x_2}{k_2} \tag{3.15}
$$

$$
\cdot (k_3)!(q-k_3)! \binom{\Delta_3}{q-k_3}\binom{\Delta_1 + x_3}{k_3}\binom{\Delta_2 - x_3}{k_3}.
$$

It can be verified that this explicit result for $G_q^1(x)$ has the properties (a)-(e) mentioned above.

The generalization to all G_q^t is the subject of the next section.

4.   CONJECTURED FORM OF THE POLYNOMIAL $G_q^t(\Delta;x)$ AND A VERIFICATION OF SOME OF ITS PROPERTIES

We give below the general polynomial that we conjecture to be the $G_q^t(\Delta;x)$ discussed in the previous sections (for t=1,2 all results given here are known to be true as discussed). We give this polynomial in a somewhat more general form than required for the theory of tensor operators, but only to the extent that the $\xi = (\xi_1, \xi_2, \xi_3)$ variables [see Eq.(4.2)] may be an arbitrary point in R^3. The relation with the polynomials of interest in the tensor operator problem is given by

$$
\xi_i = \Delta_i - t + 1, \quad i = 1,2,3. \tag{4.1}
$$

With these points in mind, we now define the polynomial $G_q^t(\xi;x)$, each t = 1, ... , q, each $\xi \in R^3$, and each $x \in R^2$ with $x_1 + x_2 + x_3 = 0$ by

$$
G_q^t \begin{pmatrix} \xi_1 & \xi_2 + x_1 & \xi_3 - x_1 \\ \xi_2 & \xi_3 + x_2 & \xi_1 - x_2 \\ \xi_3 & \xi_1 + x_3 & \xi_2 - x_3 \end{pmatrix} = G_q^t(\xi;x)
$$

$$= (-1)^{t(q-t+1)} \prod_{s=1}^{t} \frac{(q-s+1)!}{(s-1)!} \sum_{\lambda\mu\nu\rho} h(\lambda\mu\nu\rho) \frac{\prod\limits_{s=1}^{t} (\xi_1+\xi_2+\xi_3+2t-q-s)_{\rho_s}}{M(\rho)}$$

$\cdot\ F_{q-t+1,\lambda}\ (\xi_1,\xi_2+x_1,\ \xi_3-x_1)$

$\cdot\ F_{q-t+1,\mu}\ (\xi_2,\xi_3+x_2,\ \xi_1-x_2)$ (4.2)

$\cdot\ F_{q-t+1,\nu}\ (\xi_3,\xi_1+x_3,\ \xi_2-x_3),$

where the various symbols in this result have the following definitions:

(a) $\lambda = [\lambda_1,\lambda_2,\ldots,\lambda_t]$ denotes an irrep label of $U(t)$ with $\lambda_1 \geq \lambda_2 \geq \ldots \geq \lambda_t \geq 0$, each $\lambda_i =$ a non-negative integer; λ may also be regarded as the <u>shape</u> of a Young frame, $Y(\lambda)$. The symbols $\mu,\nu,\ldots$ denote irrep labels of the same type as λ.

(b) $h(\lambda\mu\nu\rho)$ denotes the number of times irrep $[q-t+1,\ldots,q-t+1]$ ($q-t+1$ repeated t times) is contained in the direct product $\lambda\times\mu\times\nu\times\rho$, and is defined to be zero if $[q-t+1,\ldots,q-t+1] \notin \lambda\times\mu\times\nu\times\rho$.

(c) $M(\lambda)$ is the measure of the Young frame $Y(\lambda)$ and has the definition

$$M(\lambda) = \prod_{s=1}^{t} (\lambda_s-s+1)! \Big/ \prod_{r<s} (\lambda_r-\lambda_s+s-r),$$ (4.3)

and $M(\mu)$, $M(\nu)$, ... have corresponding definitions.

(d) For each non-negative integer k and each Young frame $Y(\lambda)$, the function $F_{k,\lambda}(x,y,z)$ is defined for indeterminates x,y,z by

$$F_{k,\lambda}(x,y,z) = \begin{cases} \dfrac{1}{M(\lambda)} \prod\limits_{s=1}^{t} [x+t-s]_{k-\lambda_s} [y+s-1]_{\lambda_s} [z+s-1]_{\lambda_s}, \\ 0, \text{ if any } \lambda_s > k . \end{cases}$$ (4.4)

Let us next describe the properties of the polynomials (4.2) to the extent that they have been determined, noting from the outset that we have found no simple recursion relations for these polynomials for general $t(t > 1)$.

• Symmetries: From the fact that $F_{k,\lambda}(xyz)$ is invariant under interchange of y and z, and that the summation in (4.2) is invariant under the interchange of the summation labels λ, μ, and ν, it follows easily that $G_q^t(\xi;x)$ is invariant under those transformations of ξ and x corresponding to row permutations, and column 2 and column 3 interchange, in the array

$$\begin{pmatrix} \xi_1 & \xi_2+x_1 & \xi_3-x_1 \\ \xi_2 & \xi_3+x_2 & \xi_1-x_2 \\ \xi_3 & \xi_1+x_3 & \xi_2-x_3 \end{pmatrix} . \qquad (4.5)$$

Thus far we have not been able to prove that $G_q^t(\xi;x)$ is invariant under the transformations of ξ and x corresponding to transposition of the array (4.5). The proof of this transposition symmetry is required in order to establish the full determinantal symmetry. There is no reason to doubt the validity of the transposition symmetry, but it is technically difficult to prove because it involves explicit use of the condition $x_1+x_2+x_3 = 0$.

• Polynomial: $G_q^t(\xi;x)$ is clearly a polynomial of degree $2t(q-t+1)$ in ξ and in x.

• Zeros: $G_q^t(\Delta;x)$ has at least the zeros and multiplicities of the points of the weight space diagram of irrep $[q-t,0,-t+1]$. The proof of this result is quite lengthy (and is accordingly omitted), and, indeed, can be given only after we have learned how to transform the basic definition (4.2) to a new form. This, in turn, requires a generalization of the famous Saalschütz formula given in Ref. [23]. Let us note here, however, that this transformation depends essentially on the occurrence of the intertwining number $h(\lambda\mu\nu\rho)$ in Eq.(4.2); accordingly, this is indirect evidence for the correctness of this factor.

• Reduction formulas: The results for the zeros described above are restricted to the case $\Delta_i \geq q$, which, in turn, imply that $M = q+1$. This would appear to be a severe restriction on the applicability of those results. That those results include, in fact, the general case for $(\delta_1,\delta_2,\delta_3) \neq (0,0,0)$ is a consequence of the fact that the latter can be reduced to a case of a G_q^t, function that has $(\delta_1',\delta_2',\delta_3') = (0,0,0)$. This is a very significant simplification for the theory of tensor operators, since it reduces substantially the "kinds" of polynomials that occur. This reduction formula reads [see Eq.(4.1)]:

$$G_q^t \begin{pmatrix} \Delta_1-t+1 & \Delta_2-t+1+x_1 & \Delta_3-t+1-x_1 \\ \Delta_2-t+1 & \Delta_3-t+1+x_2 & \Delta_1-t+1-x_2 \\ \Delta_3-t+1 & \Delta_1-t+1+x_3 & \Delta_2-t+1-x_3 \end{pmatrix}$$

$$= (-1)^{t(q+1-M)} \prod_{s=1}^{t} \frac{(q+1-s)!}{(M-s)!}$$

(4.6)

$$\cdot \prod_{s=1}^{t} \prod_{\substack{ijk \\ cyclic}} \frac{[\Delta_i + \delta_i - s + 1]_{\delta_j + \delta_k}}{[q - s + 1]_{\delta_i}}$$

$$\cdot \prod_{s=1}^{t} \prod_{\substack{ijk \\ cyclic}} [\Delta_j + x_i - s + 1]_{\delta_i} [\Delta_k - x_i - s + 1]_{\delta_i}$$

$$\cdot G_{q'}^t \begin{pmatrix} \Delta_1'-t+1 & \Delta_2'-t+1+x_1' & \Delta_3'-t+1-x_1' \\ \Delta_2'-t+1 & \Delta_3'-t+1+x_2' & \Delta_1'-t+1-x_2' \\ \Delta_3'-t+1 & \Delta_1'-t+1+x_3' & \Delta_2'-t+1-x_3' \end{pmatrix},$$

where

$$q' = M - 1$$

$$x_i' = x_i - \delta_j + \delta_k, \quad ijk \text{ cyclic},$$

$$\Delta_i' = \Delta_i + \delta_i - \delta_j - \delta_k, \quad ijk \text{ cyclic},$$

$$\delta_i' = \max (0, q' - \Delta_i') = 0, \quad i = 1,2,3,$$

$$\Delta_1' + \Delta_2' + \Delta_3' = p + q' \quad (\text{for } \Delta_1 + \Delta_2 + \Delta_3 = p + q).$$

This remarkable formula establishes the important result:
The <u>properties of each</u> $G_q^t(\Delta;x)$ <u>polynomial for</u> $t = 1,2,\ldots,M$
<u>may be obtained through the reduction formula from those in
the special class</u> $\{G_{q'}^t(\Delta';x') | \Delta_i' \geq q', \ t = 1,2,\ldots,q'\}$.

A second kind of reduction formula for G_q^t can also be
proved and is closely related to the fact that one has a con-
jugation operation defined for tensor operators:

$$G_q^t(\xi;x) = G_q^{q-t+1}(-\xi;-x), \quad t = 1, \ldots, q. \tag{4.7}$$

It should be remarked that the proofs of both of the reduction formulas make essential use of the properties of the intertwining number $h(\lambda\mu\nu\rho)$, and this again supports the correctness of that factor in our conjectured form for G_q^t.

An interesting application of Eq.(4.7) occurs in the calculation of the minimal null space tensor operators, where one obtains the following formula (see Refs.[20,21]):

$$G_q^q \begin{pmatrix} 1 & 1+x_1 & 1-x_1 \\ 1 & 1+x_2 & 1-x_2 \\ 1 & 1+x_3 & 1-x_3 \end{pmatrix} = G_q^1 \begin{pmatrix} -1 & -1-x_1 & -1+x_1 \\ -1 & -1-x_2 & -1+x_2 \\ -1 & -1-x_3 & -1+x_3 \end{pmatrix}$$

$$= \frac{2(2q+1)!(q+1)!(q+1)!}{x_1 x_2 x_3} \sum_{i=1}^{3} \binom{x_i + q + 1}{2q + 3} \; . \tag{4.8}$$

Let us emphasize again that in establishing the symmetries of the G^t function one is led to a generalization of the Saalschütz identity and to a far-reaching generalization of the Euler hypergeometric function itself. This fact strongly hints that the extension from SU(3) to SU(n) may involve some very interesting special function generalizations.

5.   GENERALIZATION TO SU(n)

For SU(3) the multiplicity problem is resolved by the canonical labelling induced from the ordering of the characteristic null spaces of tensor operators with the same irrep and shift labels. In Refs. [24,25] this labelling was extended to the <p,q,...,q,0,...,0> tensor operators in U(n) uniquely characterized by maximal null space. To each such tensor operator there was associated an invariant polynomial denoted by

$$_\mu G_q^{(n)}(X) \equiv \; _\mu G_q^{(n)}((x_{ij} + \Delta_i))$$

$$\equiv \; _\mu G_q^{(n)}(\Delta_1, \ldots, \Delta_n; x_{12}, x_{23}, \ldots, x_{n-1,n} x_{n,1}), \tag{5.1}$$

where X is the n by n matrix [U(n) array] defined by means of:

Definition 5.2   Let $n \geq 2$.   Then

$$(X) \equiv ((X_{ij})) \equiv ((x_{ij} + \Delta_i)), \tag{5.3a}$$

where

$$\Delta_i \in R[,x_{rs},], \qquad 1 \leq i \leq n, \tag{5.3b}$$

$$x_{ij} \equiv -x_{ji}, \qquad i < j, \tag{5.3c}$$

$$x_{ii} \equiv 0, \qquad 1 \leq i \leq n, \tag{5.3d}$$

$$x_{ik} \equiv x_{ij} + x_{jk}, \; 1 \leq i,j,k \leq n. \tag{5.3e}$$

Remark: For $n = 3$, we have also denoted above $x_1 = x_{12}$, $x_2 = x_{23}$, and $x_3 = x_{31}$.

The above irrep label $<p,q,\ldots,q,0,\ldots,0>$ consists of one p, μ q's, and $n-\mu-1$ 0's.   Furthermore, q determines p since $\Delta_1 +\ldots+\Delta_n = p+\mu q$ and we are given q, $\Delta_1+\ldots+\Delta_n$, and μ.

The problems that give rise to $_\mu G_q^{(n)}(X)$ are motivated in more detail and put into a broader mathematical setting in Refs. [14,15,24].

Recently in Refs. [26-28] an alternative method for explicitly writing down $_\mu G_q^{(n)}(X)$ in polynomial form has been given. These methods provide a direct connection between the classical theory of symmetric functions as presented in Macdonald [29], and the symmetries satisfied by a family of $U(n)$ invariant polynomials even more general than $_\mu G_q^{(n)}(X)$.   Reference [26] gives a direct formulation of the invariant polynomials $_\mu G_q^{(n)}(,\Delta_i,;,x_{i,i+1},)$ characterizing $U(n)$ tensor operators $<p,q,\ldots,q,0,\ldots,0>$ in terms of the symmetric functions S_λ.

Given a partition $\lambda = (\lambda_1,\ldots,\lambda_n)$ of length $\leq n$ (the number of nonzero λ_i, denoted by $\ell(\lambda)$, is called the length of λ), the Schur functions S_λ are defined by

$$S_\lambda(x_1,\ldots,x_n) \equiv \frac{\det(x_i^{\lambda_j+n-j})_{1 \leq i,j \leq n}}{\det(x_i^{n-j})_{1 \leq i, j \leq n}}. \tag{5.4}$$

The determinant in the numerator is divisible in $Z[x_1,\ldots,x_n]$ by each of the differences $(x_i - x_j)$, $1 \leq i < j \leq n$, and hence by their product, which is the Vandermonde determinant

$$\prod_{1 \le i < j \le n} (x_i - x_j) = \det(x_i^{n-j})_{1 \le i,j \le n}$$

(5.5)

$$\equiv V_n(x_1, \ldots, x_n).$$

Thus, the above quotient of determinants is a symmetric polynomial in $x_1, \ldots, x_n$ with coefficients in Z.

After the change of variables

$$\Delta_i = \gamma_i - \delta_i \quad \text{and} \quad x_{i,i+1} = \delta_i - \delta_{i+1},$$

(5.6)

$_\mu G_q^{(n)}(,\Delta_i,;,x_{i,i+1},)$ becomes an integral linear combination of products of Schur functions $S_\alpha(,\gamma_i,) \cdot S_\beta(,\delta_i,)$ in the variables $\{\gamma_1, \ldots, \gamma_n\}$ and $\{\delta_1, \ldots, \delta_n\}$, respectively. That is, it is _directly_ proved that $_\mu G_q^{(n)}(,\Delta_i,;,x_{i,i+1},)$ is a _bisymmetric_ polynomial in the variables $\{\gamma_1, \ldots, \gamma_n\}$ and $\{\delta_1, \ldots, \delta_n\}$ with integer coefficients. This motivated the study of the yet more general bisymmetric polynomials

$$_\mu^m G_q^{(n)}(\gamma; \delta) \equiv {}_\mu^m G_q^{(n)}(\gamma_1, \ldots, \gamma_n; \delta_1, \ldots, \delta_m)$$

(5.7)

which are a common generalization of Eq.(2.2b) of Ref. [24] and Eq.(2.17) of Ref. [25] with no numerator parameters. These polynomials are given by

Definition 5.8   Given that $_\mu^m G_0^{(n)}(\gamma; \delta) \equiv 1$, we uniquely determine $_\mu^m \overline{G_q^{(n)}}(\gamma; \delta)$ by means of

$$_\mu^m G_q^{(n)}(,\gamma_i,;,\delta_i,) = \sum_{\substack{S \in I_n \\ \|S\| = \mu+1}} (-1)^{\mu+1+\Sigma(S)} \prod_{\substack{i<j \\ i \in S, j \in S^c}} (\gamma_i - \gamma_j)^{-1}$$

(5.9)

$$\cdot \prod_{\substack{i<j \\ i \in S^c, j \in S}} (\gamma_i - \gamma_j)^{-1} \cdot \prod_{i=1}^{n} \prod_{\ell=1}^{m} (\gamma_i - \delta_\ell) \cdot {}_\mu^m G_{q-1}^{(n)}(,\gamma_i - \chi(i \in S),;,\delta_i,),$$

where $S \subset I_n$ is a $(\mu+1)$-element subset of $\{1,2,\ldots n\}$, $\Sigma(S)$ denotes the sum of the elements in S, and $\chi(A)$ is 1 if statement A is true and 0 otherwise.

Remark.   We have $m = j-n$ with $j \ge n$ where j is the number of denominator parameters in Eq.(2.17) of Ref. [25]. Only m of these parameters actually appear in Eq. (5.9). The case that is of most interest (in physics) is $m = n$.

At this point it is convenient to establish some notation. Fix μ and assume that $m - n \equiv \nu$ is a constant. Denote the sets of variables $\{\gamma_1,\ldots,\gamma_{\mu+1}\}$, $\{\gamma_{\mu+2},\ldots,\gamma_n\}$, $\{\gamma_1,\ldots,\gamma_n\}$, and $\{\delta_1,\ldots,\delta_m\}$ by A, B, E, and F, respectively. If $Z = m_1 + \ldots + m_n$, with m_i monomials, then

$$S_\lambda(Z) \equiv S_\lambda(m_1 + \ldots + m_n) \equiv S_\lambda(x_1,\ldots,x_n)\big|_{x_i = m_i} . \quad (5.10a)$$

On the other hand, $S_\lambda(-(Z))$ is defined symbolically by

$$S_\lambda(-(Z)) \equiv S_\lambda(-(m_1+\ldots+m_n)) \equiv (-1)^{|\lambda|} S_{\lambda'}(Z), \quad (5.10b)$$

where $|\lambda| = \lambda_1+\ldots+\lambda_n$ is the sum of the parts of λ and λ' is the conjugate partition to λ. That is, $\lambda' = (\lambda'_1,\lambda'_2,\ldots,\lambda'_{(\lambda_1)})$, with $\lambda'_i = \|\{j \,|\, \lambda_j \geq i\}\|$. For example $(5,2,1)$ is the conjugate partition of $(3,2,1,1,1)$.

If $\rho = (\rho_1,\ldots,\rho_k)$, where ρ_i are integers then denote by $(X)^\rho$ the monomial $x_1^{\rho_1}\ldots x_k^{\rho_k}$ where $X = \{x_1,\ldots,x_k\}$. Let the permutations $w \in S_k$ act on subscripts. For example, $w(X)^\rho = x_{w(1)}^{\rho_1}\ldots x_{w(k)}^{\rho_k}$. Finally, let δ_k be the partition $(k-1,k-2,\ldots 0)$ and m_ℓ the partition consisting of ℓ parts equal to m.

One of the main results in Ref. [26] is the following fundamental

Theorem 5.11. Let $^m_\mu G^{(n)}_q(E;F)$ be defined as in Eq.(5.9). We then have

$$^m_\mu G^{(n)}_q(E;F) = \frac{(-1)^{\binom{\mu+1}{2}}}{V_n(E)} \sum_{\substack{\lambda=(\lambda_1,\ldots,\lambda_{\mu+1}) \\ \lambda_1 \leq m}} S_\lambda(-F)$$

$$\cdot \left\{ \sum_{w \in S_n} \epsilon(w) \cdot w[(A)^{m^{\mu+1}+\delta_{\mu+1}-(\lambda_{\mu+1},\ldots,\lambda_1)} (B)^{\delta_{n-(\mu+1)}} \right.$$

$$\cdot \left. {}^m_\mu G^{(n)}_{q-1}(\gamma_1 - 1,\ldots,\gamma_{\mu+1} - 1, B; F)] \right\} , \quad (5.12)$$

where, without loss of generality, w acts only on $\gamma_1, \ldots, \gamma_{\mu+1}$ when applied to $^m_\mu G^{(n)}_{q-1}$.

Starting with (5.12) and making direct use of the new symmetries discovered in Ref. [26], it is shown in Ref. [30] that the bisymmetric polynomials $^m_\mu G^{(n)}_q(E;F)$ are an integral linear combination of Schur functions $S_\lambda(E-F)$ in the symbol E-F, where E-F denotes the difference of the two sets of variables E and F. Making use of properties of skew Schur functions $S_{\lambda/\rho}$ and $S_\lambda(E-F)$ one puts together an umbral calculus for $^m_\mu G^{(n)}_q(E;F)$. That is, working entirely with polynomials, one uniquely determines $^m_\mu G^{(n)}_q(E;F)$ from $^m_\mu G^{(n)}_{q-1}(E;F)$ and combinatorial rules [such as the Littlewood-Richardson rule (see Ref. [29]) involving Ferrers diagrams, that is, partitions]. The deepest part of this umbral calculus is a summation theorem, involving many different aspects of the theory of Schur functions, which essentially reduces the double sum in Eq.(5.12) to a single term. Although simple to apply, this summation theorem is rather technical. Therefore, we will just give the final formulas for $^m_\mu G^{(n)}_1(E;F)$ and $^n_1 G^{(n)}_2(E;F)$, which are a direct result of the umbral calculus and the work in Ref. [26], and state a general theorem from Ref.[30] that illustrates how the structure of $^m_\mu G^{(n)}_q(E;F)$ "stabilizes" as ν increases while m-n = ν remains fixed.

Before giving these formulas we need some more notation. Let λ, ρ be partitions. The skew Schur function $S_{\lambda/\rho}$ is defined by

$$S_{\lambda/\rho}(E) = \sum_{\nu \subset \lambda} c^\lambda_{\rho\nu} \cdot S_\nu(E), \qquad (5.13)$$

where $\nu \subset \lambda$ means $\nu_i \le \lambda_i$ for all $i \ge 1$, and the integers $c^\lambda_{\rho\nu}$ are the Littlewood-Richardson rule coefficients determined by ·

$$S_\rho(E) \cdot S_\nu(E) = \sum_\lambda c^\lambda_{\rho\nu} S_\lambda(E). \qquad (5.14)$$

Note that $S_{\lambda/0} = S_\lambda$, where 0 denotes the zero partition. Also, $c^\lambda_{\rho\nu} = 0$ unless $|\lambda| = |\rho| + |\nu|$, so that $S_{\lambda/\rho}$ is homogeneous of degree $|\lambda| - |\rho|$, and is zero if $|\lambda| < |\rho|$. In fact, $S_{\lambda/\rho} = 0$ unless $\rho \subset \lambda$. The skew Schur function $S_{\lambda/\rho}$ can also be written as the determinant

$$S_{\lambda/\rho}(E) = \det(h_{\lambda_i-\rho_j-i+j}(E))_{1 \le i,j \le n}, \qquad (5.15)$$

where $n \ge \ell(\lambda)$, and h_ℓ is the ℓth homogeneous symmetric function of the variables E.

Given the sets of variables E and F, let E+F denote the set union $\{\gamma_1,\ldots,\gamma_n,\delta_1,\ldots,\delta_m\}$. There is then the following classical result

$$S_\lambda(E+F) = \sum_{\rho \subset \lambda} S_{\lambda/\rho}(E)S_\rho(F). \tag{5.16}$$

It is immediate from Eqs.(5.10b) and (5.16) that

$$S_\lambda(E-F) = \sum_{\rho \subset \lambda} S_{\lambda/\rho}(E)S_\rho(-F) \tag{5.17}$$

$$= \sum_{\rho \subset \lambda} (-1)^{|\rho|} S_{\lambda/\rho}(E) \cdot S_{\rho'}(F), \tag{5.18}$$

where ρ' is the conjugate partition to ρ.

The above classical formulas can be found in Chapter I of Macdonald [29]. Eqs.(5.13)-(5.18) are crucial to the analysis in Ref. [30].

We now state a fundamental result of Ref. [30].

Theorem 5.19. Fix $\mu \geq 0$ and $k = m-n$. There are integers $a_{\lambda,q}$, λ a partition, q a non-negative integer such that

$$_\mu^m G_q^{(n)}(E;F) = \sum_{\substack{|\lambda| \leq (\mu+1)(\mu+1+k)q \\ \lambda \subset [(\mu+1+k)q]^{(\mu+1)q}}} a_{\lambda,q} S_\lambda(E-F). \tag{5.20}$$

If $n \geq (\mu+1)q$ and $m \geq (\mu+1+k)q$ then $a_{\lambda,q}$ is independent of n. More generally, if $S_\lambda(E-F) \neq 0$, then $a_{\lambda,q}$ is independent of n.

A similar "stabilization" theorem involving $S_\alpha(E) \cdot S_\beta(F)$ instead of $S_\lambda(E-F)$ was proven in Ref.[26]. Just as in Ref. [26], it follows from Eq. (5.20) that the $n = (\mu+1)q$ and $m = (\mu+1+m-n)q$ case of this relation gives the correct formula for $_\mu^{m+\ell} G_q^{(n+\ell)}(E;F)$, for all integers ℓ, when the sets E and F contain $(n+\ell)$ and $(m+\ell)$ variables, respectively.

Theorem 1.22 of Ref. [26] expresses $_\mu^m G_1^{(n)}(E;F)$ as a sum of products of Schur functions $S_\alpha(E) \cdot S_\beta(F)$. Using Eq.(5.18) to rewrite this sum, immediately gives

$$_\mu^m G_1^{(n)}(E;F) = (-1)^{\binom{\mu+1}{2}} S_{(\mu+1+m-n)^{\mu+1}}(E-F), \tag{5.21}$$

where $(\mu+1+m-n)^{\mu+1}$ denotes the partition consisting of $(\mu+1)$ parts equal to $(\mu+1+m-n)$.

In Ref. [26] it took three pages to write ${}_1^n G_2^{(n)}(E;F)$ as a sum of products of Schur functions. Starting with Eq.(5.21) and making use of the umbral calculus developed in Ref. [30], it is not hard to see that

$$
\begin{aligned}
{}_1^n G_2^{(n)}(E;F) \; = \;\; & \{S_{2^2}(D)\}^2 + \{-[2S_{4,3}(D) + 2S_{2^3,1}(D) \\
& + 3S_{3,2^2}(D) + 3S_{3^2,1}(D) + S_{4,3,1}(D) + S_{3,2,1^2}(D)] \\
& + [S_{4,2}(D) + S_{2^2,1^2}(D) + 3S_{2^3}(D) + 3S_{3^2}(D) + 3S_{3,2,1}(D)] \\
& - [2S_{3,2}(D) + 2S_{2^2,1}(D)] + S_{2^2}(D)\} \; ,
\end{aligned}
\tag{5.22}
$$

where D denotes the set difference E-F.

Further applications of Theorem 5.19 and the umbral calculus, as well as detailed proofs, can be found in Ref.[30].

We conclude roughly where we began: special functions are interesting in theoretical physics because they arise not only from the matrix elements of symmetry group representations, but also from the associated algebras of tensor operators, the symmetry itself still being fundamental.

NOTES

[*] A joint report based on three invited papers [1-3] presented at the Mathematisches Forschungsinstitut Oberwolfach, 13-19 March 1983.

[†] Research support in part by the National Science Foundation, the U.S. Department of Energy, and a Sloan Fellowship (S.C.M.)

[1] Department of Physics, Duke University, Durham, North Carolina 27706, USA.

[2] Department of Mathematics, Texas A & M University, College Station, Texas 77843, USA.

[3] Los Alamos National Laboratory, University of California, Los Alamos, New Mexico, 87545, USA.

FOOTNOTE

[1] For t = 1, the nine entries in the array A may be written as $p_{j3}^{f} - p_{k3}^{i}$ (j,k = 1,2,3), where $p_{k3}^{i} = \lambda_k + 3 - k$ and $p_{j3}^{f} = p_{j3}^{i} + \Delta_j$. For the generalization to SU(n) discussed in Section 5, it is useful to rearrange the elements in array A into an array X in which the entry appearing in row j and column k is $X_{jk} = p_{j3}^{f} - p_{k3}^{i}$. The symmetries of G_q^t may then equally well be expressed as the determinantal symmetries of the array X. [One sees now that X is also invariant under constant shift of $(p_{13}^{f}, p_{23}^{f}, p_{33}^{f})$ and $(p_{13}^{i}, p_{23}^{i}, p_{33}^{i})$.]

REFERENCES

[1] L.C. Biedenharn, M.A. Lohe and J.D. Louck, "Special functions and group theory in theoretical physics," (unpublished).

[2] J.D. Louck, L.C. Biedenharn and M.A. Lohe, "Canonical SU(3) Wigner coefficients and special functions," (unpublished).

[3] S.C. Milne and R.A. Gustafson, "An umbral calculus for polynomials characterizing U(n) tensor operators," (to appear in Advances in Math.; see Ref. [30]).

[4] R. Askey, Orthogonal Polynomials and Special Functions. (Regional Conference Series in Applied Mathematics, Vol. 21). Society for Industrial and Applied Mathematics, 1975.

[5] E.P. Wigner, Application of Group Theory to the Special Functions of Mathematical Physics. (Unpublished lecture notes). Princeton University, Princeton, New Jersey, 1955.

[6] J. Talman, Special Functions: A Group Theoretic Approach. W.A. Benjamin, New York, 1968.

[7] N. Vilenkin, Special Functions and the Theory of Group
    Representations. (transl. from the Russian; Amer. Math.
    Soc. Transl., Vol. 22). Amer. Math. Soc., Providence,
    R.I., 1968.

[8] S. Helgason, Differential Geometry and Symmetric Spaces.
    Academic Press, New York, 1962.

[9] L.C. Biedenharn and P.J. Brussaard, Coulomb Excitation.
    Clarendon, Oxford, 1965.

[10] W. Miller, Jr., Symmetry and Separation of Variables.
     Encyclopedia of Mathematics and Its Applications, Vol. 4:
     G.-C. Rota, Ed.). Addison-Wesley, 1977.

[11] B. Kaufman, "Special functions of mathematical physics
     from the viewpoint of Lie algebra," J. Math Phys. $\underline{7}$(1966),
     447.

[12] C. Truesdell, An Essay Toward a Unified Theory of Special
     Functions. (Ann. of Math. Studies, No. 18). Princeton
     Univ. Press, Princeton, New Jersey, 1948.

[13] L. Weisner, "Group-theoretic origin of certain generating
     functions," Pacific J. Math. $\underline{5}$(1955), 1033-1039.

[14] L.C. Biedenharn and J.D. Louck, Angular Momentum in
     Quantum Physics. (Encyclopedia of Mathematics and Its
     Applications, Vol. 8: G.-C. Rota, Ed.). Addison-Wesley,
     1981.

[15] L.C. Biedenharn and J.D. Louck, The Racah-Wigner Algebra.
     in Quantum Theory. (Encyclopedia of Mathematics and Its
     Applications, Vol. 9: G.-C. Rota, Ed.). Addison-Wesley,
     1981.

[16] G.E. Baird and L.C. Biedenharn, "On the representations
     of semisimple Lie groups, II," J. Math. Phys. $\underline{4}$(1963),
     1449-1466; "III, The explicit conjugation operation for
     SU(n)." ibid $\underline{5}$(1964),1723-1730; "IV, A canonical classi-
     fication for tensor operators in SU_3," ibid $\underline{5}$(1965),
     1730-1747.

[17] D.E. Flath and L.C. Biedenharn, "Beyond the enveloping
     algebra of $\mathfrak{sl}_3$," (preprint, submitted for publication,1982).

[18] L.C. Biedenharn, J.D. Louck, E. Chacon, and M. Ciftan, "On the structure of the canonical tensor operators in the unitary groups. I. An extension of the pattern calculus rules and the canonical splitting in U(3)," J. Math. Phys. 13(1972),1957-1984.

[19] L.C. Biedenharn and J.D. Louck, "On the structure of the canonical tensor operators in the unitary groups. II. The tensor operators in U(3) characterized by maximal null space," J. Math Phys. 13(1972),1985-2001.

[20] J.D. Louck, M.A. Lohe, and L.C. Biedenharn, "Structure of the canonical U(3) Racah functions and the U(3):U(2) projective functions," J.Math. Phys. 16(1975),2408-2426.

[21] M.A. Lohe, L.C. Biedenharn, and J.D. Louck, "Structural properties of the self-conjugate SU(3) tensor operators," J. Math Phys. 18(1977),1883-1891.

[22] J.A. Castilho Alcaras, L. C. Biedenharn, K.T. Hecht, and G. Neely, "On the 27-plet unitary symmetry operator," Annals of Physics (N.Y.) 60(1970),85-147.

[23] J.D. Louck and L.C. Biedenharn, "A generalization of the Gauss hypergeometric series," J. Math. Analysis and Appl. 59(1977),423-431.

[24] L.C. Biedenharn, W. Holman III, and S. Milne, "The invariant polynomials characterizing U(n) tensor operators <p,q,...,q,0,...,0> having maximal null space," Advances in Applied Math., 1(1980),390-472.

[25] S.C. Milne, "Hypergeometric series well-poised in SU(n) and a generalization of Biedenharn's G functions," Advances in Math. 36(1980),169-211.

[26] R.A. Gustafson and S.C. Milne, "Schur functions and the invariant polynomials characterizing U(n) tensor operators," Advances in Applied Math. (in press).

[27] R.A. Gustafson and S.C. Milne, "Schur functions, Good's identity, and hypergeometric series well poised in SU(n)," Advances in Math. 48(1983),177-188.

[28] R.A. Gustafson and S.C. Milne, "A new symmetry for
     Biedenharn's G-functions and classical hypergeometric
     series," (in preparation).

[29] I.G. Macdonald, <u>Symmetric</u> <u>Functions</u> <u>and</u> <u>Hall</u> <u>Polynomials</u>.
     Oxford Univ. Press, 1979.

[30] L.C. Biedenharn, R.A. Gustafson, and S.C. Milne, "An
     umbral calculus for polynomials characterizing U(n)
     tensor operators," Advances in Math. (in press).

George E. Andrews and Enrico Onofri

LATTICE GAUGE THEORY, ORTHOGONAL POLYNOMIALS

AND q-HYPERGEOMETRIC FUNCTIONS

0. ABSTRACT

A simple closed form for a particular instance of Wilson's loop variables is derived both via group theory and via q-hypergeometric series. Several facets of these diverse approches are explored.

1. INTRODUCTION

This joint work has its genesis in the study of the following seemingly bizarre identity

$$\sum_{j\geq 0} \frac{(q^s;q)_j(q^{-s};q)_j q^{(N+1)j}}{(q;q)_j^2}$$

$$= \frac{(-1)^{s-1}(1-q^s)}{(q;q)_N} \sum_{r=0}^{N} (-1)^r(N-r)(1-q^{N-r})$$

$$\sum_{n=0}^{r} \begin{bmatrix} N \\ n \end{bmatrix}_q \begin{bmatrix} N \\ r-n \end{bmatrix}_q q^{n+\binom{2n-r+s}{2}} \qquad (1.1)$$

where
$$\begin{bmatrix} A \\ B \end{bmatrix}_q = \frac{(1-q^A)(1-q^{A-1})\ldots(1-q^{A-B+1})}{(1-q^B)(1-q^{B-1})\ldots(1-q)} \qquad (1.2)$$

for $B \geq 0$ (vanishing for $B<0$, $A>0$) and

$$(A;q)_n = (1-Aq^{n-1})(1-Aq^{n-2})\ldots(1-A) \qquad (1.3)$$

Several initial observations should be made. For the

R. A. Askey et al. (eds.), Special Functions: Group Theoretical Aspects and Applications, 163–188.
© 1984 by D. Reidel Publishing Company.

mathematician, the left side of (1.1) (actually a slight va-
riant of it) provides the spectral values of certain orthogonal polynomials (the little q-Jacobi polynomials); consequently (1.1) provides some wild transformation of these spectral values. For the physicist the right side of (1.1) arises naturally in representing certain Wilson's loop variables
while the left hand side provides a much more tractable closed form for these objects. These observations are, in fact,
what led to this collaboration of a mathematician and a physicist. In Section 2 we shall present a brief survey of the
aspects of lattice gauge theory which led one of us [13] to
conjecture (1.1). In Sections 3 and 4 we prove (1.1) by pro-
viding two different evaluations of the integral representing
Wilson's loop variable. In Sections 5 and 6 we analyze the
problem from the viewpoint of basic hypergeometric series,
and we provide a second proof utilizing primarily classical
q-series results. The tecniques required in Sections 5 and
6 suggest further applications which are briefly explored in
Section 7, where still another proof is sketched, based on
the orthogonality property of Rogers-Szegö polynomials.

While the identity (1.1) originally arose in attempting
to obtain a simple closed formula for the Wilson loop variables, we feel that the interactions and directions suggested
by this study are also of substantial interest. This is espe-
cially so in light of the recent interest in the Selberg integral [19] , another N-dimensional integral with kernel
$\left|\Delta(e^{i\theta})\right|^{2k}$.

2. LATTICE GAUGE THEORY IN TWO-DIMENSIONS

We shall provide only a brief sketch of lattice gauge theory.
Extensive accounts are available in some recent review papers
[5], [8]. Here it will suffice to recall that the two-dimensional theory is exactly soluble; i.e. it reduces to the problem of evaluating integrals over the unitary group. Let us
consider a plane square lattice

$$L = \left\{ x = n_1 ae_1 + n_2 ae_2 \mid (n_1, n_2) \in \mathbf{Z}^2 \right\} ,$$

where a is the lattice spacing which we can put equal to one.
We then consider gauge variables $U_{ij} \in G$ where (i,j) denotes
any pair of nearest neighbour points in L and G is any (com-
pact) group. Applications are most often concerned with G=
SU(3); we shall more generally consider G=U(N), the group of
N × N unitary matrices, or G = SU(N), its unimodular subgroup.
The U_{ij} are random variables (subject to the condition $U_{ij} U_{ji} = 1$)
their probability distribution being defined as follows.

Let P be any elementary square in the lattice (called a
"plaquette") and let $U_{\partial P}$ be the (ordered) product of the variables U_{ij} along the boundary of P (starting from any point
of P). Let Φ be any positive symmetric class function on G
($\Phi(U) = \Phi(VUV^{-1}) = \Phi(U^{-1})$). The measure

$$\mu_\Lambda = \prod_{<ij>} [dU_{ij}] \prod_{P \subset \Lambda} \Phi(U_{\partial P}) \qquad (2.1)$$

(where $<ij>$ and P are contained in some finite square $\Lambda \subset L$
and [dU] is the Haar measure) is taken as the unnormalized
probability measure. There are some restrictions on the choice of μ which are dictated by the heuristic continuum limit
$(a \to 0)$: namely, in this limit, μ should reproduce the classical Lagrangian of Yang-Mills fields, i.e.

$$\Phi \underset{a \to 0}{\sim} \exp\left\{-\frac{1}{g^2} \int d^4x \sum_{\mu\nu} \text{tr } F_{\mu\nu}^2\right\} \qquad (2.2)$$

where

$$U_{\partial P} = \exp\left\{ia^2 F_{\partial P} + O(a^3)\right\}$$

Even with this stipulation there is still great latitude in
the choice of Φ. The original choice [24] was the following:

$$\Phi(U) = \exp \frac{1}{g^2} \left\{ \chi_f(U) + \chi_f(U^{-1}) \right\} \qquad (2.3)$$

where χ_f is the fundamental character of SU(N) and g is a nonvanishing real parameter called the "coupling constant". Other choices will be introduced subsequently.

The fundamental property of the probability measure is its gauge invariance under the transformations

$$U_{ij} \rightarrow U_{ij} = \gamma_i \, U_{ij} \, \gamma_j^{-1} \qquad (2.4)$$

where γ is any map $L \rightarrow G$. This is a local symmetry in contrast with the "global" or "rigid" symmetry of other lattice models such as those of Ising or Heisenberg.

The central object of gauge theory is the so-called "Wilson loop" W_Γ. Take any simple oriented closed contour $\Gamma \subset \Lambda$ and define $U_\Gamma = \prod_{<ij>\subset\Gamma} U_{ij}$ (again the initial point is immaterial since we are interested only in the class of U_Γ). The trace $\chi_f(U_\Gamma)$ is gauge-invariant (i.e. it is invariant under the local transformations (2.4)), and one defines W_Γ as

$$W_\Gamma = \frac{1}{d_f} \frac{\int \mu \, \chi_f(U_\Gamma)}{\int \mu} = \frac{1}{d_f} < \chi_f(U_\Gamma)> \qquad (2.5)$$

where $d_f = \chi_f(I)$ with I the identity matrix. One is interested in the asymptotic behaviour of W_Γ as the size of Γ goes to infinity; it is known that for sufficiently large g

$$W_\Gamma \sim \exp(-\sigma A_\Gamma) \qquad (2.6)$$

where A_Γ is the area of Γ and σ is some positive constant [14]. Noteworthy about W_Γ is that its asymptotic behaviour is known to be related to the force which a Yang-Mills field generates between a quark-antiquark pair. In particular, the asymptotic behaviour (2.6) implies that a constant force permanently binds together the two quarks. This property has been verified in the full four-dimensional theory for g sufficiently large. Furthermore there is a strong evidence from numerical experiments that this still holds down to g=0 (in

dimension not exceeding four) although some kind of phase
transition occurring around g=1 prevents to extend the ana-
lytical calculation to small g.

In two dimensions the theory is essentially trivial because one always finds [12]

$$W_\Gamma = (W)^{A_\Gamma/a^2} \tag{2.7}$$

where

$$W = \frac{1}{d_f} \frac{[dU]\, \Phi(U)\, _f(U)}{[dU]\, \Phi(U)} = a_f/a_0 \tag{2.8}$$

having introduced the character expansion of $\Phi(U)$:

$$\Phi(U) = a_0 + d_f a_f (\chi_f(U) + \chi_f(U^{-1})) + \ldots \tag{2.9}$$

In this case the asymptotic behaviour (2.6) always satisfies
$\sigma a^2 = -\ln(W) > 0$.

More generally one considers the multiple loop variables

$$W_n(g,N) = \frac{1}{N} < Tr(U^n) > = \frac{\int [dU]\, \Phi(U) Tr(U^n)}{N \int [dU]\, \Phi(U)}, \tag{2.10}$$

where $U \in SU(N)$ is identified with its fundamental matrix
representative. These variables have no direct physical meaning, but they arise naturally when one tries to calculate
W_Γ by some recursive method (the so-called "loop equations"
[10], [16]).

Significant interest is attached to the asymptotic behaviour of such integrals as N goes to infinity [7]. In a
variety of models this limit can be found in closed form and
gives a good approximation to the actual value for finite N.
To work out such an asymptotic expansion for W_n, a method
was introduced [6] based on the orthogonal polynomials constructed out of $\Phi(U)$. In the next section, we shall introduce a particular choice of $\Phi(U)$ which leads to the Rogers-
Szego polynomials and allows us to calculate W_n in terms of
q-binomial coefficients. An independent calculation of W_n can
be obtained in terms of the character expansion of Φ. This
provides us with our first proof of (1.1) and Theorem 2.

From a physical point of view these results are interesting primarily because they provide an example where the expansion in the parameter 1/N can be studied in full detail. The convergence properties of this kind of expansion, however, are strongly dependent on the choice of the measure; a general mathematical approach to this subject is still lacking.

3. THE FIRST EXPANSION OF $W_n(q,N)$

Let us now introduce a specific choice of Φ:

$$\Phi(U) = \det\left[\vartheta_3(U|q)\right], \qquad (3.1)$$

where $\vartheta_3(z) = \sum_{n=-\infty}^{\infty} q^{n^2} z^n$ is one of Jacobi's ϑ-functions; q is related to the usual coupling constant by $q = \exp(-g^2/4) \in (0,1)$. Jacobi's transformation shows that (2.2) is actually satisfied. By introducing polar coordinates on the group $U(N)$, $U = S\,\text{diag}\left[\exp(i\varphi_1),\ldots,\exp(i\varphi_N)\right]S^\dagger$, we obtain

$$W_n(q,N) = \frac{Z^{-1}}{N}\int\frac{d\varphi_1}{2\pi}\ldots\int\frac{d\varphi_N}{2\pi}\left|\Delta(e^{i\varphi})\right|^2$$

$$\prod_{j=1}^{N}\vartheta_3(e^{i\varphi_j}|q)\sum_{m=1}^{N}e^{in\varphi_m} \qquad (3.2)$$

where $Z(q,N)$ is chosen in such a way that $W_0 = 1$ and $\Delta(e^{i\varphi}) = \prod_{1\leq k<j\leq N}(\exp(i\varphi_j) - \exp(i\varphi_k))$ is Vandermonde's determinant.

Now, if $P_j(z)$ is any family of monic polynomials (with P_j of degree j), then

$$\Delta(e^{i\varphi}) = \det\left\|P_{j-1}(e^{i\varphi_k})\right\|_{j,k=1,\ldots,N} \qquad (3.3)$$

Moreover, if the P_j are orthogonal with respect to the measure

$$\vartheta_3(e^{i\varphi}|q)\,d\varphi$$

on the unit circle, then the integral (3.2) reduces to

$$W_n(q,N) = \frac{1}{N}\sum_{j=0}^{N-1}\frac{\langle P_j, z^n P_j\rangle}{\|P_j\|^2} \qquad (3.4)$$

Letting $v_j(z) = P_j(z)/\|P_j\|$, we have

$$W_n(q,N) = \frac{1}{N} \sum_{j=0}^{N-1} \int d\omega \; \vartheta_3(e^{i\omega}|q) e^{in\varphi} \left| v_j(e^{i\varphi}) \right|^2$$

$$= \int d\varphi \; e^{in\varphi} \; \rho_N(\varphi,q) \qquad (3.5)$$

where ρ_N represents the probability density for the eigenvalues φ of U, namely

$$\rho_N(\varphi,q) = \frac{1}{N} \sum_{j=0}^{N-1} \left| v_j(e^{i\varphi}) \right|^2 \vartheta_3(e^{i\varphi}|q) . \qquad (3.6)$$

Now, ρ_N can actually be evaluated in the standard way through the use of the Christhoffel-Darboux formula [21; pp.42-44]. The polynomials $v_j(z)$ are slightly transformed Rogers-Szego polynomials [1; pp.49-52], and are explicitly given by

$$v_j(z) = (q^2;q^2)_j^{-\frac{1}{2}} \sum_{k=0}^{j} \begin{bmatrix} j \\ k \end{bmatrix}_{q^2} (-q)^{j-k} z^k \qquad (3.7)$$

where $((q^2;q^2)_j$ and $\begin{bmatrix} j \\ k \end{bmatrix}_q$ are defined in (1.2) and (1.3).

The recurrence relation

$$z \, v_j(z) = (1-q^{2j+2})^{\frac{1}{2}} \, v_{j+1}(z) + q \, v_j(z)$$

$$- q(1-q^{2j})^{\frac{1}{2}} \, z \, v_{j-1}(z) \qquad (3.8)$$

can be effectively applied to obtain the matrix elements $\langle v_j, z^n \, v_j \rangle$, at least for small n. One finds, for instance,

$$\langle v_j, z \, v_j \rangle = q^{2j+1} \qquad (3.9a)$$

$$\langle v_j, z^2 v_j \rangle = q^{4j} \begin{bmatrix} 3 \\ 1 \end{bmatrix}_{q^2} - q^{2j} \begin{bmatrix} 2 \\ 1 \end{bmatrix}_{q^2} \qquad (3.9b)$$

$$\langle v_j, z^3 v_j \rangle = q^{6j-3} \begin{bmatrix} 5 \\ 2 \end{bmatrix}_{q^2} - q^{4j-3} \begin{bmatrix} 4 \\ 1 \end{bmatrix}_{q^2} \begin{bmatrix} 3 \\ 1 \end{bmatrix}_{q^2} +$$

$$+ \; q^{2j-1} \begin{bmatrix} 3 \\ 1 \end{bmatrix}_{q^2} \tag{3.9c}$$

but a general expression for $<v_j, z^n v_j>$ seemed, at first sight, hard to obtain directly from the recurrence relation (this circumstance will be remedied in Theorem 2; see also the comments in Section 8).

We shall now apply the Christoffel-Darboux [21,pp 42-44] formula to obtain $\rho_N(\varphi,q)$:

$$\rho_N(\varphi,q) = \frac{1}{N} \vartheta_3(e^{i\varphi}|q) \left. \frac{v_N^*(\zeta) v_N^*(z) - \overline{v_N(\zeta)} \, v_N(z)}{1 - z\overline{\zeta}} \right|_{\zeta = z = e^{i\varphi}} \tag{3.10}$$

After inserting the explicit form of $v_N(z)$ and of $v_N^*(z) = z^N v_N(z^{-1})$ one easily finds

$$\rho_N(\varphi,q) = \frac{1}{N} \sum_{0 \le j,k \le N} (q^2;q^2)_N^{-1} \begin{bmatrix} N \\ j \end{bmatrix}_{q^2} \begin{bmatrix} N \\ k \end{bmatrix}_{q^2}$$

$$(-q)^{j+k}(N-j-k) \; e^{i(j-k)\varphi} \; \vartheta_3(e^{i\varphi}|q) \tag{3.11}$$

Hence, by (3.5)

$$W_n(q,N) = \frac{1}{N} (q^2;q^2)_N^{-1} \sum_{r=0}^{N} (N-r)(-q)^r (1-q^{2(N-r)})$$

$$\sum_{j=0}^{r} \begin{bmatrix} N \\ j \end{bmatrix}_{q^2} \begin{bmatrix} N \\ r-j \end{bmatrix}_{q^2} q^{(2j-r+n)^2} \tag{3.12}$$

This last expression does provide a closed form for $W_n(q,N)$; indeed after q is replaced by q^2 it is (after multiplication by an elementary factor) the right hand side of (1.1). However (3.12) is too complicated to allow an analysis for $N \to \infty$ at $q^N = \exp(-\lambda/4)$ fixed, which was the original motivation of our study. Even for small n, where we can easily get W_n by summing the matrix elements $<v_j, z^n v_j>$, it is not obvious at all that the two expressions coincide; indeed this latter method of evaluation yields

$$W_0(q,N) = 1 \tag{3.13a}$$

$$W_1(q,N) = \frac{q}{N} \begin{bmatrix} N \\ 1 \end{bmatrix}_q^2 \tag{3.13b}$$

$$W_2(q,N) = \frac{1}{N} \begin{bmatrix} 3 \\ 1 \end{bmatrix}_q^2 \begin{bmatrix} N \\ 1 \end{bmatrix}_q^4 - \frac{1}{N} \begin{bmatrix} 2 \\ 1 \end{bmatrix}_q^2 \begin{bmatrix} N \\ 1 \end{bmatrix}_q^2 \tag{3.13c}$$

We shall show in the next two sections how one finds the most useful closed form representation of $W_n(q,N)$. In Section 4 our proof relies on the character expansion of $\Phi(U)$. In Section 5 we study the right hand side of (3.12) as a multiple q-hypergeometric series expansion; this latter approach leads to some analytic devices with further interesting applications considered in Section 6 and 7. A third method, which was actually the first to be tried, consists of studying the matrix $<v_j, z^n v_k>$ in the 1/N-expansion; the first two terms of the expansion were derived by one of us in [13] and led to the conjecture that (1.1) was valid. It turned out, when this paper was in its final form, that this path can be pursued to the end and it gives still another way to calculate $W_n(q,N)$. This will be sketched in Section 8: essentially it turns out that the result is equivalent to a property of Rogers-Szego polynomials which can be found in [25];

4. THE SECOND EXPANSION OF $W_n(q,N)$ VIA GROUP THEORY

It is well-known [9], [22] that any combination of traces of a matrix and of its powers of the form

$$S_\alpha = (\text{tr } U)^{\alpha_1} (\text{tr } U^2)^{\alpha_2} \ldots (\text{tr } U^r)^{\alpha_r}$$

can be expanded into characters $\chi_\lambda(U)$ of GL(N,C), the coefficients being precisely the characters of the symmetric group S_n ($n = \alpha_1 + \alpha_2 + \ldots + \alpha_r$) on n objects [9] :

$$S_\alpha = \sum \chi^{(\lambda)}_{(1^{\alpha_1}, 2^{\alpha_2}, \ldots, r^{\alpha_r})} \chi_\lambda(U) \tag{4.1}$$

Here $(\alpha_1, \ldots, \alpha_r)$ gives the cycle structure of the class $C_\alpha \subset S_n$

and the partition $\{\lambda\}$ identifies the irreducible representation. It is also known [15; p. 75] that the character of a full cycle is always 1, 0 or -1, namely:

$$X^{\lambda}_{(n)} = \begin{cases} (-1)^r & \lambda = (n-r, \; 1^r) \\ 0 & \text{otherwise} \end{cases} \tag{4.2}$$

Hence we have

$$W_n(q,N) = \frac{1}{N} \frac{\int [dU] \Phi(U) \sum_{r \geq 0} (-1)^r X_{(n-r,1^r)}(U)}{\int [dU] \; \Phi(U)} \tag{4.3}$$

We shall now determine the character expansion of $\Phi(U)$ which will allow us to calculate W_n. We have

<u>Theorem 1</u>. The character expansion of $\Phi(U) = \det \left[\vartheta_3(U|q) \right]$ is given by

$$\frac{\Phi(U)}{\int [dU] \; \Phi(U)} = \sum_{m_1 \geq m_2 \geq \dots \geq m_N} q^{\sum m_i^2}$$

$$\prod_{1 \leq j < k \leq N} \left(\frac{1 - q^{2(m_j - m_k - j + k)}}{1 - q^{2(k-j)}} \right) X_{(m_1, \dots, m_n)}(U) \tag{4.4}$$

<u>Corollary 1</u>. The coefficient of $X_{(n-r,1^r)}$ in the character expansion of $\Phi(U)$ is given by

$$< X_{(n-r,1^r)} > = q^{(n-r)^2 + r} \begin{bmatrix} N+n-r-1 \\ n \end{bmatrix}_{q^2} \begin{bmatrix} n-1 \\ r \end{bmatrix}_{q^2} \tag{4.5}$$

<u>Theorem 2</u>.

$$W_n(q,N) = \frac{1}{N} \sum_{r \geq 0} (-1)^r q^{(n-r)^2 + r} \begin{bmatrix} N+n-r-1 \\ n \end{bmatrix}_{q^2} \begin{bmatrix} n-1 \\ r \end{bmatrix}_{q^2}$$

$$= - \frac{1}{N} \frac{(-q)^n}{1 - q^{2n}} \; {}_2\phi_1 \left(\begin{array}{c} q^{2n}, q^{-2n} \; ; \; q^2, \; q^{2N+2} \\ q^2 \end{array} \right) \tag{4.6}$$

where in the standard notation $[20; \, p.90]$:

$$
{}_r\phi_s \left(\begin{array}{c} a_1, \; a_2, \ldots \; a_r \; ; \; q, \; t \\ b_1, \; b_2, \ldots, b_s \end{array} \right) =
$$

$$
\sum_{j \geq 0} \frac{(a_1;q)_j (a_2;q)_j \ldots (a_r;q)_j}{(q;q)_j (b_1;q)_j (b_2;q)_j \ldots (b_s;q)_j} \, t^j \tag{4.7}
$$

Corollary 2. The diagonal matrix elements $<v_j, z^n v_j>$ of z^n in the basis of Rogers-Szego polynomials are given by:

$$
<v_j, z^n v_j> = (-1)^{n+1} q^{2j+2-n} \begin{bmatrix} n \\ 1 \end{bmatrix}_{q^2}
$$

$$
{}_2\phi_1 \left(\begin{array}{c} q^{2n+2}, \; q^{2-2n} \; ; \; q^2, \; q^{2j+2} \\ q^4 \end{array} \right) \tag{4.8}
$$

Proof of Theorem 1. Let $\Phi(U) = \sum_\lambda a_\lambda \chi_\lambda(U)$; then by Weyl's character formula,

$$
a_\lambda = \int_0^{2\pi} \ldots \int_0^{2\pi} \frac{d\varphi_1 \ldots d\varphi_N}{(2\pi)^N \, N!} \, |\Delta(e^{i\varphi})|^2 \Phi(\mathrm{diag}[e^{i\varphi_1}, \ldots, e^{i\varphi_N}])
$$

$$
\chi_\lambda(\varphi_1, \ldots, \varphi_N)
$$

$$
= \int_0^{2\pi} \ldots \int_0^{2\pi} \frac{d\varphi_1 \ldots d\varphi_N}{(2\pi)^N \, N!} \, \det \| e^{i(N-j)\varphi_k} \| \det \| e^{-i\ell_j \varphi_k} \|
$$

$$
\prod_{j=1}^N \vartheta_3(e^{i\varphi_j} | q)
$$

$$(\ell_j = m_j + N - j)$$

$$= \frac{1}{N!} \sum_{p,q \in S_N} \varepsilon_p \varepsilon_q \prod_{j=1}^{N} \int \frac{d\varphi_j}{2\pi} \sum_{r=-\infty}^{\infty} q^{r^2} e^{ir\varphi_{j_e}} e^{-i\ell_{p_j}\varphi_j + i(N-q_j)\varphi_j}$$

$$= \frac{1}{N!} \sum_{p,q \in S_N} \varepsilon_p \varepsilon_q \prod_{j=1}^{N} q^{(\ell_{p_j} - N + q_j)^2}$$

(now set $q_j \to j$)

$$= \frac{1}{N!} \sum_{p,q \in S_N} \varepsilon_p \varepsilon_q \prod_{j=1}^{N} q^{(\ell_{pq^{-1}_j} - N + j)^2}$$

$$= \sum_{p \in S_N} \varepsilon_p \prod_{j=1}^{N} q^{(\ell_{p_j} - N + j)^2}$$

$$= \det \left\| q^{(\ell_j - N + k)^2} \right\|_{j,k=1,\dots,N} \tag{4.9}$$

Since the identity representation corresponds to $\ell_j = N-j$, we get

$$a_\lambda / a_0 = \frac{\det \left\| q^{(m_j - j + k)^2} \right\|}{\det \left\| q^{(k-j)^2} \right\|} \tag{4.10}$$

These determinants are easily evaluated once they are reduced to the Vandermonde determinant. Notice also that the expression a_λ/a_0 coincides with a <u>specialization</u> of χ_λ, namely

$$a_\lambda / a_0 = q^{2c_\lambda^{(2)}} \chi_\lambda(T_q) \tag{4.11}$$

where $c_\lambda^{(2)}$ is the quadratic Casimir invariant and $T \in GL(N,C)_q$ has eigenvalues

$$\left[q^{N-1}, q^{N-3}, \ldots, q^{N-2}, \ldots, q^{1-N} \right] .$$

Also of interest is the choice T = identity, which leads to the "heat kernel" on the group (i.e. $\Phi(U)[dU]$ represents in this case the heat distribution on the group at the time t ($q=\exp-\frac{1}{2}t$) if the initial distribution at t=0 is the Dirac measure $\delta(U)$ centred at the identity).

Proof of Corollary 1. This result is easily obtained by inserting ($m_1 = n-r$, $m_2 = \ldots = m_{r+1} = 1$, $m_{r+2} = \ldots = m_N = 0$) into equation (4.4). The easiest way to perform the calculation is by realizing that, apart from trivial factors, a_λ/a_0 is formally identical to the <u>dimension</u> χ_λ(identity) after the substitution $n \mapsto [n] = 1 - q^{2n}$ in every integer factor; hence we can find a_λ/a_0 by the well-known formula involving the "hook graph" [15]:

$$a_\lambda/a_0 = q^{\sum m_j^2} \begin{vmatrix} [N] & [N+1] & \ldots & [N+n-r-1] \\ [N-1] & & & \\ \vdots & & & \\ [N-r] & & & \end{vmatrix} \div \begin{vmatrix} [n] & [n-r-1] & \ldots & [1] \\ [r] & & & \\ \vdots & & & \\ [1] & & & \end{vmatrix} \qquad (4.12)$$

where we have to multiply together all the factors inside each set of curly brackets which are arranged according to the Young diagram of $(n-r, 1^r)$. We also remark that we have checked that as $q \to 1$ $\Phi(U)[dU]/\int \Phi(U)[dU]$ converges to the distribution $\delta(U) = \sum_\lambda \chi_\lambda$(identity) $\chi_\lambda(U)$.

Proof of Theorem 2. Combining now the expansion of $\mathrm{tr}(U^n)$ into characters with the result of Corollary 1, we obtain

$$W_n(q,N) = \frac{1}{N} \sum_{r \geq 0} (-1)^r q^{(n-r)^2 + r} \begin{bmatrix} N+n-r-1 \\ n \end{bmatrix}_2 \begin{bmatrix} n-1 \\ r \end{bmatrix}_2 \qquad (4.13)$$

which is perhaps the best possible result for computational purposes, since it involves at most n or N terms whichever is

smallest, but id does not clearly exhibit the asymptotic behaviour as $N \to \infty$. In order to obtain the representation of W_n
stated in Theorem 2 we first transform the expression in
(4.13) into q-hypergeometric series notation utilizing the
standard identity [20; p.241, eq.II.9]:

$$(q^2;q^2)_{n-k} = (-1)^k q^{k(k+1)-2k(n+1)} (q^2;q^2)_n / (q^{-2n};q^2)_k$$

(4.14)

This yields for $W_n(q,N)$ the value

$$\frac{(-1)^{n+1}}{N} q^n \begin{bmatrix} N \\ n \end{bmatrix}_{q^2} \, {}_2\phi_1\left(\begin{matrix} q^{2N+2}, q^{2-2n}; q^2, q^{2n} \\ q^{2(N-n+1)} \end{matrix} \right)$$

(4.15)

To this ${}_2\phi_1$-function we apply the Heine's transformation [1;
p.19, Cor.2.3] :

$$ {}_2\phi_1\left(\begin{matrix} a,b; \ q,t \\ c \end{matrix} \right) = \frac{(b;q)_\infty (at;q)_\infty}{(c;q)_\infty (t;q)_\infty} \, {}_2\phi_1\left(\begin{matrix} c/b,t; \ q,b \\ at \end{matrix} \right) $$ (4.16)

with the replacement of q by q^2 and the substitutions $a=q^{2n-2}$,
$b=q^{2N+2}$, $c=q^{2(N-n+1)}$, $t=q^{2n}$, and we obtain the final result,
Equation(4.6). To be totally safe, we should assume initially
in the use of (4.16) that N is nonnegative and nonintegral;
we then appeal to continuity to extend our result to integral
nonnegative N.

Proof of Corollary 2. From equation (3.4) we know that

$$\frac{1}{N} \langle \mathcal{V}_N, z^n \mathcal{V}_N \rangle = W_n(q,N+1) - W_n(q,N)$$ (4.17)

hence by Theorem 2,

$$\langle \mathcal{V}_j, z^n \mathcal{V}_j \rangle = \frac{(-q)^n}{1-q^{2n}} \sum_{r \geq 0} \frac{(q^{2n};q^2)_r (q^{-2n};q^2)_r}{(q^2;q^2)_r^2}$$

$$\times q^{2r(j+1)} (1- q^{2r})$$ (4.18)

$$= (-q)^n (1-q^{-2n}) \sum_{r \geq 0} \frac{(q^{2n+2};q^2)_r (q^{2-2n};q^2)_r q^{2(j+1)(r+1)}}{(q^2;q^2)_r (q^2;q^2)_{r+1}}$$

$$= (\text{equation } (4.8))$$

$$= \sum_{r=1}^{n} (-1)^{n+r} \begin{bmatrix} n \\ r \end{bmatrix}_{q^2} \begin{bmatrix} n+r-1 \\ n \end{bmatrix}_{q^2} q^{r(r+1)-n(2r-1)+2jr}. \qquad (4.19)$$

For $n > j$ it is perhaps convenient to transform this last expression into a sum involving $(j+1)$-terms, namely

$$\langle v_j, z^n v_j \rangle = \sum_{r \geq 0} (-1)^r \begin{bmatrix} j+n-r-1 \\ j \end{bmatrix}_{q^2} \begin{bmatrix} j \\ r \end{bmatrix}_{q^2} q^{(n-r)^2+2j-r} \qquad (4.20)$$

which is readily obtained from (4.13).

5. RESULTS FROM BASIC HYPERGEOMETRIC FUNCTIONS

We begin with two elementary lemmas.

Lemma 1. Let η denote the operator defined by $\eta f(x) = f(xq)$. Let $D = d/dx$; assume $g(1/x) = g(xq)$. Then

$$\left[D(1-\eta)g(x) \right]_{x=1} = 2\left[Dg(x) \right]_{x=1} = 2g'(1)$$

Proof. $\left[D(1-\eta)g(x) \right]_{x=1}$

$$= \left[D(g(x) - g(xq)) \right]_{x=1}$$

$$= \left[D(g(x) - g(1/x)) \right]_{x=1}$$

$$= \left[g'(x) + g'(1/x)/x^2 \right]_{x=1} = 2g'(1)$$

Lemma 2. $\left[D(ax;q)_j (aqx^{-1};q)_j \right]_{x=1} = -a(1-q^j)(aq;q)_{j-1}^2$

Proof. We proceed by mathematical induction on j. For j=0 both sides are 0. For j=1

$$\left[D(ax;q)_1(aqx^{-1};q)_1\right]_{x=1} = \left[D(1-ax)(1-aqx^{-1})\right]_{x=1}$$

$$= \left[-a + aqx^{-2}\right]_{x=1}$$

$$= -a(1-q)$$

in agreement with our assertion. Now we a-sume that the formula is true up to a particular j. Then

$$\left[D(ax;q)_{j+1}(aqx^{-1};q)_{j+1}\right]_{x=1}$$

$$=\left[D\{(1-axq^j)(1-ax^{-1}q^{j+1})\}\{(ax;q)_j(aqx^{-1};q)_j\}\right]_{x=1}$$

$$= (1-aq^j)(1-aq^{j+1})\left[D(ax;q)_j(aqx^{-1};q)_j\right]_{x=1}$$

$$+ (a;q)_j(aq;q)_j(-aq^j+aq^{j+1})$$

$$= (1-aq^j)(1-aq^{j+1})(-a(1-q^j)(aq;q)_{j-1}^2$$

$$+(a;q)_j(aq;q)_j(-aq^j+aq^{j+1})$$

(by the induction hypothesis)

$$= -a(aq;q)_{j-1}(aq;q)_j((1-aq^{j+1})(1-q^j) + q^j(1-q)(1-a))$$

$$= -a(aq;q)_{j-1}(aq;q)_j(1+aq^{2j+1}-q^{j+1}-aq^j)$$

$$= -a(aq;q)_{j-1}(aq;q)_j(1-q^{j+1})(1-aq^j)$$

$$= -a(1-q^{j+1})(aq;q)_j^2$$

and Lemma 2 thus follows by mathematical induction.

Apart from these two curious lemmas, there are several formulas from the literature that we require: the q-Binomial theorem $[1; \text{ p.36, eq.}(3.3.6)]$

$$(z;q)_M = \sum_{j=0}^{M} \begin{bmatrix} M \\ j \end{bmatrix}_q (-1)^j z^j q^{j(j-1)/2} \qquad (5.1)$$

Heine's analog of Euler's transformation $[1; \text{ p.39, eq.}(3.3.13)]$

$$_2\phi_1\begin{pmatrix} a,b; \ q,t \\ c \end{pmatrix} = \frac{(abt/c;q)_\infty}{(t;q)_\infty} \ _2\phi_1\begin{pmatrix} c/a, \ c/b; \ q, \ abt/c \\ c \end{pmatrix} \qquad (5.2)$$

Series reversal $[20; \text{ p.190}]$ provides basic idea

$$_2\phi_1\begin{pmatrix} q^{-n},B; \ q,T \\ C \end{pmatrix} \qquad (5.3)$$

$$= \frac{(B;q)_n}{(C;q)_n} (-1)^n \ q^{-n(n+1)/2} T^n \ _2\phi_1\begin{pmatrix} q^{-n},q^{1-n}/C;q,\dfrac{Cq^{n+1}}{BT} \\ q^{1-n}/B \end{pmatrix}$$

Jackson's transformation $[18; \text{ p.175, eq.}(10.2)]$, $a= q^{-n}$, $b=B$,

$c= fCq''/TB$, then $f \to 0$

$$_2\phi_1\begin{pmatrix} q^{-n},B; \ q, \ T \\ C \end{pmatrix} = \frac{(C/B;q)_n}{(C;q)_n} \ _3\phi_1\begin{pmatrix} q^{-n},B,BTq^{-n}/C; \ q,q \\ B \ q^{1-n}/C \end{pmatrix} \qquad (5.4)$$

6. THE SECOND EXPANSION OF $W_n(q,N)$ VIA q-HYPERGEOMETRIC SERIES

We shall prove (1.1) directly. Theorem 2 follows from (1.1) by applying (1.1) to (3.12).

Proof of (1.1). We begin by noting that if the sum for $0 \le r \le N$ in (1.1) is replaced by a sum for $0 \le r \le 2N$, the right hand side is merely doubled since the summand is zero for r=N and it is invariant under the successive transformations $r \to 2n-r$, then $n \to N -r+n$. Hence

$$\sum_{r=0}^{N} (-1)^r (N-r)(1-q^{N-r}) \sum_{n=0}^{r} \begin{bmatrix} N \\ n \end{bmatrix}_q \begin{bmatrix} N \\ r-n \end{bmatrix}_q q^{n+\left(\frac{2n-r+s}{2}\right)}$$

$$= \tfrac{1}{2} \left[D(1-\eta) \sum_{r=0}^{2N} \sum_{n=0}^{r} \begin{bmatrix} N \\ n \end{bmatrix}_q \begin{bmatrix} N \\ r-n \end{bmatrix}_q x^{N-r} (-1)^r q^{n+\binom{2n-r+s}{2}} \right]_{x=1}$$

$$= \tfrac{1}{2} \left[D(1-\eta) \sum_{n=0}^{N} \sum_{r=0}^{N} (-1)^{r+n} \begin{bmatrix} N \\ n \end{bmatrix}_q \begin{bmatrix} N \\ r \end{bmatrix}_q x^{N-r-n} q^{n+\binom{n-r+s}{2}} \right]_{x=1}$$

$$= \tfrac{1}{2} \left[D(1-\eta) \sum_{r=0}^{N} (-1)^r \begin{bmatrix} N \\ r \end{bmatrix}_q x^{N-r} q^{\binom{s-r}{2}} (x^{-1}q^{s-r+1};q)_N \right]_{x=1}$$

(by (5.1))

$$= \tfrac{1}{2} \left[D(1-\eta) \sum_{r=0}^{N} (-1)^r \begin{bmatrix} N \\ r \end{bmatrix}_q x^{N-r} {}_{-}q^{\binom{s-r}{2}} \frac{(x^{-1}q^{s-r+1};q)_{N+r}}{(x^{-1}q^{s+N-r+1};q)_r} \right]_{x=1}$$

$$= \tfrac{1}{2} \left[D(1-\eta) (\frac{q^{s+1}}{x};q)_N \sum_{r=0}^{N} (-1)^r \begin{bmatrix} N \\ r \end{bmatrix}_q x^{N-r} q^{\binom{s-r}{2}} \right.$$
$$\left. \frac{(x^{-1}q^{s-r+1};q)_r}{(x^{-1}q^{s+N-r+1};q)_r} \right]_{x=1}$$

$$= \tfrac{1}{2} \left[D(1-\eta) (x^{-1}q^{s+1};q)_N \sum_{r=0}^{N} (-1)^r \begin{bmatrix} N \\ r \end{bmatrix}_q x^{N-r} q^{\binom{s-r}{2}-rN} \right.$$
$$\left. \frac{(xq^{-s};q)_r}{(xq^{-s-N};q)_r} \right]_{x=1}$$

$$= \tfrac{1}{2} \left[D(1-\eta) x^N q^{\binom{s}{2}} (x^{-1}q^{s+1};q)_N \right.$$
$$\left. \sum_{r=0}^{N} \frac{(q^{-N};q)_r (x/q^s;q)_r q^{r-rs} x^{-r}}{(q;q)_r (xq^{-s-N};q)_r} \right]_{x=1}$$

$$= \tfrac{1}{2}\left[D(1-\eta)x^N q^{\binom{s}{2}}(x^{-1}q^{s+1};q)_N \; {}_2\phi_1\left(\begin{matrix} q^{-N},xq^{-s};q,q^{1-s}/x \\ xq^{-s-N}\end{matrix}\right)\right]_{x=1}$$

$$= \tfrac{1}{2}\left[D(1-\eta)x^N q^{\binom{s}{2}}(x^{-1}q^{s+1};q)_N \frac{(q^{-N};q)_N}{(xq^{-s-N};q)_N}\right.$$

$$\left.{}_3\phi_1\left(\begin{matrix} q^{-N},xq^{-s},q^{1-s}x^{-1};\;q,q \\ q\end{matrix}\right)\right]_{x=1}$$

(by 5.4)
$$= \tfrac{1}{2}\left[D(1-\eta)q^{\binom{s}{2}+sN}(q;q)_N \; {}_3\phi_1\left(\begin{matrix} q^{-N},xq^{-s},q^{1-s};q,q \\ q\end{matrix}\right)\right]_{x=1}$$

$$= \left[Dq^{\binom{s}{2}+sN}(q;q)_N \; {}_3\phi_1\left(\begin{matrix} q^{-N},xq^{-s},q^{1-s}x^{-1};q,q \\ q\end{matrix}\right)\right]_{x=1}$$

(by Lemma 1)
$$- (q;q)_N q^{\binom{s}{2}+sN} \sum_{j\geq 0} \frac{(q^{-N};q)_j(-q^{-s}(1-q^j)(q^{1-s};q)^2_{j-1}q^j}{(q;q)^2_j}$$

(by Lemma 2)
$$= (1-q^N)(q^2;q)_{N-1}q^{\binom{s-1}{2}+N(s-1)} \; {}_3\phi_1\left(\begin{matrix} q^{1-N},q^{1-s},q^{1-s};q,q \\ q^2\end{matrix}\right)$$

$$= q^{\binom{s-1}{2}+N(s-1)}(1-q^N)(q^2;q)_{N-1}\frac{(q^{1-2s};q)_{s-1}}{(q^{-s};q)_{s-1}}$$

$$\qquad {}_2\phi_1\left(\begin{matrix} q^{1-s},q^{1-s};q,q^{-N} \\ q^{1-2s}\end{matrix}\right)$$

$$= q^{\binom{s-1}{2}+N(s-1)}(1-q^N)(q^2;q)_{N-1}\frac{(q^{1-s};q)_{s-1}}{(q^{1-2s};q)_{s-1}}$$

$$\frac{(q^{1-2s};q)_{s-1}}{(q^{-s};q)_{s-1}} (-1)^{s-1} q^{-\binom{s}{2}-N(s-1)} {}_2\phi_1\left(\begin{array}{c} q^{1-s},q^{s+1} \\ q \end{array} ;q,q^N \right)$$

$$= \frac{(-1)^{s-1}(q;q)_N(1-q^N)}{(1-q^s)} {}_2\phi_1\left(\begin{array}{c} q^{1-s},q^{s+1} \\ q \end{array} ;q,q^N \right)$$

$$= \frac{(-1)^{s-1}(q;q)_N}{(1-q^s)} {}_2\phi_1\left(\begin{array}{c} q^{-s},q^s \\ q \end{array} ;q,q^{N+1} \right) \qquad \text{(by 5.2)}.$$

This concludes the proof of (1.1) and with it of Theorem 2.

7. APPLICATIONS OF LEMMA 2

The simplicity and elegance of the elementary Lemma 2 quite
surprised us in light of the fact that we had not seen it before. The rule for the differentiation of quotients allows
one to establish immediately that

$$\left[D \frac{(xb;q)_j(x^{-1}bq;q)_j}{(xa;q)_j(x^{-1}aq;q)_j} \right]_{x=1} = \frac{(a-b)(bq;q)^2_{j-1}(1-q^j)(1-abq^j)}{(a;q)^2_{j+1}} \qquad (7.1)$$

from Lemma 2. One can now apply (7.1) in numerous instances
to obtain new and rather surprising identities. We shall list
only a few to indicate the possibilities.

First let $j \rightarrow \infty$ in (7.1) and recall that [1; p.17, eq.
(2.2.1)]

$$\sum_{n\geq0} \frac{(A;q)_n X^n}{(q;q)_n} = \frac{(AX;q)_\infty}{(X;q)_\infty}$$

Hence

$$\sum_{n,m\geq0} \frac{(b/a;q)_n(b/a;q)_m a^{n+m}(n-m)}{(q;q)_n(q;q)_m} = \frac{(a-b)(bq;q)^2_\infty}{(a;q)^2_\infty} \qquad (7.2)$$

Alternative expressions precisely like the one being differentiated in (7.1) arise in identities for very well-poised basic
hypergeometric series. For example, an identity of F.H.Jackson

[20; p.96, eq.(3.3.14), corrected] may be formulated as

$$
{}_6\phi_5 \left(\begin{array}{c} ab,\ q\sqrt{ab},\ -q\sqrt{ab},\ xb,\ x^{-1}qb,\ q^{-N};\ q, aq^{N}/b \\[4pt] \sqrt{ab},\ -\sqrt{ab},\ aqx^{-1},\ xa,\ abq^{N+1} \end{array} \right)
$$

$$
= \frac{(abq;q)_N (a/b;q)_N}{(aqx^{-1};q)_N (ax;q)_N} \tag{7.3}
$$

Unfortunately application of (7.1) to this identity produces a special case of the same result with N replaced by N-1.

Finally we consider an application to a fascinating family of orthogonal polynomials introduced by J.Wilson in his Ph.D. thesis [23] and first presented in the literature in [4]. The Wilson polynomials are defined by

$$
w_n(a,b,c,d;x) = {}_4\phi_3 \left(\begin{array}{c} q^{-n},\ abq^{n+1},\ x, x^{-1}qcd;\ q,q \\[4pt] aq,\ bdq,\ cq \end{array} \right) \tag{7.4}
$$

If we replace x by $x\sqrt{cd}$, then we may apply (7.1) to obtain

$$
w'_n(a,b,c,d;\sqrt{cd}) =
$$

$$
- \frac{(1- q^{-n})(1-abq^{n+1})q}{(1-aq)(1-bdq)(1-cq)}\ w_{n-1}(aq,bq,cq,d;q\sqrt{cd}) \ .
$$

These few results indicate how useful (7.1) might be in applications. While neither (7.2) nor (7.5) is deep, nonetheless the possibilities for evaluating derivatives related to basic hypergeometric series may well prove valuable in other applications just as it has in treating (1.1).

8. CONCLUSIONS.

We conclude with several observations about the results we have obtained. First of all, it was pointed out to us that the matrix elements $\langle \mathcal{v}_j, z^n \mathcal{v}_j \rangle$ can be obtained in a simple way starting from a formula which gives the orthogonality of Szego polynomials (R.Askey gave this formula in his introdu-

ction to paper 26-6 of Szego's Collected Papers). The point
is that there is no serious difficulty in evaluating the matrix elements of z^n starting from the recurrence relation.
Let the linear operator A be defined as follows

$$A \mathcal{V}_j(z) = (1 - q^{2j})^{\frac{1}{2}} \mathcal{V}_{j-1}(z) \tag{8.1}$$

Let A^* be its adjoint. Then, equation (3.8) can be rewritten
as

$$Z = (q + A^*)(1 + qA)^{-1} \tag{8.2}$$

(where Z denotes the multiplication operator $Z\psi(z) = z\psi(z)$).
The matrix elements of Z^n are then calculated by noticing the
identity

$$\left[(q + A^*)(1 + qA)^{-1}\right]^n =$$

$$\prod_{j=1}^{n} (q^{2j-1} + A^*) \prod_{k=0}^{n} (1 + q^{2k-1}A)^{-1} \tag{8.3}$$

By standard (q-binomial) expansions one gets

$$z^n = \sum_{j=0}^{n} \begin{bmatrix} n \\ j \end{bmatrix}_{q^2} q^{(n-j)^2} A^{*j} \sum_{k=0}^{\infty} \begin{bmatrix} n+k-1 \\ k \end{bmatrix}_{q^2} (-qA)^k \tag{8.4}$$

Having "disentangled" A's from A^*'s , it is now immediate to
calculate the matrix elements. To get $W_n(q,N)$ one has still
to work with several "basic" transformations, but the essential step is done. To prove (8.3) start from

$$(1 + xA)^{-1}(x + A^*) = (xq^2 + A^*)(1 + xq^2 A)^{-1}$$

which is an immediate consequence of the commutation properties of A and A^*.

As a second remark, we note that the expressions in equation (4.8)

$$\frac{(1-q^2)(-1)^{n+1} q^{n-2j-2} \langle \mathcal{V}_j, z^n \mathcal{V}_j \rangle}{(1 - q^{2n})}$$

$$= {}_2\phi_1 \left(\begin{array}{c} q^{2-2n}, q^{2n+2} \; ; \; q^2, q^{2j+2} \\ q^4 \end{array} \right) \qquad (8.5)$$

$$= p_{n-1}(q^{2j}; q^2, 1 \, | \, q^2) \qquad (\text{by } [2; \text{p.11}, \text{eq.}(3.1)], [3])$$

are the spectral values of the little q-Jacobi polynomials $p_{n-1}(x; q^2, 1 \, | \, q^2)$. Consequently one could establish Theorem 2 merely by proving the following identity:

$$\sum_{j=0}^{\infty} q^{j(2m+2)} <\upsilon_j, z^{n+1}\upsilon_j> = \begin{cases} 0 \text{ if } n>m \\ \neq 0 \text{ otherw.} \end{cases} \qquad (8.6)$$

This is because this identity is only fulfilled (up to normalizing factor) by the little q-Jacobi polynomials [2], [3]; it is the uniquely defining orthogonality relation. In fact, starting from (8.4), it is not hard to show that

$$\sum_{j=0}^{\infty} x^j <\upsilon_j, z^n\upsilon_j> = q^{n^2}(x; q^2)_{n+1}^{-1}(xq^{2-2n}; q^2)_n \qquad (8.7)$$

a formula which was first derived by D.Stanton.

As a third remark, consider the insertion of $T = \text{id}$ in to the character expansion (4.11), yielding the "heat kernel" measure. This does not admit a treatment in terms of orthogonal polynomials since, unlike $\Phi(U)$, it is not factorized as a function of $(\varphi_1, \ldots, \varphi_N)$. Apparently, the only way to calculate $W_n(q,N)$ in this case is through the character expansion. This gives

$$W_n(q,N)_{\text{H.K.}} = q^{n(N+1-n)} {}_2F_1(N+1, 1-n; 2; 1-q^{2n}) \qquad (8.8)$$

$$= \frac{(-1)^{n+1}}{N} q^{n(N+1-n)} {}_2F_1(N+1, 1-n; N-n+1; q^n)$$

which is formally very similar to equation (4.15) wherein ordinary binomial coefficients and hypergeometric functions now appear. The reason for this strong similarity is purely geometrical: $\Phi(U)$ is easily recognized as being constructed from the heat kernel on the tangent space at the identity. As $q \to 1$ the two measures converge to the same Gaussian distribution centered at the origin. The asymptotic behaviour

$$W_n(e^{-\lambda/2N},N) \xrightarrow[N \to \infty]{} e^{-\frac{1}{2}n\lambda} \, {}_1F_1(1-n;2;n\lambda) \qquad (8.9)$$

for the H.K. measure was anticipated in [17].

Finally we refer to the well-known relation

$$\det(1 - zU) = \sum_{j=0}^{N} (-z)^j \chi_{(1^j)}(U) \qquad (8.10)$$

Taking the average over our measure $\Phi(U)[dU]$, we find

$$\langle \det(1- zU)\rangle = \sum_{j=0}^{N} (-z)^j q^j \begin{bmatrix} N \\ j \end{bmatrix}_{q^2} = (q^2;q^2)_N^{\frac{1}{2}} \, v_N^*(z) \quad (8.11)$$

(here $v_N^*(z) = z^N v_N(1/z)$, with v_N denoting the Rogers-Szego polynomials as before). This phenomenon is true in general: take any family of orthogonal polynomials on the unit circle $p_n(z)$ with respect to the measure $f(\exp i\varphi)d\varphi$ and let $\mu(dU) = \det \|f(U)\| \, [dU]$ be the measure on $U(N)$; then

$$\langle \det(zI-U)\rangle = \int \mu(dU)\det(zI-U)/\mu[U(N)] =$$

$$= p_N(z) \qquad (8.12)$$

It follows from a general property of such orthogonal polynomials [21] that $\langle \det(zI-U)\rangle$ does not vanish for $|z| > 1$. What about a general positive measure $\mu[dU]$ not necessarily factorized as in the examples above? Is it still true that $\langle \det(zI-U)\rangle$ does not vanish for $|z| > 1$? An explicit example is provided by the heat kernel measure, where direct inspection of the first few polynomials suggests that the property is verified.

THE PENNSYLVANIA STATE UNIVERSITY   ISTITUTO DI FISICA
UNIVERSITY PARK, PENNSYLVANIA 16802   UNIVERSITA' DI PARMA
U.S.A.   43100 PARMA
                                            ITALY

REFERENCES

1. Andrews, G.E.: The Theory of Partitions, Encyclopedia of Mathematics and its Applications, Vol.2, G.C.Rota ed., Addison-Wesley, Reading, 1976.
2. Andrews, G.E. and Askey, R.: Enumeration of Partitions: The role of Eulerian series and q-orthogonal polynomials, Higher Combinatorics, M.Aigner ed., D.Reidel, Dordrecht, Boston, 1977, pp.3-26.
3. Andrews, G.E. and Askey, R.: Some basic hypergeometric analogues of the classical polynomials and applications, (to appear).
4. Askey, R. and Wilson, J.: A set of orthogonalpolynomials that generalize the Racah coefficients or 6-j symbols, S.I.A.M. J.Math.Anal., 10 (1979), 1008-1016.
5. Drouffe,J.M. and Itzykson, C.: Lattice Gauge Fields, Phys. Reports, 38C (1978), 133-175.
6. Goldschmidt, Y.Y.: 1/N expansion in two-dimensional lattice gauge theory, J.Math.Phys.,21 (1980), 1842-1850.
7. Gross, D.J. and Witten, E.: Possible third order phase transition in the large-N lattice gauge theory, Phys.Rev. D21 (1980), 446-453.
8. Kogut, J.B.: An introduction to lattice gauge theory and spin systems, Rev.Mod.Phys., 51 (1979), 659-713.
9. Littlewood, D.E.: The Theory of Group Characters, Oxford University Press, London, 1940.
10. Makeenko, Yu.M. and Migdal, A.A.: Exact equation for the loop average in multicolor QCD, Phys.Letters, 88B (1979), 135-137.
11. Mehta, M.L.: Random Matrices, Academic Press, NewYork, 1967
12. Migdal, A.A.: Recursion equations in gauge field theories, Sov.Phys.-J.E.T.P., 42(3) (1976), 413-418.
13. Onofri, E.: SU(N) lattice gauge theory with Villain's action, Nuovo Cimento 66A, (1981) 293-
14. Osterwalder, K. and Seiler, E.: Gauge fileld theories on a lattice, Ann.of Physics (N.Y.), 110 (1978), 440-471
15. Robinson, G.deB.: Representation Theory of the Symmetric Group, University of Toronto Press, Toronto, 1961.

16. Rossi, P.: A solution of Wilson's loop equation in lattice QCD2, Phys.Letters, 92B (1980), 321-323.
17. Rossi, P.: Continuum QCD2 from a fixed-point lattice action, Ann.of Physics (N.Y.), 132 (1981), 463-481.
18. Sears, D.B.: On the transformation theory of basic hypergeometric functions, Proc.London Math.Soc. (2), 53 (1951), 158-180.
19. Selberg, A.: Bemerkninger om et multipelt integral, Norsk Matematisk Tidsskrift, 26 (1944), 71-78.
20. Slater, L.J.: Generalized Hypergeometric Functions, Cambridge University Press, London 1966.
21. Szego, G.: Orthogonal Polynomials, Vol.23, 3rd Ed., Amer. Math.Soc.Colloq.Publ., Providence 1966.
22. Weyl, H.: Classical Groups, 2nd Ed., Princeton University Press, Princeton 1946.
23. Wilson, J.: Hypergeometric series recurrence relations and some new orthogonal functions, Ph.D. thesis, University of Wisconsin, Madison 1978.
24. Wilson, K.G.: Confinement of quarks, Phys.Rev., D10 (1974) 2445-2459.
25. Askey, R.: Collected Papers of Gabor Szego, Birkhauser Boston (1982).

R.W. Beals, P.C. Greiner and J. Vauthier

THE LAGUERRE CALCULUS ON THE HEISENBERG GROUP

1. INTRODUCTION. The purpose of this article is to derive
a multiplicative symbolic calculus for left-invariant convolution operators on the Heisenberg group. We let H_n denote the n-th Heisenberg group with underlying manifold

$$\mathbb{R}^{2n+1} = \{(x_0, x')\} = \{(x_0, x_1, \ldots, x_{2n})\} , \tag{1.1}$$

and with the group law

$$xy = (x_0, x')(y_0, y')$$

$$= \left(x_0 + y_0 + \frac{1}{2} \sum_{j=1}^{n} a_j [x_j y_{j+n} - x_{j+n} y_j], \; x'+y'\right) . \tag{1.2}$$

We assume that $a_j, j=1,\ldots,n$ are positive numbers.
(1.2) should be looked upon as a non-commutative analogue of
Euclidean translation on $\mathbb{R}^{2n+1}$. We note that $\mathbb{R}^{2n+1} = \mathbb{R} \times \mathbb{C}^n$ and writing

$$z_j = x_j + i x_{j+n} , \qquad w_j = y_j + i y_{j+n} \tag{1.3}$$

the group law (1.2) can be written in the following symbolic
form

$$(x_0, z)(y_0, w) = \left(x_0 + y_0 - \frac{1}{2} \text{Im} \sum_{j=1}^{n} a_j z_j \bar{w}_j, \; z+w\right) . \tag{1.4}$$

Here (x_0, z) stands for $(x_0, z, \bar{z})$. H_n has $(0,0)$ for
its unit and

$$(x_0, x')^{-1} = (-x_0, -x') . \tag{1.5}$$

Given functions $\phi, \psi \in C_0^{\infty}(H_n)$ the Heisenberg convolut-

189

R. A. Askey et al. (eds.), Special Functions: Group Theoretical Aspects and Applications, 189–216.
© 1984 by D. Reidel Publishing Company.

ion (H-convolution) is given by

$$\phi *_H \psi (x) = \int_{H_n} \phi(y^{-1}x)\psi(y)\, dH_n(y) \ ,\tag{1.6}$$

where

$$dH_n(y) = dy_0 \prod_{j=1}^{n} \left(\frac{a_j}{4}\, dy_j dy_{j+n}\right) \ .\tag{1.7}$$

Set

$$T_{x_1}\phi(x_2) = \phi(x_1 x_2) \ ,\tag{1.8}$$

i.e. T_x is left-translation with respect to x . Then

$$\int_{H_n} \phi(y^{-1}x_1)\psi(x_2 y)\, dH_n(y)\tag{1.9}$$

$$= \int_{H_n} \phi(y^{-1}[x_2 x_1])\psi(y)\, dH_n(y) \ ,$$

since $dH_n(x_2^{-1}y) = dH_n(y)$ follows easily from the definition of the Heisenberg translation. Thus we have

$$\left(\phi *_H (T_{x_2}\psi)\right)(x_1) = \left(T_{x_2}(\phi *_H \psi)\right)(x_1) \ .\tag{1.10}$$

In other words the Heisenberg convolution commutes with left-translations on H_n , or, the H-convolution product is left-invariant. (1.10) also implies that the H-convolution product is associative:

$$\phi *_H (\psi *_H \chi) = (\phi *_H \psi) *_H \chi \ .\tag{1.11}$$

Now we are ready to introduce Principal Value (or PV) convolution operators on H_n . These are the analogues of the Mikhlin-Calderon-Zygmund PV convolution operators on $\mathbb{R}^n$. Let

$$r(x_0, x') = (r^2 x_0, rx')\tag{1.12}$$

define the Heisenberg dilation. *F is said to be H-homogeneous of degree $\cdot m$ on H_n if*

$$f(r^2 x_0, rx') = r^m f(x_0, x') \, , \quad r > 0 \, . \tag{1.13}$$

We also define a norm on H_n :

$$\left| (x_0, x') \right|_H = \left(x_0^2 + \left[\sum_{j=1}^n \frac{a_j}{4} (x_j^2 + x_{j+n}^2) \right]^2 \right)^{1/4} \, , \tag{1.14}$$

which is H-homogeneous of degree one. This also yields a distance function, namely the distance $d(x,y)$ *of the points* $x,y \in H_n$ *is*

$$d(x,y) = d(y^{-1} x, 0) = \left| y^{-1} x \right|_H \, . \tag{1.15}$$

d *is left-invariant in the sense that* $d(x,y)$ *remains unchanged when both* x *and* y *are left-translated by a fixed vector in* H_n *. Furthermore* d *satisfies the triangle inequality*

$$d(x,z) \le d(x,y) + d(y,z) \, , \quad x,y,z \in H_n \, . \tag{1.16}$$

Suppose $F \in C^\infty(H_n \backslash 0)$ *is H-homogeneous of degree* γ *. Then* F *is integrable near the origin if* $\gamma > -2n-2$ *- see* [4].

We are mainly concerned with convolution operators on H_n which are induced by H-homogeneous functions of the critical degree: $-2n-2$.

1.17 Definition. *Let* $F \in C^\infty(H_n \backslash 0)$ *, H-homogeneous of degree* $-2n-2$ *.* F *is said to have vanishing Principal Value if*

$$\int_{|x|_H = 1} f(x) d\sigma(x) = 0 \, , \tag{1.18}$$

where $d\sigma(x)$ *is the induced measure on the unit Heisenberg sphere.*

The following result justifies the notion of vanishing Principal Value - see [4].

1.19 Proposition. *Let* $F \in C^\infty(H_n \backslash 0)$ *, H-homogeneous of de-*

gree -2n-2 with zero Principal Value. Then F induces a Principal Value (or PV) convolution operator on functions $\phi \in C_0^\infty(H_n)$ as follows:

$$F*_H\phi(x) = \lim_{\varepsilon \to 0} \int_{d(x,y)>\varepsilon} F(y^{-1}x)\phi(y)dH_n(y) \ . \qquad (1.20)$$

The operator F given by (1.20) can be extended to a bounded operator:

$$F : L^2(H_n) \to L^2(H_n) \ . \qquad (1.21)$$

In particular PV convolution operators can be composed. Furthermore, their composition yields another PV convolution operator. We shall denote PV convolution operators by capital letters F,G,... and their composition simply by $F*_HG$.

The best known examples of left-invariant PV convolution operators on H_n are induced by the "Cauchy-Szegö" kernels $S_\pm(y^{-1}x)$, where

$$S_\pm(x) = \frac{n!}{(2\pi)^{n+1}} \frac{a_1a_2\cdots a_n}{\left(\sum_{j=1}^{n} \frac{a_j}{4}[x_j^2 + x_{j+n}^2] \mp ix_0\right)^{n+1}} \ . \qquad (1.22)$$

$S_\pm$ are projections, so

$$S_\pm *_H S_\pm = S_\pm \quad \text{or} \quad S_\pm^2 = S_\pm \ . \qquad (1.23)$$

It is useful to derive (1.23) by an explicit calculation, see [7], because it yields the first clue to finding the multiplicative symbolic (tensor) calculus on H_n .

2. THE LAGUERRE CALCULUS ON H_1 . To simplify the discussion we shall derive the calculus on H_1 first. This was already done in [7] although there the Heisenberg translation was normalized differently. We set $a_1 = 4$. Then the group law is

$$(x_0,x')(y_0,y') = (x_0+y_0+2[x_1y_2-x_2y_1] , x'+y') , \qquad (2.1)$$

where $x' = (x_1, x_2)$ and $y' = (y_1, y_2)$. The key to understanding the convolution on H_1 is the careful study of the
Cauchy-Szegö kernels $S_{\pm}(y^{-1}x)$, where

$$S_{\pm}(x) = \frac{1}{\pi^2 (|x'|^2 \mp ix_0)^2} \qquad (2.2)$$

$S_{\pm}$ turn out to be the simplest of a large number of
basic convolution operators on H_1, which are induced by
Laguerre functions. We define the generalized Laguerre polynomials, $L_k^{(p)}$, $p, k = 0, 1, 2, \ldots$ by the usual generating function formula

$$\sum_{k=0}^{\infty} L_k^{(p)}(x) z^k = \frac{1}{(1-z)^{p+1}} e^{-\frac{xz}{1-z}}. \qquad (2.3)$$

Then

$$\ell_k^{(p)}(x) = \left[\frac{\Gamma(k+1)}{\Gamma(k+p+1)} \right]^{\frac{1}{2}} x^{p/2} L_k^{(p)}(x) e^{-x/2} \qquad (2.4)$$

are known as the Laguerre functions, where $x \geq 0$ and
$p, k = 0, 1, 2, \ldots$ For each fixed $p = 0, 1, 2, \ldots$

$$\ell_0^{(p)}(x), \ \ell_1^{(p)}(x), \ \ell_2^{(p)}(x), \ldots \qquad (2.5)$$

form a complete orthonormal set of functions in $L^2(0, \infty)$ –
consult [17].

We let $\tilde{\phi}(\tau, x')$ *denote the partial Fourier transform
of* $\phi \in C_0^\infty(\mathbb{R}^3)$:

$$\tilde{\phi}(\tau, x') = \int_{-\infty}^{\infty} e^{-i\tau x_0} \phi(x_0, x') dx_0. \qquad (2.6)$$

2.7 Definition. *We define the functions* $L_k^{(p)}(x)$, $x \in H_1$,
$\pm p, k = 0, 1, 2, \ldots$ *via their partial Fourier transform in* x_0:

$$\tilde{L}_k^{(p)}(\tau, x') \qquad (2.8)$$

$$= \frac{2|\tau|}{\pi} (sgn\ p)^p \ell_k^{(|p|)} \{|\sqrt{2|\tau|}\ x'|^2\} e^{ip\theta},$$

where

$$x_1 + ix_2 = |x'| e^{i\theta} .$$
(2.9)

We note that the exponential Laguerre functions

$$\{\tilde{L}_k^{(p)} (\frac{1}{2}, x'), \; x' \in \mathbb{R}^2, \; \pm p, k = 0,1,2,\ldots\}$$
(2.10)

form a complete orthonormal set of functions in $L^2(\mathbb{R}^2)$.

The functions $L_k^{(p)} (x_0, x')$, $\pm p, k = 0,1,2,\ldots$ *induce the generalized Cauchy-Szegö kernels on* H_1 . *In particular*

$$\tilde{S}_\pm = \begin{cases} \tilde{L}_0^{(0)} & \text{if} \; \tau \gtrless 0 , \\ 0 & \text{if} \; \tau \lessgtr 0 . \end{cases}$$
(2.11)

Let $F \in C^\infty(H_n \backslash 0)$ be homogeneous of degree -4, the critical degree on H_1 , with vanishing Principal Value and consider

$$\tilde{F}(\tau, x') = \int_{-\infty}^{\infty} e^{-i\tau x_0} F(x_0, x') dx_0 .$$
(2.12)

Then

$$\tilde{F}(\frac{\tau}{\lambda^2}, \lambda x') = \lambda^{-2} \tilde{F}(\tau, x') , \; \lambda > 0 .$$
(2.13)

In particular we set $\lambda = \sqrt{2|\tau|}$. Then

$$\tilde{F}(\tau, x') = 2|\tau| \tilde{F}(\frac{1}{2} \text{sgn} \; \tau, \; \sqrt{2|\tau|} x') .$$
(2.14)

We define

$$\tilde{F}_\pm (\tau, x') = \begin{cases} \tilde{F}(\tau, x') & \text{if} \; \tau \gtrless 0 \\ 0 & \text{if} \; \tau \lessgtr 0 \end{cases} .$$
(2.15)

Then

$$\tilde{F}_{\pm}(\tau,x') = \begin{cases} 2|\tau|\tilde{F}(\pm\frac{1}{2}, \sqrt{2|\tau|}x') & \text{if } \tau \gtrless 0, \\ 0 & \text{if } \tau \lessgtr 0. \end{cases} \tag{2.16}$$

<u>2.17 Proposition</u>. *Let* F *and* G *define PV convolution operators on* H_1. *Set*

$$F_{\pm}(x_0,x') = \frac{1}{2\pi}\int_{-\infty}^{\infty} e^{ix_0\tau} \tilde{F}_{\pm}(\tau,x')d\tau, \tag{2.18}$$

where $\tilde{F}_{\pm}$ *are given by (2.15). Then*

$$F = F_+ + F_- \tag{2.19}$$

and

$$F_+ *_H G_- = F_- *_H G_+ = 0. \tag{2.20}$$

<u>Proof</u>. (2.19) is a consequence of the definition of $\tilde{F}_+$. (2.20) follows from the following formula for the Heisenberg convolution:

$$\phi *_H \psi(x) = \frac{1}{2\pi}\int_{-\infty}^{\infty} e^{ix_0\tau} (\tilde{\phi} *_\tau \tilde{\psi})(\tau,x')d\tau, \tag{2.21}$$

where the twisted convolution $\tilde{\phi} *_\tau \tilde{\psi}$ *is given by*

$$\tilde{\phi} *_\tau \tilde{\psi}(\tau,x') \tag{2.22}$$

$$= \int_{-\infty}^{\infty} e^{2i\tau(x_1y_2-x_2y_1)}\tilde{\phi}(\tau,x'-y')\tilde{\psi}(\tau,y')dy'$$

with

$$dy' = dy_1 dy_2. \tag{2.23}$$

This proves Proposition *2.16*.

Taking the partial Fourier transform of both sides of (2.21) we have

$$(\phi *_H \psi)^{\sim} (\tau, x') = \tilde{\phi} *_\tau \tilde{\psi}(\tau, x') \; . \tag{2.24}$$

We write (2.19) and (2.20) as

$$F = F_+ \oplus F_- \; . \tag{2.25}$$

Let F and G induce PV convolution operators on H_1. Using (2.24), i.e.

$$(F *_H G)^{\sim} = \tilde{F} *_\tau \tilde{G} \; , \tag{2.26}$$

we have

$$(F *_H G)_\pm = (F_\pm *_H G_\pm)_\pm = F_\pm *_H G_\pm \; . \tag{2.27}$$

Now (2.20) implies

$$F *_H G = (F_+ *_H G_+) \oplus (F_- *_H G_-) \; . \tag{2.28}$$

Applying the partial Fourier transform we find

$$(F *_H G)^{\sim} = \tilde{F} *_\tau \tilde{G} \tag{2.29}$$

$$= (\tilde{F}_+ *_\tau \tilde{G}_+) \oplus (\tilde{F}_- *_\tau \tilde{G}_-) \; .$$

Consequently, to obtain $F *_H G$ it suffices to find

$$\tilde{F}_\pm *_\tau \tilde{G}_\pm \; . \tag{2.30}$$

To find $\tilde{F}_+ *_\tau \tilde{G}_+$ we shall write $\tilde{F}_+$ and $\tilde{G}_\pm$ as a series of exponential Laguerre functions

$$\tilde{L}_k^{(p)} (\tau, x') = \tilde{L}_{k;+}^{(p)}(\tau, x') \oplus \tilde{L}_{k;-}^{(p)}(\tau, x') \; , \tag{2.31}$$

where

$$\tilde{L}_{k;\pm}^{(p)}(\tau, x') = \begin{cases} \tilde{L}_k^{(p)} (\tau, x') & \text{if } \tau \gtrless 0 \; , \\ 0 & \text{if } \tau \lessgtr 0 \; . \end{cases} \tag{2.32}$$

We note that

$$\tilde{L}_k^{(p)}(\tau, x') = \tilde{L}_k^{(p)}(|\tau|, x') \ . \tag{2.33}$$

The twisted convolution of two exponential Laguerre functions was found in Theorem 5.1 of [7]. Namely

2.34 <u>Theorem</u>. *Let* $p, k, q, m = 1, 2, \ldots$ *Then*

$$\tilde{L}_{p\vee k-1}^{(p-k)} * |\tau| \tilde{L}_{q\vee m-1}^{(q-m)} = \delta_k^{(q)} \tilde{L}_{p\vee m-1}^{(p-m)} \ , \tag{2.35}$$

where $a \vee b = min(a, b)$ *and* $\delta_m^{(q)}$ *denotes the Kronecker delta,*
i.e. $= 1$ *if* $q = m$ *and vanishes otherwise.*

Thus the twisted convolution of two exponential Laguerre
functions is another exponential Laguerre function. This
surprising result justifies the use of the Laguerre series
on H_1 in analogy with Mikhlin's use of the Fourier series
on the complex plane – see [14].

Formula (2.35) is reminiscent of matrix multiplication.
To expoit this similarity we introduce the positive Laguerre
matrix.

We define the positive Laguerre matrix

$$L_+\left(L_{p\vee k-1}^{(p-k)}\right) \tag{2.36}$$

of $L_{p\vee k-1}^{(p-k)}$ *to be the infinite matrix with a one at the in-*
tersection of the p-th row with k-th column and zeros every-
where else. Then Theorem 2.34 takes on the following form

$$L_+\left(L_k^{(p)} *_H L_m^{(q)}\right) = L_+\left(L_k^{(p)}\right) L_+\left(L_m^{(q)}\right) \ . \tag{2.37}$$

Next we define the positive Laguerre matrix $L_+(F)$ for
F , where F induces a PV convolution operator on H_1 . We
start with expanding $\tilde{F}_+(\tau, x')$ in a series of exponential
Laguerre functions:

$$\tilde{F}_+(\tau, x') = \sum_{p, k=1}^{\infty} F_{+;k}^{(p)} \tilde{L}_{p\vee k-1}^{(p-k)}(\tau, x') \ . \tag{2.38}$$

According to (2.8) and (2.16) this is equivalent to

$$\tilde{F}(\tfrac{1}{2}, \sqrt{2|\tau|}x') = \frac{1}{\pi} \sum_{p,k=1}^{\infty} F_{+;k}^{(p)} \tag{2.39}$$

$$[\text{sgn}(p-k)]^{p-k} \ell_{p\vee k-1}^{(|p-k|)} (2|\tau||x'|^2) e^{i(p-k)\theta}$$

or

$$\tilde{F}(\tfrac{1}{2}, x') = \frac{1}{\pi} \sum_{p,k=1}^{\infty} F_{+;k}^{(p)} \tag{2.40}$$

$$[\text{sgn}(p-k)]^{p-k} \ell_{p\vee k-1}^{(|p-k|)} (|x'|^2) e^{i(p-k)\theta}$$

where

$$x_1 + ix_2 = |x'|e^{i\theta} . \tag{2.41}$$

Since $\ell_k^{(p)}, k=0,1,2,\dots$ are orthonormal on $(0,\infty)$ the coefficients $F_{+;k}^{(p)}$ can be found by

$$[\text{sgn}(p-k)]^{(p-k)} F_{+;k}^{(p)} \tag{2.42}$$

$$= \int_{\mathbb{R}^2} \tilde{F}(\tfrac{1}{2}, x') \ell_{p\vee k-1}^{(|p-k|)} (|x'|^2) e^{-i(p-k)\theta} dx' .$$

2.43 Definition. *Let* F *induce a PV convolution operator on* H_1 . *We define its positive Laguerre matrix by*

$$L_+(F) = \sum_{p,k=1}^{\infty} F_{+;k}^{(p)} L_+\left(L_{p\vee k-1}^{(p-k)}\right) . \tag{2.44}$$

In other words $L_+(F)$ *is the infinite matrix whose* (p,k)-*th term is* $F_{+;k}^{(p)}$.

An immediate consequence of (2.37) and (2.44) is

2.45 Theorem. *Let* F *and* G *induce PV convolution operators on* H_1 . *Then*

$$L_+(F *_H G) = L_+(F) L_+(G) . \tag{2.46}$$

2.47 Remark. We note that (2.40) can be written in the equivalent form

$$\tilde{F}(\tfrac{1}{2},x') = \frac{1}{\pi} \sum_{p=-\infty}^{\infty} \sum_{k=0}^{\infty} F \frac{(\tfrac{1}{2}[|p|+p]+k+1)}{+;\tfrac{1}{2}[|p|-p]+k+1} \tag{2.48}$$

$$(\text{sgn } p)^p \ell_k^{(|p|)} (|x'|^2) e^{ip\theta} .$$

Such an expansion always exists since

$$\frac{1}{\pi} \ell_k^{(|p|)} (|x'|^2) e^{ip\theta} , \quad \pm p, k = 0,1,2,\dots \tag{2.49}$$

is a complete orthonormal set of functions in $L^2(\mathbb{R}^2)$. This justifies the expansion (2.38).

We still need the equivalent calculus for F_- . Again we have the expansion

$$\tilde{F}_-(\tau,x') = \sum_{p,k=1}^{\infty} F_{-;k}^{(p)} \tilde{L}_{p\vee k-1}^{(p-k)} (\tau,x') , \quad \tau < 0 , \tag{2.50}$$

where

$$[\text{sgn}(p-k)]^{p-k} F_{-;k}^{(p)} \tag{2.51}$$

$$= \int_{\mathbb{R}^2} \tilde{F}(-\tfrac{1}{2},x') \ell_{p\vee k-1}^{(|p-k|)} (|x'|^2) e^{-i(p-k)\theta} dx' ,$$

for $p,k = 1,2,\dots$ According to Lemma 6.1 of [7] the twisted convolution commutes the exponential Laguerre functions when $\tau < 0$. More precisely we have

2.52 Lemma. *For* $\tau \in \mathbb{R}$

$$\tilde{L}_k^{(p)} *_\tau \tilde{L}_m^{(q)} = \tilde{L}_m^{(q)} *_{-\tau} \tilde{L}_k^{(p)} . \tag{2.53}$$

Let F and G induce PV convolution operators on H_1 . Then

$$\widetilde{F}_- \underset{\tau}{*} \widetilde{G}_- = \widetilde{F}_- *_{-|\tau|} \widetilde{G}_- \tag{2.54}$$

$$= \sum_{p,k=1}^{\infty} F_{-;k}^{(p)} \widetilde{L}_{p\vee k-1}^{(p-k)} *_{-|\tau|} \sum_{q,m=1}^{\infty} G_{-;m}^{(q)} \widetilde{L}_{q\vee m-1}^{(q-m)}$$

$$= \sum_{p,k,q,m=1}^{\infty} F_{-;k}^{(p)} G_{-;m}^{(q)} \widetilde{L}_{p\vee k-1}^{(p-k)} *_{-|\tau|} \widetilde{L}_{q\vee m-1}^{(q-m)}$$

$$= \sum_{p,k,q,m=1}^{\infty} F_{-;k}^{(p)} G_{-;m}^{(q)} \widetilde{L}_{q\vee m-1}^{(q-m)} *_{|\tau|} \widetilde{L}_{p\vee k-1}^{(p-k)}$$

$$= \sum_{p,k,q,m=1}^{\infty} F_{-;k}^{(p)} G_{-;m}^{(q)} \delta_m^{(p)} \widetilde{L}_{q\vee k-1}^{(q-k)}$$

$$= \sum_{k,q=1}^{\infty} \left(\sum_{\ell=1}^{\infty} F_{-;k}^{(\ell)} G_{-;\ell}^{(q)} \right) \widetilde{L}_{q\vee k-1}^{(q-k)} .$$

On the other hand

$$\widetilde{F}_- \underset{\tau}{*} \widetilde{G}_- = (F_- \underset{H}{*} G_-)^{\sim} = (F *_H G)_-^{\sim} \tag{2.55}$$

$$= \sum_{k,q=1}^{\infty} (F *_H G)_{-;k}^{(q)} \widetilde{L}_{q\vee k-1}^{(q-k)} ,$$

where $\tau < 0$. Thus

$$\sum_{\ell=1}^{\infty} F_{-;k}^{(\ell)} G_{-;\ell}^{(q)} = (F *_H G)_{-;k}^{(q)} . \tag{2.56}$$

<u>2.57 Definition.</u> *We set*

$$L_-(F) = \left(F_{-;k}^{(p)} \right)^t , \tag{2.58}$$

i.e. the negative Laguerre matrix $L_-(F)$ of F is the transpose of the infinite matrix whose (p,k)-th term is $F_{-;k}^{(p)}$

For example

$$L_-(L_k^{(p)}) = [L_+(L_k^{(p)})]^t . \tag{2.59}$$

Now (2.56) is simply

$$L_-(F*_H G) = L_-(F)L_-(G) .$$ (2.60)

2.61 Definition. *Let F induce a PV convolution operator on H_1 . We define the Laguerre matrix $L(F)$ of F by*

$$L(F) = L_+(F) \oplus L_-(F) .$$ (2.62)

Finally, our calculations can be collected in the follow-form.

2.63 Theorem. *Let F and G induce PV convolution operators on H_1 . Then*

$$L(F*_H G) = L(F)L(G) .$$ (2.64)

2.65 Corollary. *Let I_H denote the PV convolution operator on H_1 which induces identity operator on $C_0^\infty(H_1)$. Then*

$$L_\pm(I_H) = (\delta_k^{(p)}) .$$ (2.66)

Thus the Laguerre matrix plays the same role for PV convolution operators on H_1 which is assigned to the Fourier transform, i.e. the classical symbol, of Mikhlin-Calderon-Zygmund operators on $\mathbb{R}^n$. Analogously we refer to Theorem 2.63 as the *Laguerre symbolic calculus on H_1* .

3. THE LAGUERRE CALCULUS ON H_n . We recall the group law (1.2)

$$(x_0,x')(y_0,y')$$

$$= (x_0 + y_0 + \frac{1}{2} \sum_{j=1}^n a_j(x_j y_{j+n} - x_{j+n} y_j), \ x'+y') ,$$

where $x' = (x_1,\ldots,x_{2n})$, $y' = (y_1,\ldots,y_{2n})$ and $a_j > 0$, $j = 1,\ldots,n$. We normalized the measure dH_n by

$$dH_n(x) = dx_0 \ \prod_{j=1}^n \left(\frac{a_j}{4} dx'_j\right) ,$$ (3.1)

where

$$dx'_j = dx_j dx_{j+n} \; . \tag{3.2}$$

Again, we introduce the *partial Fourier transform*

$$\tilde{\phi}(\tau, x') = \int_{-\infty}^{\infty} e^{-i\tau x_0} \phi(x_0, x') dx_0 \; . \tag{3.3}$$

We define the n-dimensional version of the exponential Laguerre functions by the n-fold product

$$\prod_{j=1}^{n} \tilde{L}_{k_j}^{(p_j)} (\tau, \frac{\sqrt{a_j}}{2} x'_j) \; , \tag{3.4}$$

where $\pm p_j$, $k_j = 0,1,2,\ldots$, $j = 1,\ldots,n$ and

$$x'_j = (x_j, x_{j+n}) \; . \tag{3.5}$$

The $\tilde{L}_k^{(p)}$ are given by (2.8).

The twisted convolution on H_n takes the following form:

$$\tilde{\phi} *_\tau \tilde{\psi}(\tau, x') \tag{3.6}$$

$$= \int_{\mathbb{R}^{2n}} e^{2i\tau \sum_{j=1}^{n} \frac{a_j}{4}(x_j y_{j+n} - x_{j+n} y_j)}$$

$$\tilde{\phi}(\tau, x'-y') \tilde{\psi}(\tau, y') \prod_{j=1}^{n} (\frac{a_j}{4} dy'_j) \; .$$

In particular we have

$$\phi *_H \psi(x) = \int_{H_n} \phi(y^{-1}x) \psi(y) dH_n(y) \tag{3.7}$$

$$= \frac{1}{2\pi} \int_{-\infty}^{\infty} e^{ix_0 \tau} (\tilde{\phi} *_\tau \tilde{\psi})(\tau, x') d\tau \; .$$

Next we compose the n-fold exponential Laguerre functions

$$\left(\prod_{j=1}^{n} \tilde{L}_{k_j}^{(p_j)}\right) *_{\tau} \left(\prod_{j=1}^{n} \tilde{L}_{m_j}^{(q_j)}\right)(\tau,x') \tag{3.8}$$

$$= \prod_{j=1}^{n} \int_{\mathbb{R}^2} e^{2i\tau[\frac{a_j}{4}(x_j y_{j+n} - x_{j+n} y_j)]}$$

$$\tilde{L}_{k_j}^{(p_j)}\left(\tau,\frac{\sqrt{a_j}}{2}x_j' - \frac{\sqrt{a_j}}{2}y_j'\right)\tilde{L}_{m_j}^{(q_j)}\left(\tau,\frac{\sqrt{a_j}}{2}y_j'\right)\frac{a_j}{4}\,dy_j'$$

$$= \prod_{j=1}^{n} \left(\tilde{L}_{k_j}^{(p_j)} *_{\tau} L_{m_j}^{(q_j)}\right)(\tau,x') \ .$$

Consequently Theorem 2.34 implies

3.9   Theorem. *Let* $p_j, k_j, q_j, m_j = 1,2,\dots$ *for* $j = 1,\dots,n$.
Then

$$\left(\prod_{j=1}^{n} \tilde{L}_{p_j\sqrt{k_j}-1}^{(p_j-k_j)}\right) *_{|\tau|} \left(\prod_{j=1}^{n} \tilde{L}_{q_j\sqrt{m_j}-1}^{(q_j-m_j)}\right)(\tau,x') \tag{3.10}$$

$$= \prod_{j=1}^{n} \delta_{k_j}^{(q_j)} \tilde{L}_{p_j\sqrt{m_j}-1}^{(p_j-m_j)}\left(\tau,\frac{\sqrt{a_j}}{2}x_j'\right) \ .$$

Let F induce a PV convolution operator on H_n i.e.
$F \in C^{\infty}(H_n\backslash 0)$ is homogeneous of degree $-2n-2$ with vanishing PV.   For $\lambda > 0$

$$\tilde{F}\left(\frac{\tau}{\lambda^2},\lambda x'\right) = \int_{-\infty}^{\infty} e^{-i(\tau/\lambda^2)x_0} F(x_0,\lambda x')\,dx_0 \tag{3.11}$$

$$= \lambda^2 \int_{-\infty}^{\infty} e^{-i\tau y_0} F(\tau^2 y_0,\lambda x')\,dy_0$$

$$= \lambda^{-2n}\tilde{F}(\tau,x') \ .$$

Setting $\lambda = \sqrt{2|\tau|}$ we have

$$\widetilde{F}(\tau,x') = (2|\tau|)^n \widetilde{F}(\tfrac{1}{2} \operatorname{sgn} \tau, \sqrt{2|\tau|}\, x') \tag{3.12}$$

$$= \widetilde{F}_+(\tau,x') + \widetilde{F}_-(\tau,x')$$

where

$$\widetilde{F}_\pm(\tau,x') = \begin{cases} (2|\tau|)^n \widetilde{F}(\pm\tfrac{1}{2}, \sqrt{2|\tau|}x') & \text{if } \tau \gtrless 0, \\ 0 & \text{if } \tau \lessgtr 0. \end{cases} \tag{3.13}$$

As in (2.38) we expand $\widetilde{F}_\pm$ in a Laguerre series:

$$\widetilde{F}_\pm(\tau,x') = \sum_{p_j,k_j=1}^{\infty} F_{\pm;k_1,\ldots,k_n}^{(p_1,\ldots,p_n)} \tag{3.14}$$

$$\prod_{j=1}^{n} \widetilde{L}_{p_j\vee k_j-1}^{(p_j-k_j)}(\tau, \frac{\sqrt{a_j}}{2} x_j') \ .$$

To find the coefficients $F_{\pm;k_1,\ldots,k_n}^{(p_1,\ldots,p_n)}$ we rewrite (3.14) in the following form

$$\widetilde{F}(\pm\tfrac{1}{2}, \sqrt{2|\tau|}x') \tag{3.15}$$

$$= \frac{1}{\pi^n} \sum_{p_j,k_j=1}^{\infty} F_{\pm;k_1,\ldots,k_n}^{(p_1,\ldots,p_n)} \prod_{j=1}^{n} \left\{ [\operatorname{sgn}(p_j-k_j)]^{p_j-k_j} \right.$$

$$\left. \ell_{p_j\vee k_j-1}^{(|p_j-k_j|)} \left(|\sqrt{2|\tau|}\, \frac{\sqrt{a_j}}{2} x_j'|^2\right) e^{i(p_j-k_j)\theta} \right\}$$

or

$$\widetilde{F}(\pm\tfrac{1}{2}, \frac{2}{\sqrt{a_j}} x_1',\ldots) = \frac{1}{\pi^n} \sum_{p_j,k_j=1}^{\infty} F_{\pm;k_1,\ldots,k_n}^{(p_1,\ldots,p_n)} \tag{3.16}$$

$$\prod_{j=1}^{n} \left\{ [\operatorname{sgn}(p_j-k_j)]^{p_j-k_j} \ell_{p_j\vee k_j-1}^{(|p_j-k_j|)}(|x'|^2) e^{i(p_j-k_j)\theta_j} \right\} ,$$

where

$$x_j + ix_{j+n} = |x_j'|e^{i\theta_j} \, , \; j = 1,\ldots,n \; . \tag{3.17}$$

Now (3.16) and the orthonormality of $\ell_k^{(p)}$, $k = 0,1,2,\ldots$ on $(0,\infty)$ yield the coefficients:

$$F_{\pm;k_1,\ldots,k_n}^{(p_1,\ldots,p_n)} \prod_{j=1}^{n} [sgn(p_j-k_j)]^{p_j-k_j} \tag{3.18}$$

$$= \int_{\mathbb{R}^{2n}} F(\pm\tfrac{1}{2},x') \prod_{j=1}^{n} \left[\ell_{p_j\forall k_j-1}^{(|p_j-k_j|)}(|x'|^2)e^{-i(p_j-k_j)\theta_j} \right] dx' \; .$$

<u>3.19 Definition</u>. *Let F induce a PV convolution operator on H_n. We define its positive Laguerre tensor, $L_+(F)$, by*

$$L_+(F) = \left(F_{+;k_1,\ldots,k_n}^{(p_1,\ldots,p_n)}\right) , \tag{3.20}$$

where $p_j,k_j = 1,2,\ldots$ for $j = 1,2,\ldots,n$.

Then we have the analogue of Theorem 2.45:

<u>3.21 Theorem</u>. *Let F and G induce PV convolution operators on H_n. Then*

$$L_+(F *_H G) = L_+(F)L_+(G) , \tag{3.22}$$

where the composition on the right hand side denotes tensor contraction.

We recall the notion of tensor contraction:

Let

$$U = \left(U_{k_1,\ldots,k_n}^{(p_1,\ldots,p_n)}\right) , \quad V = \left(V_{m_1,\ldots,m_n}^{(q_1,\ldots,q_n)}\right) \tag{3.23}$$

denote two infinite (n,n)-tensors. Their composition, UV, is defined to be

$$UV = W = \left(W^{(p_1,\ldots,p_n)}_{m_1,\ldots,m_n}\right),$$

(3.24)

where

$$W^{(p_1,\ldots,p_n)}_{m_1,\ldots,m_n} = \sum_{k_1,\ldots,k_n=1}^{\infty} U^{(p_1,\ldots,p_n)}_{k_1,\ldots,k_n} V^{(k_1,\ldots,k_n)}_{m_1,\ldots,m_n}.$$

(3.25)

The tensor W *is the contraction of the tensors* U *and* V.

Tensor contraction is not commutative. This is to be expected since it imitates the convolution on H_n. We note that the "Laguerre matrix symbol" on H_1 is simply a $(1,1)$-tensor.

To obtain the calculus for $\tau < 0$ we need the notion of the transpose of an (n,n)-tensor.

We say that

$$U^t = \left(U^{t;(p_1,\ldots,p_n)}_{k_1,\ldots,k_n}\right) = \left(U^{(p_1,\ldots,p_n)}_{k_1,\ldots,k_n}\right)^t$$

(3.26)

is the transpose of

$$U = \left(U^{(p_1,\ldots,p_n)}_{k_1,\ldots,k_n}\right)$$

(3.27)

if

$$U^{t;(p_1,\ldots,p_n)}_{k_1,\ldots,k_n} = U^{(k_1,\ldots,k_n)}_{p_1,\ldots,p_n}.$$

(3.28)

3.29 Definition. *Let F induce a PV convolution operator on H_n. We define its negative Laguerre tensor, $L_-(F)$, by*

$$L_-(F) = \left(F^{(p_1,\ldots,p_n)}_{-;k_1,\ldots,k_n}\right)^t.$$

(3.30)

Now the n-fold version of (2.60) is the following formula:

Let F and G induce PV convolution operators on H_n. Then

$$L_-(F*_H G) = L_-(F)L_-(G) \ . \tag{3.31}$$

Finally, we collect our results.

3.32 Definition. Let F induce a PV convolution operator on H_n. We define its Laguerre tensor, $L(F)$, as follows:

$$L(F) = L_+(F) \oplus L_-(F) \ . \tag{3.33}$$

3.34 Theorem. (The Laguerre calculus) Let F and G induce PV convolution operators on H_n. Then

$$L(F*_H G) = L(F)L(G) \ . \tag{3.35}$$

We note that

$$L(F)L(G) = L_+(F)L_+(G) \oplus L_-(F)L_-(G) \ . \tag{3.36}$$

3.37 Corollary. Let I_H denote the PV convolution operator on H_n which induces the identity operator on $C_0^\infty(H_n)$. Then

$$L_\pm(I_H) = \left(\delta_{k_1}^{(p_1)} \cdots \delta_{k_n}^{(p_n)}\right) \ . \tag{3.38}$$

4. AN EXAMPLE ON H_n. Let

$$X_0 = \frac{\partial}{\partial x_0} \ , \tag{4.1}$$

$$X_j = \frac{\partial}{\partial x_j} - \frac{1}{2} a_j x_{j+n} \frac{\partial}{\partial x_0} \ , \quad j = 1,\ldots,n \ , \tag{4.2}$$

$$X_j = \frac{\partial}{\partial x_j} + \frac{1}{2} a_j x_{j-n} \frac{\partial}{\partial x_0} \ , \quad j = n+1,\ldots,2n \tag{4.3}$$

denote a basis for the Lie algebra of left-invariant vector-fields on H_n. Using the Laguerre calculus of Chapter 3 we shall invert the differential operator

$$\Box_\lambda = -\sum_{j=1}^{2n} X_j^2 - i\lambda X_0 \ , \quad \lambda \in \mathbb{C} \ , \tag{4.4}$$

whenever possible. Since $\Box_\lambda$ is left-invariant on H_n there exists a function K_λ, H-homogeneous (deg-2n), which induces the convolution operator $\Box_\lambda^{-1}$.

At first sight differential operators do not look like convolution operators. To remedy this apparent dissimilarity we shall use the Euclidean Fourier transform to obtain a common formalism. We set

$$\hat{\phi}(\tau,\xi) = \int_{\mathbb{R}^{2n}} e^{-i<\xi,x'>} \tilde{\phi}(\tau,x') \prod_{j=1}^{n} \left(\frac{a_j}{4} \, dx_j'\right) \ , \tag{4.5}$$

where $\phi \in C_0^\infty(H_n)$ and $<\xi,x'> = \xi_1 x_1 + \ldots + \xi_{2n} x_{2n}$. Its inverse is

$$\tilde{\phi}(\tau,x') = \frac{1}{(2\pi)^{2n}} \int_{\mathbb{R}^{2n}} e^{i<x',\xi>} \hat{\phi}(\tau,\xi) \prod_{j=1}^{n} \left(\frac{4}{a_j} \, d\xi_j'\right) \ , \tag{4.6}$$

where $\xi_j' = (\xi_j, \xi_{j+n})$ and $d\xi_j' = d\xi_j d\xi_{j+n}$. As an example consider

$$i^{-1} X_j \phi(x) = \frac{1}{(2\pi)^{2n+1}} \int_{\mathbb{R}^{2n+1}} e^{ix_0\tau + i<x',\xi>} \tag{4.7}$$

$$\sigma(i^{-1} X_j)(x;\tau,\xi) \hat{\phi}(\tau,\xi) dH_n'(\tau,\xi) \ ,$$

where we set

$$dH_n'(\tau,\xi) = d\tau \prod_{j=1}^{n} \left(\frac{4}{a_j} \, d\xi_j d\xi_{j+n}\right) \ , \tag{4.8}$$

$$\sigma(i^{-1} X_0) = \tau \ , \tag{4.9}$$

$$\sigma(i^{-1} X_j) = \xi_j - \frac{1}{2} a_j x_{j+n} \tau \ , \quad j = 1,\ldots,n \ , \tag{4.10}$$

$$\sigma(i^{-1} X_j) = \xi_j + \frac{1}{2} a_j x_{j-n} \tau \ , \quad j = n+1,\ldots,2n \ . \tag{4.11}$$

$\sigma(i^{-1}X_j)$ is called the Fourier symbol of $i^{-1}X_j$. Next a simple calculation yields

4.12 **Proposition.** *Let* $\phi, \psi \in C_0^\infty(H_n)$. *Then*

$$\phi *_H \psi(x) = (2\pi)^{-2n-1} \int_{H_n'} e^{ix_0\tau + i<x',\xi>} \tag{4.13}$$

$$\hat{\phi}\left(\sigma(i^{-1}X)(x;\tau,\xi)\right)\hat{\psi}(\tau,\xi)\,dH_n'(\tau,\xi) ,$$

where

$$\sigma(i^{-1}X) = \left(\sigma(i^{-1}X_0), \sigma(i^{-1}X_1), \ldots, \sigma(i^{-1}X_{2n})\right) . \tag{4.14}$$

We calculate $\square_\lambda$ under the Fourier integral:

$$\square_\lambda \phi(x) = (2\pi)^{-2n-1} \int_{H_n'} e^{ix_0\tau + i<x',\xi>} \tag{4.15}$$

$$\left(\sum_{j=1}^{2n}\sigma(i^{-1}X_j)^2 + \lambda\tau\right)\hat{\phi}(\tau,\xi)\,dH_n'(\tau,\xi)$$

$$= (2\pi)^{-2n-1} \int_{H_n'} e^{ix_0\tau + i<x',\xi>}$$

$$\hat{\square}_\lambda\left(\sigma(i^{-1}X)(x;\tau,\xi)\right)\hat{\phi}(\tau,\xi)\,dH_n'(\tau,\xi) ,$$

where we set

$$\hat{\square}_\lambda(\tau,\xi) = \sum_{j=1}^{2n}\xi_j^2 + \lambda\tau . \tag{4.16}$$

We note that

$$\hat{\square}_\lambda(\tau,\xi) = \sigma(\square_\lambda)(0;\tau,\xi) , \tag{4.17}$$

where

$$\sigma(\square_\lambda) = \hat{\square}_\lambda\left(\sigma(i^{-1}x)\right) . \tag{4.18}$$

Comparing (4.13) and (4.15) we see that under the Fourier integral $\Box_\lambda$ does behave like a convolution operator on H_n . Thus we are justified to use the Laguerre calculus to invert $\Box_\lambda$. To do this first we need to compute the Laguerre tensor, $L(F)$, from $\hat{F}$. We start with the *Fourier transform of the n-fold exponential Laguerre functions. Namely*

$$\prod_{j=1}^{n} \hat{L}_{k_j}^{(p_j)} \left(\tau, \frac{2}{\sqrt{a_j}} \xi_j'\right) = \prod_{j=1}^{n} \hat{L}_{k_j}^{(p_j)} \left(\frac{1}{2}, \frac{2}{\sqrt{a_j}} \frac{\xi_j'}{\sqrt{2|\tau|}}\right) , \qquad (4.19)$$

where

$$\hat{L}_k^{(p)} \left(\frac{1}{2}, \xi'\right) = 2(-i)^p (-1)^k \ell_k^{(|p|)} (|\xi'|^2) e^{ip\theta} \qquad (4.20)$$

with

$$\xi_j + i\xi_{j+n} = |\xi_j'| e^{i\theta} . \qquad (4.21)$$

Let F induce a PV convolution operator on H_n . Then

$$\hat{F}(r^2\tau, r\xi) = \hat{F}(\tau, \xi) , \quad r > 0 , \qquad (4.22)$$

i.e. $\hat{F}$ is H'-homogeneous of degree zero. Therefore we may write

$$\hat{F}_\pm(\tau, \xi) = \begin{cases} \hat{F}\left(\pm\frac{1}{2}, \dfrac{\xi}{\sqrt{2|\tau|}}\right) , & \tau \gtrless 0 , \\[2ex] 0 , & \tau \lessgtr 0 , \end{cases} \qquad (4.23)$$

and the Fourier transform of (3.14) is

$$\hat{F}_\pm(\tau, \xi) = \sum_{p_j, k_j = 1}^{\infty} F_{\pm; k_1, \ldots, k_n}^{(p_1, \ldots, p_n)} \qquad (4.24)$$

$$\prod_{j=1}^{n} \hat{L}_{p_j \vee k_j - 1}^{(p_j - k_j)} \left(\tau, \frac{2}{\sqrt{a_j}} \xi_j'\right) .$$

We rewrite this in the following form:

$$\hat{F}(\pm\frac{1}{2}, \frac{\sqrt{a_1}}{2}\xi'_1, \ldots) = \sum_{p_j, k_j = 1}^{\infty} F_{\pm; k_1, \ldots, k_n}^{(p_1, \ldots, p_n)}$$

$$\prod_{j=1}^{n} \hat{L}_{p_j \vee k_j - 1}^{(p_j - k_j)} (\frac{1}{2}, \xi'_j) \quad . \tag{4.25}$$

4.26 Proposition. *Let* F *induce a PV convolution operator on* H_n. *Then*

$$(4\pi)^n F_{\pm; k_1, \ldots, k_n}^{(p_1, \ldots, p_n)} \tag{4.27}$$

$$= \int_{\mathbb{R}^{2n}} F(\pm\frac{1}{2}, \frac{\sqrt{a_1}}{2}\xi'_1, \ldots) \prod_{j=1}^{n} \hat{L}_{k_j \vee p_j - 1}^{(k_j - p_j)} (\frac{1}{2}, \xi'_j) d\xi \quad .$$

We return to $\hat{\square}_\lambda$.

$$\hat{\square}_\lambda = \sum_{j=1}^{n} (\xi_j^2 + \xi_{j+n}^2) + \lambda\tau \tag{4.28}$$

$$= 2|\tau|\left\{ \sum_{j=1}^{n} \frac{a_j}{4} \left| \frac{2}{\sqrt{a_j}} \frac{\xi'_j}{\sqrt{2|\tau|}} \right|^2 + \frac{1}{2}(\operatorname{sgn} \tau)\lambda \right\} \quad .$$

We note that $(2|\tau|)^{-1}\hat{\square}_\lambda$ is H'-homogeneous of degree zero. Using (4.25) we expand it in a Laguerre series:

$$\sum_{j=1}^{n} \frac{a_j}{4} |\xi'_j|^2 \pm \frac{1}{2}\lambda \tag{4.29}$$

$$= \sum_{p_j, k_j = 1}^{\infty} (\frac{1}{2|\tau|}\hat{\square}_\lambda)_{\pm; k_1, \ldots, k_n}^{(p_1, \ldots, p_n)} \prod_{j=1}^{n} \hat{L}_{p_j \vee k_j - 1}^{(p_j - k_j)} (\frac{1}{2}, \xi'_j)$$

$$= \sum_{k_1, \ldots, k_n = 1}^{\infty} (\frac{1}{2|\tau|}\hat{\square}_\lambda)_{\pm; k_1, \ldots, k_n}^{(k_1, \ldots, k_n)} \prod_{j=1}^{n} \hat{L}_{k_j - 1}^{(0)} (\frac{1}{2}, \xi'_j) \quad ,$$

since the left-hand side depends on $|\xi_1'|^2,\ldots,|\xi_n'|^2$ only. To find the coefficients we start with

$$|\mu|^2 = \sum_{k=0}^{\infty} 2(2k+1)\hat{L}_k^{(0)}(\tfrac{1}{2},\mu) , \qquad (4.30)$$

$\mu = (\mu_1,\mu_2)$. Then

$$\sum_{j=1}^{\infty} \frac{a_j}{4} |\xi_j'|^2 \pm \tfrac{1}{2}\lambda \qquad (4.31)$$

$$= \sum_{k_1,\ldots,k_n=1}^{\infty} \frac{1}{2}\Big(\sum_{j=1}^{n}(2k-1)a_j\pm\lambda\Big) \prod_{j=1}^{n}\hat{L}_{k_j-1}^{(0)}(\tfrac{1}{2},\xi_j') .$$

Consequently, with a slight abuse of notation, we have

$$\Big(L_{\pm}\Big[\frac{1}{2|\tau|}\hat{\Box}_\lambda\Big]\Big)_{k_1,\ldots,k_n}^{(p_1,\ldots,p_n)} \qquad (4.32)$$

$$= \frac{1}{2}\Big(\sum_{j=1}^{n}(2k_j-1)a_j\pm\lambda\Big)\delta_{k_1}^{(p_1)}\cdots\delta_{k_n}^{(p_n)} .$$

Therefore, Corollary 3.37 implies that $(2|\tau|)^{-1}\hat{\Box}_\lambda$ is invertible as long as

$$\pm\lambda \neq \sum_{j=1}^{n}(2k_j-1)a_j , \qquad (4.33)$$

$k_1,\ldots,k_n = 1,2,3,\ldots$ If (4.33) holds

$$L_{\pm}\Big(\Big[\frac{1}{2|\tau|}\hat{\Box}_\lambda\Big]^{-1}\Big) \qquad (4.34)$$

$$= \Big(\Big[\frac{1}{2}\sum_{j=1}^{n}(2k_j-1)a_j\pm\lambda\Big]^{-1}\delta_{k_1}^{(p_1)}\cdots\delta_{k_n}^{(p_n)}\Big) ,$$

and the Laguerre calculus yields $\hat{\Box}_\lambda^{-1}$:

$$\left(\Box_\lambda^{-1}\right)^\wedge = \sum_{k_1,\ldots,k_n=0}^{\infty} \left(\sum_{j=1}^{n}(2k_j+1)a_j\,|\tau|+\lambda\tau\right)^{-1} \tag{4.35}$$

$$\prod_{j=1}^{n} \hat{L}_{k_j}^{(0)}\left(\frac{1}{2},\ \frac{2}{\sqrt{a_j}}\ \frac{\xi_j'}{\sqrt{2|\tau|}}\right) \ .$$

This can be summed as follows.

$$\left(\Box_\lambda^{-1}\right)^\wedge(\tau,\xi) = \sum_{k_1,\ldots,k_n=0}^{\infty} \int_0^\infty e^{-\left[\sum_{j=1}^{n}(2k_j+1)a_j|\tau|+\lambda\tau\right]s}\, ds$$

$$\prod_{j=1}^{n} \hat{L}_{k_j}^{(0)}\left(\frac{1}{2},\ \frac{2}{\sqrt{a_j}}\ \frac{\xi_j'}{\sqrt{2|\tau|}}\right) \tag{4.36}$$

$$= \int_0^\infty e^{-\lambda\tau s}\, G(\tau,\xi;s)\, ds \ ,$$

with

$$G = \sum_{k_1,\ldots,k_n=0}^{\infty} e^{-\sum_{j=1}^{n}(2k_j+1)a_j|\tau|s} \tag{4.37}$$

$$\prod_{j=1}^{n} \hat{L}_{k_j}^{(0)}\left(\frac{1}{2},\ \frac{2}{\sqrt{a_j}}\ \frac{\xi_j'}{\sqrt{2|\tau|}}\right)$$

$$= \prod_{j=1}^{n} g(p_j,q_j) \ ,$$

where we set

$$p_j = a_j\,|\tau|s \ , \tag{4.38}$$

$$q_j = 2\ \frac{\xi_j^2 + \xi_{j+n}^2}{a_j|\tau|} \tag{4.39}$$

and

$$g(p,q) = 2e^{-p} \sum_{k=0}^{\infty} (-e^{-2p})^k L_k^{(0)}(q) e^{-q/2} \qquad (4.40)$$

$$= \frac{1}{\cosh p} e^{-\frac{q}{2} \tanh p} .$$

Here we used (2.3) to sum the series (4.40). Therefore

$$G(\tau,\xi;s) = \left(\prod_{j=1}^{n} \cosh(a_j|\tau|s) \right)^{-1} \qquad (4.41)$$

$$\exp\left(- \sum_{j=1}^{n} \frac{\xi_j^2 + \xi_{j+n}^2}{a_j|\tau|} \tanh(a_j|\tau|s) \right) .$$

(4.41) occurs in [2]. We collect these calculations in

4.42   Theorem. *Assume*

$$\pm\lambda \neq \sum_{j=1}^{n} (2k_j-1)a_j , \qquad (4.43)$$

$k_1,\ldots,k_n = 1,2,3,\ldots$ *Then* $\square_\lambda$ *is invertible.* $\square_\lambda^{-1}$ *is in-*
duced by a convolution operator on H_n *whose Laguerre tensor*
can be obtained from

$$L_{\pm}\left(\left[\frac{1}{2|\tau|} \hat{\square}_\lambda \right]^{-1} \right) \qquad (4.44)$$

$$= \left(2 \left[\sum_{j=1}^{n} (2k_j-1)a_j \pm \lambda \right]^{-1} \delta_{k_1}^{(p_1)} \ldots \delta_{k_n}^{(p_n)} \right) .$$

Consequently

$$\left(\square_\lambda^{-1} \right)^{\hat{}} = \sum_{k_1,\ldots,k_n=1}^{\infty} \left(\sum_{j=1}^{n} (2k_j-1)a_j |\tau|+\lambda\tau \right)^{-1} \qquad (4.45)$$

$$\prod_{j=1}^{n} \hat{L}_{k_j-1}^{(0)} \left(\frac{1}{2}, \frac{2}{\sqrt{a_j}} \frac{\xi_j'}{\sqrt{2|\tau|}} \right) .$$

If

$$- \sum_{j=1}^{n} a_j < Re \; \lambda < \sum_{j=1}^{n} a_j \; , \tag{4.46}$$

(4.45) can be summed to

$$\left(\Box_\lambda^{-1}\right)^{\hat{}} = \int_0^\infty e^{-\lambda\tau s} G(\tau, \xi; s) ds \; , \tag{4.47}$$

where $G(\tau, \xi; s)$ is given by (4.41).

4.48 <u>Remarks</u>. For many derivations and proofs missing from this article and for a great deal of further work on this subject the reader should consult [2], [7] and [10]. In particular, the idea to use the Laguerre series on H_1 in analogy with Mikhlin's use of the Fourier series on C was suggested by some work in [10]. The central result, Theorem 2.34, was derived in [7]. Finally, both [2] and [7] contain further results we have not been able to mention in this short article.

Yale University
The University of Toronto
Universitè de Paris VI

This work has been supported in part by the National Science Foundation of U.S.A. and by the National Science and Engineering Research Council of Canada under Grant No.A3017.

REFERENCES.

[1]   Beals, R.W. and Greiner, P.C., "Pseudo-differential operators associated to hyperplane bundles", Bull. Sem. Mat. Torino, pp.7-40, 1983.

[2]   Beals, R.W. and Greiner, P.C., "Non-elliptic differential operators of type $\Box_b$", (in preparation).

[3]   Folland, G.B., "A fundamental solution for a subelliptic operator", Bull. Amer. Math. Soc. <u>79</u> (1973), pp.373-376.

[4]   Folland, G.B. and Stein, E.M., "Estimates for the $\bar{\partial}_b$-complex and analysis on the Heisenberg group", Comm. Pure Appl. Math. <u>27</u> (1974), pp.429-522.

[5]   Geller, D., "Fourier analysis on the Heisenberg group. I. Schwartz space", J. Func. Analysis <u>36</u> (1980), pp. 205-254.

216   R. W. BEALS ET AL.

[6]   Geller, D., "Local solvability and homogeneous distribu-
      tions on the Heisenberg group", Comm. PDE, 5 (5) (1980),
      pp.475-560.
[7]   Greiner, P.C., "On the Laguerre calculus of left-invari-
      ant convolution (pseudo-differential) operators on the
      Heisenberg group", Seminaire Goulaouic-Meyer-Schwartz,
      1980-81, Exposè no.XI, pp.1-39.
[8]   Greiner, P.C., Kohn, J.J. and Stein, E.M., "Necessary
      and sufficient conditions for the solvability of the
      Lewy equation", Proc. Nat. Acad. of Sciences, U.S.A.,
      72 (1975), pp.3287-3289.
[9]   Greiner, P.C. and Stein, E.M., "Estimates for the $\bar{\partial}$-
      Neumann problem", Math. Notes Series, no.19, Princeton
      Univ. Press, Princeton, N.J. 1977.
[10] Greiner, P.C., and Stein, E.M., "On the solvability of
      some differential operators of type $\Box_b$", Proc. of the
      Seminar on Several Complex Variables, Cortona, Italy,
      1976-1977, pp.106-165.
[11] Koranyi, A. and Vagi, S., "Singular integrals in homo-
      geneous spaces and some problems of classical analysis",
      Ann. Scuola Norm. Sup. Pisa 25 (1971), pp.575-648.
[12] Lewy, H., "An example of a smooth linear partial diff-
      erential equation without solution", Ann. of Math., 66
      (1957), pp.155-158.
[13] Mauceri, G., "The Weyl transform and bounded operators
      on $L^p(\mathbb{R}^n)$", Report no.54 of the Math. Inst. of the
      Univ. of Genova, 1980.
[14] Mikhlin, S.G., "Multidimensional singular integrals and
      integral equations", Pergamon Press, 1965.
[15] Nagel, A. and Stein, E.M., "Lectures on pseudo-differen-
      tial operators", Math. Notes Series, no.24, Princeton
      Univ. Press, Princeton, N.J. 1979.
[16] Seeley, R.T., "Elliptic Singular Integral Equations",
      Amer. Math. Soc. Proc. Symp. Pure Math. 10 (1967), pp.
      308-315.
[17] Szegö, G., "Orthogonal polynomials", Amer. Math. Soc.
      Colloquium Publ., V. 23, Amer. Math. Soc., Providence,
      R.I., 1939.

Walter Schempp

RADAR AMBIGUITY FUNCTIONS, NILPOTENT HAR-
MONIC ANALYSIS, AND HOLOMORPHIC THETA
SERIES

1. INTRODUCTION

As is well known, radar (=abbreviation of RAdio
Detection And Ranging) systems are a device for
discovering distant objects that are stationary or
moving such as ships, aeroplanes, and satellites.
Besides the detection of the presence of a remote
target, the purpose of a radar system is basically
to extract information of interest (such as range,
relative velocity, etc.) about the target. The radar transmitter generates electromagnetic energy
of a few centimeters' wavelength in the form of
pulses of large amplitude and brief duration which
are emitted periodically through an antenna that
produces a narrow beam of radiation. Any object
located in the path of the propagating beam scat-
ters the radiation in all directions and a small
portion of the scattered radiation excites the re-
ceiving antenna. It can be achieved by means of
modern electronical equipments that the radar system uses a common antenna for both transmission
and reception: In an elementary form of a radar
system a duplexer enables the radar antenna to
operate in the transmission mode as well as in the
reception mode. The reflected signal energy picked
up by the radar antenna (operating in the reception
mode) is led to a receiver, amplified, and then
applied to the vertical deflection plates of a
cathode-ray oscilloscope to detect the presence of
the radar target and estimate its parameters.

In studying radar signal pulses in an ideali-
zed situation let $\mathscr{S}(\mathbb{R})$ denote the Schwartz-Bruhat
subspace of the complex Hilbert space $L^2(\mathbb{R})$ formed

R. A. Askey et al. (eds.), Special Functions: Group Theoretical Aspects and Applications, 217–260.
© 1984 by D. Reidel Publishing Company.

by the complex-valued infinitely differentiable
functions on the real line $\mathbb{R}$, rapidly decaying at
infinity. Consider an amplitude modulated radar
signal pulse described by the complex waveform

$$t \longrightarrow f(t) e^{2\pi i \omega t}$$

where $f \in \mathscr{S}(\mathbb{R})$ is the transmitted signal envelope
function and ω denotes the constant radar carrier
frequency. The envelope f is a relatively slowly
varying time function in contrast with the mono-

chromatic radar carrier. Its squared L^2-norm

$$||f||^2 = <f|f> = \int_{\mathbb{R}} |f(t)|^2 dt$$

gives the total input signal energy which is double
the actual signal energy. If we assume that the received signal is reflected from a stationary target then the echo signal is delayed by time and the
carrier frequency of the transmitted and the reflected signals coincide. To measure the distance
of the radar target it is necessary to measure the
time x at which the echo from it arrives at the receiver. If time is counted from the transmission
of the radar pulse, the distance is

$$\frac{1}{2}cx,$$

where c denotes the velocity of electromagnetic radiation. In the case when the radar target is not
stationary but is moving toward or away from the
antenna, the carrier frequency ω of the echo signal
also differs from that of the transmitted pulse because of the Doppler effect. If the Doppler frequency shift y can be measured, the observer can
calculate the component v of the target velocity
in the direction of the radar antenna, obtaining
valuable information for tracking the target effi-
ciently. For instance, if we pick the transmitted
frequency as our basic reference frequency, a target moving radially at a rate of v = 500 m.p.h.
generates for a carrier frequency ω = 3000 MHz a
Doppler frequency shift

$$y = 2\frac{v}{c} = 4500 \text{ Hz}$$

which is an appreciable fraction of the one-mega-
hertz bandwidth typical of a radar pulse. By drop-
ping an inessential phase factor, the auto-corre-
lation function of the transmitted and reflected
signal takes the symmetric form

$$H(f;x,y) = \int_{\mathbb{R}} f(t+\tfrac{1}{2}x)\,\overline{f}(t-\tfrac{1}{2}x)\,e^{2\pi iyt}dt.$$

The function $H(f;.,.)$ was introduced in signal
theory by J. Ville in his fundamental 1948 paper
[26] while its importance with respect to radar
analysis and design was pointed out by Woodward
[30]. In the language of electrical engineering it
is commonly referred to as the radar auto-ambiguity function for target range (delay time) and
range rate (Doppler frequency shift) with respect
to the signal envelope function $f \in \mathscr{S}(\mathbb{R})$ [14-20]
whereas the Fourier transform of $H(f;.,.)$ repre-
sents the Wigner quasi-probability distribution
function [21-23] which plays an important rôle in
quantum mechanics and has received recently a good
deal of attention in nuclear physics (oral communications by H.-J. Krappe, Berlin, and P. Kramer,
Tübingen). If $g \in \mathscr{S}(\mathbb{R})$ denotes the envelope of
a second radar signal pulse, a slightly more gene-
ral notion is the mixed radar ambiguity (or radar
cross-ambiguity) function given by

$$H(f,g;x,y) = \int_{\mathbb{R}} f(t+\tfrac{1}{2}x)\,\overline{g}(t-\tfrac{1}{2}x)\,e^{2\pi iyt}dt;$$

cf. de Bruijn [7]. The function $H(f,g;.,.)$ and its
extension to envelopes $f,g \in L^2(\mathbb{R})$ are at the basis in modern radar technology for the systematic
search for the optimal waveform in order to dis-
tinguish two echo signals by means of their arri-
val times x and their Doppler shifts y of their
carrier frequencies from a common reference value.
Therefore the understanding of these functions
and their fundamental properties is of particular

importance for the analysis and the waveform design of radar systems.

Radar observations can be interpreted as a
physical experiment for determining simultaneously the target coodinates (range) and radial component of the relative velocity (range rate) of the
remote target at a given instant of time. The radar uncertainty principle states that there exists
a "unique" uncertainty (ambiguity) in determining
the target range and range rate simultaneously
which cannot be decreased by changing the signal
parameters. Consequently there is an analogy with
the Heisenberg uncertainty principle of quantum
mechanics. Quantum mechanics stands here for the
quantum-mechanical description, at a given instant
of time, of a non-relativistic particle with one
degree of freedom. Roughly speaking not all physical quantities observed in any realizable experiment (even in principle only) can be determined
according to the Heisenberg uncertainty principle
with an arbitrarily high accuracy. Since the mu-
tually exclusive (conjugate, non-commuting) quantities the measurement errors of which are inter-
related can be embodied into the notion of Heisenberg nilpotent group, the analogy between the radar uncertainty principle on the one hand and the
Heisenberg uncertainty principle of quantum physics on the other hand suggests an application of
the harmonic analysis of the real Heisenberg group
$\tilde{A}(\mathbb{R})$ to radar analysis. It is the purpose of the
present paper to support the usefulness of this
intuitive idea and to point out some of its immediate consequences by working systematically from
a group representational viewpoint. Taking into
account the analogy of the two uncertainty prin-
ciples mentioned above we will use the languages
of electrical engineering and elementary quantum
mechanics simultaneously.

The paper is organized as follows. In Section
2, we briefly review some aspects of nilpotent
harmonic analysis in a geometric language. Of course, the central point due to Kirillov [10] is that
for a connected, simply connected nilpotent Lie

group the unitary dual can be parametrized by the coadjoint orbits in the real vector space dual of its Lie algebra. The Kirillov correspondence is obtained by selecting a polarization for an element in the orbit and then forming an appropriate unitarily induced group representation. In Section 4, we determine via a representation-theoretic characterization of the radar cross-ambiguity function the group of signal energy preserving invariants of the radar ambiguity surface by an application of I. Segal's metaplectic formula. Moreover, based on this geometric result we determine explicitly as a special case the radially symmetric radar auto-ambiguity functions by means of the logarithm of the isotropic cross-section to the center $\tilde{Z}$ in $\tilde{A}(\mathbb{R})$. Finally, in Section 7, we turn from the complex wave model to the lattice model which also has applications in signal theory (sampling theorem of digital signal processing) as well as in quantum mechanics (dynamics of Bloch electrons in crystalline solids in the presence of a constant electric or magnetic field). Transferring from $\tilde{A}(\mathbb{R})$ to the Heisenberg compact nilmanifold $D\backslash\tilde{A}(\mathbb{R})$ which is a principal circle bundle over the compact two-di-

mensional torus $\mathbb{T}^2$, we get an identity for radar cross-ambiguity functions with respect to harmonic oscillator wave functions as their envelopes. The transferral will be processed by an application of the Weil-Brezin isomorphism. It follows an identity for sums of the evaluations of Laguerre-Weber functions of different orders at quadratic lattice points. These "geometrically" established formulae are closely related to some identities of first order classical Jacobi theta functions which are deeply connected with the nil-theta functions living in a natural way on the Heisenberg compact nilmanifold $D\backslash\tilde{A}(\mathbb{R})$.

ACKNOWLEDGMENTS.

The author is grateful to Professor Lawrence C. Biedenharn (Duke University) for inviting him to present some of the results previously announced in [14-18], [21-23] as an invited lecture at the

special session "Lie groups and generalized classical special functions" of the 797th Meeting of
the American Mathematical Society at the University of Maryland (College Park, MD) on October 30-31,
1982. Moreover he is happy to express his gratitu-
de to Professor Luigi Gatteschi for inviting him
to work at the Istituto di Calcoli Numerici of the
Università di Torino and to Professors Edwin He-
witt and Mourad E.H. Ismail for their invitations
to lecture on the subject at the University of
Alaska (Fairbanks, AK) and Arizona State University
(Tempe, AZ), respectively. Finally he should like
to thank the Deutsche Forschungsgemeinschaft for
financial support.

2. THE REAL HEISENBERG NILPOTENT GROUP

In this paper we look upon the real (time-frequen-
cy resp. phase) plane $\mathbb{R} \oplus \mathbb{R}$ as the vector space
of all column vectors

$$v = \begin{bmatrix} x \\ y \end{bmatrix} \qquad (x, y \in \mathbb{R})$$

and define the standard symplectic form A on
$\mathbb{R} \oplus \mathbb{R}$ according to the prescription

$$A(v_1, v_2) = A\left(\begin{bmatrix} x_1 \\ y_1 \end{bmatrix}, \begin{bmatrix} x_2 \\ y_2 \end{bmatrix}\right) = x_1 y_2 - x_2 y_1 = \det \begin{pmatrix} x_1 & x_2 \\ y_1 & y_2 \end{pmatrix}$$

The automorphism J of $\mathbb{R} \oplus \mathbb{R}$ which admits the matrix

$$\begin{pmatrix} 0 & -1 \\ +1 & 0 \end{pmatrix}$$

with respect to the canonical basis satisfies
$J^2 = -id_{\mathbb{R} \oplus \mathbb{R}}$. We have

$$A(v_1, Jv_2) = \langle v_1 | v_2 \rangle, \quad A(Jv_1, Jv_2) = A(v_1, v_2)$$

for all vectors v_1, v_2 in $\mathbb{R} \oplus \mathbb{R}$ where $<.|.>$ denotes the standard scalar product of $\mathbb{R}^2$. In particular $A(v, Jv) > 0$ unless $v = 0$. Since A is $\mathbb{Z}$-valued on the standard lattice $\mathbb{Z} \oplus \mathbb{Z}$ in $\mathbb{R} \oplus \mathbb{R}$ it defines a Riemann form on $\mathbb{R} \oplus \mathbb{R}$ relative to $\mathbb{Z} \oplus \mathbb{Z}$ and $\mathbb{R} \oplus \mathbb{R}$ equipped with the complex structure J can be identified with the complex plane $\mathbb{C}$. Thus we have the data of an abelian variety. Obviously

$$B(v_1, v_2) = A(v_1, Jv_2) + iA(v_1, v_2)$$

is a positive definite Hermitian form on $\mathbb{R} \oplus \mathbb{R}$.

Consider the trivial principal fiber bundle $(\mathbb{R}^2 \times \mathbb{C}^\times, pr_1, \mathbb{R}^2)$ over the plane $\mathbb{R}^2 = \mathbb{R} \oplus \mathbb{R}$ with (multiplicative) group $\mathbb{C}^\times = \mathbb{C} - \{0\}$. On $\mathbb{R}^2 \times \mathbb{C}^\times$ define the multiplication law

$$(v_1, z_1) \cdot (v_2, z_2) = (v_1 + v_2, z_1 z_2 e(\tfrac{1}{2i}B(v_1, v_2)))$$

where $e: t \longrightarrow e^{2\pi it}$ denotes the exponential map of the Lie algebra $\mathbb{R}$ of $\mathbb{T}$ onto $\mathbb{T}$. Under this multiplication, $\mathbb{R}^2 \times \mathbb{C}^\times$ is a connected nilpotent real Lie group which is called the Heisenberg nilpotent group. We define a reduction of the structure group $\mathbb{C}^\times$ to the subgroup $\mathbb{T}$ on the principal fiber bundle $(\mathbb{R}^2 \times \mathbb{C}^\times, pr_1, \mathbb{R}^2)$ which is canonically determined by the Riemann form A and the complex structure J on $\mathbb{R}^2$. For this purpose let $A(\mathbb{R})$ denote the subset of $\mathbb{R}^2 \times \mathbb{C}^\times$ consisting of the elements of the form

$$(v, \zeta e(\tfrac{1}{4i}B(v,v))) \qquad (\zeta \in \mathbb{T}).$$

If such an element is denoted by the symbol

$(v, \zeta) \in \mathbb{R}^2 \times \mathbb{T}$ then we obtain as multiplication rule of $A(\mathbb{R})$

$$(v_1, \zeta_1)(v_2, \zeta_2) = (v_1 + v_2, \zeta_1 \zeta_2 e(\tfrac{1}{2} A(v_1, v_2))).$$

In view of the identity

$$(v, \zeta)^{-1} = (-v, \overline{\zeta})$$

it is evident that $A(\mathbb{R})$ forms a subgroup of $\mathbb{R}^2 \times \mathbb{C}^\times$. It is called the reduced Heisenberg nilpotent group. We may regard $(A(\mathbb{R}), pr_1, \mathbb{R}^2)$ as a sub-bundle of $(\mathbb{R}^2 \times \mathbb{C}^\times, pr_1, \mathbb{R}^2)$ with the compact structure group $\mathbb{T}$. It is easily verified that the universal covering group $\tilde{A}(\mathbb{R})$ of $A(\mathbb{R})$ is modeled on $\mathbb{R}^2 \times \mathbb{R}$ with multiplication law

$$(v_1, s_1) \cdot (v_2, s_2) = (v_1 + v_2, s_1 + s_2 + \tfrac{1}{2} A(v_1, v_2)).$$

Computations involving the reduced Heisenberg nilpotent group $A(\mathbb{R})$ and its universal covering group $\tilde{A}(\mathbb{R})$ are usually easier in their basic presentations (exponential coordinates) as indicated above than in their dual pairing presentations (canonical coordinates) given by

$$\left(\begin{bmatrix} x_1 \\ y_1 \end{bmatrix}, \zeta_1 \right) \cdot \left(\begin{bmatrix} x_2 \\ y_2 \end{bmatrix}, \zeta_2 \right) = \left(\begin{bmatrix} x_1 + x_2 \\ y_1 + y_2 \end{bmatrix}, \zeta_1 \zeta_2 e(x_1 y_2) \right),$$

and

$$\left(\begin{bmatrix} x_1 \\ y_1 \end{bmatrix}, s_1 \right) \cdot \left(\begin{bmatrix} x_2 \\ y_2 \end{bmatrix}, s_2 \right) = \left(\begin{bmatrix} x_1 + x_2 \\ y_1 + y_2 \end{bmatrix}, s_1 + s_2 + x_1 y_2 \right),$$

respectively. From the dual pairing presentation however it becomes obvious that $\tilde{A}(\mathbb{R})$ is isomorphic to the subgroup of $SL(3, \mathbb{R})$ formed by the unipotent matrices

$$
\begin{pmatrix} 1 & x & s \\ 0 & 1 & y \\ 0 & 0 & 1 \end{pmatrix} .
$$

We will return to the dual pairing presentation of $\tilde{A}(\mathbb{R})$ in Section 5.

The Heisenberg groups $A(\mathbb{R})$ and $\tilde{A}(\mathbb{R})$ are locally compact real connected two-step nilpotent Lie groups with centers

$$
Z = \{(\begin{bmatrix} 0 \\ 0 \end{bmatrix}, \zeta) \mid \zeta \in \mathbb{T}\} = [A(\mathbb{R}), A(\mathbb{R})]
$$

and

$$
\tilde{Z} = \{(\begin{bmatrix} 0 \\ 0 \end{bmatrix}, s) \mid s \in \mathbb{R}\} = [\tilde{A}(\mathbb{R}), \tilde{A}(\mathbb{R})],
$$

respectively. The two-step nilpotency can be easily checked by examining the central subgroups. Obviously the center Z of $A(\mathbb{R})$ which is isomorphic to the one-dimensional torus $\mathbb{T}$ also forms the maximal compact subgroup of $A(\mathbb{R})$. The simply connected Lie group $\tilde{A}(\mathbb{R})$ has the same Lie algebra $\mathcal{M}$ as does $A(\mathbb{R})$ and since $\tilde{A}(\mathbb{R})$ is nilpotent the exponential mapping

$$
\exp: \mathcal{M} \to \tilde{A}(\mathbb{R})
$$

is a global diffeomorphism which maps the center $\mathfrak{z}$ of the Heisenberg nilpotent Lie algebra $\mathcal{M}$ onto the one-dimensional center Z of $A(\mathbb{R})$. Observe that $\mathcal{M}$ is spanned by the vectors $\{X, Y, T\}$ over $\mathbb{R}$ such that

$$
[X, Y] = T, \quad [X, T] = [Y, T] = 0
$$

and $\mathfrak{z} = \mathbb{R}T$. Clearly

$$
\exp X = (\begin{bmatrix} 1 \\ 0 \end{bmatrix}, 0), \quad \exp Y = (\begin{bmatrix} 0 \\ 1 \end{bmatrix}, 0), \quad \exp T = (\begin{bmatrix} 0 \\ 0 \end{bmatrix}, 1).
$$

In the following we will use the coordinates $(\begin{bmatrix} p \\ q \end{bmatrix}, c)$ relative to the standard basis $\{X, Y, T\}$ to coordinatize $\mathcal{M}$. Then we obtain

$$\exp\left(\begin{bmatrix} p \\ q \end{bmatrix}, c\right) = \left(\begin{bmatrix} p \\ q \end{bmatrix}, c + \frac{1}{2}pq\right)$$

and conversely

$$\log\left(\begin{bmatrix} x \\ y \end{bmatrix}, s\right) = \left(\begin{bmatrix} x \\ y \end{bmatrix}, s - \frac{1}{2}xy\right).$$

The Lie bracket operation in $\mathcal{M}$ takes the form

$$[(w_1, c_1), (w_2, c_2)] = \left(\begin{bmatrix} 0 \\ 0 \end{bmatrix}, A(w_1, w_2)\right).$$

Since the symplectic group $\mathrm{Sp}(1, \mathbb{R}) = \mathrm{SL}(2, \mathbb{R})$ consists of the linear automorphisms of $\mathbb{R} \oplus \mathbb{R}$ which preserve the symplectic form A it can be naturally identified with the group of automorphisms of the Heisenberg Lie algebra $\mathcal{M}$ leaving the one-dimensional center $\mathcal{Z}$ pointwise fixed.

According to the Kirillov theory the isomorphy classes of the irreducible unitary continuous linear representations of $\tilde{A}(\mathbb{R})$ are in natural bijection with the orbits of $\tilde{A}(\mathbb{R})$ (or $A(\mathbb{R})$, since the center acts trivially) in the real vector space dual $\mathcal{M}^*$ of $\mathcal{M}$ under the contragredient action Coad. In other words, the Kirillov correspondence maps bijectively and functorially the unitary dual $\hat{\tilde{A}}(\mathbb{R})$ of $\tilde{A}(\mathbb{R})$ onto the orbit space $\mathcal{M}^*/\mathrm{Coad}(\tilde{A}(\mathbb{R}))$.

The adjoint action of $\tilde{A}(\mathbb{R})$ on $\mathcal{M}$ is given by

$$\mathrm{Ad}(v, s)(w, c) = (w, c + A(v, w)).$$

Let $\{X^*, Y^*, T^*\}$ denote the dual basis of $\{X, Y, T\}$ in $\mathcal{M}^*$. Then the contragredient action of $\tilde{A}(\mathbb{R})$ on $\mathcal{M}^*$ is given by

$$\text{Coad}\left(\begin{bmatrix} x \\ y \end{bmatrix}, s\right)(\alpha X^* + \beta Y^* + \lambda T^*) = (\alpha + \lambda y) X^* + (\beta - \lambda x) Y^* + \lambda T^*.$$

In particular, the $\text{Coad}(\tilde{A}(\mathbb{R}))$-orbits in $\mathcal{M}^*$ are the affine planes $\mathbb{R} X^* + \mathbb{R} Y^* + \lambda T^* = \mathfrak{z}^\perp + \lambda T^*$ parametrized by $\lambda \in \mathbb{R}^\times$ and single points in the plane $\mathfrak{z}^\perp$ spanned by $\{X^*, Y^*\}$ in $\mathcal{M}^*$ in the case $\lambda = 0$. The orbit space $\mathcal{M}^*/\text{Coad}(\tilde{A}(\mathbb{R}))$ with the "catastrophic plane $\lambda = 0$" therefore has the following simple form:

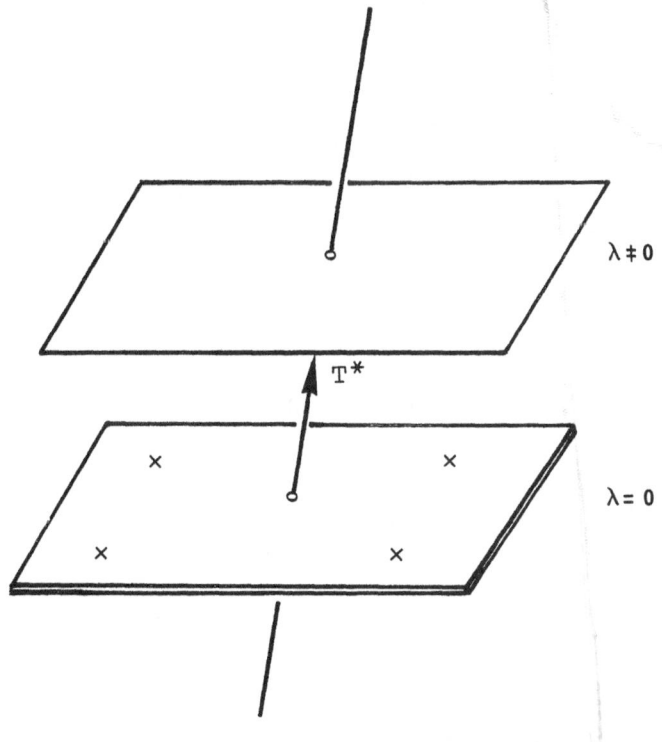

Among the two types of $\text{Coad}(\tilde{A}(\mathbb{R}))$-orbits occurring in $\mathcal{M}^*$ we shall ignore the "degenerate"

zero-dimensional orbits because in this case $\mathcal{M}$
is the unique polarization of $\mathfrak{m}$ and therefore the
Kirillov corresponding irreducible unitary linear
representations of $\tilde{A}(\mathbb{R})$ are one-dimensional and
trivial on the center $\tilde{Z}$. They take the form

$$U_0^{(\alpha,\beta)} : \left(\begin{bmatrix} x \\ y \end{bmatrix}, s \right) \longrightarrow e(\alpha x + \beta y) \, id_{\mathbb{C}} \qquad ((\alpha,\beta) \in \mathbb{R} \oplus \mathbb{R}).$$

The trivial one-dimensional identity representation $U_0^{(0,0)}$ corresponds to the origin of the catastrophic plane $\mathcal{Z}^{\perp}$. The differentiated version of
these representations, again denoted $U_0^{(\alpha,\beta)}$,
takes the form

$$U_0^{(\alpha,\beta)}(X) = 2\pi i \alpha, \quad U_0^{(\alpha,\beta)}(Y) = 2\pi i \beta, \quad U_0^{(\alpha,\beta)}(T) = 0.$$

In the "generic" case every polarization of $\mathfrak{m}$ is
of the form $L \oplus \mathbb{R}T$ where L denotes a Lagrangian
line relative to the symplectic form A in the plane W spanned by $\{X,Y\}$. It should be observed that
the notions of Lagrangian line in

$$W = \log \mathbb{R} \oplus \log \mathbb{R}$$

and polarization of $\mathfrak{m} = W \oplus \mathfrak{z}$ are in bijective
correspondence and that the polarizations corres-
ponding to a given element in a $\text{Coad}(\tilde{A}(\mathbb{R}))$-orbit
are nonunique and noncanonical.

Choose $L = \mathbb{R}Y$ and restrict the unitary character

$$\chi_\lambda : \tilde{A}(\mathbb{R}) \ni \left(\begin{bmatrix} x \\ y \end{bmatrix}, s \right) \longrightarrow e(\lambda s) \in \mathbb{T} \qquad (\lambda \in \mathbb{R}^\times)$$

to the maximal normal abelian subgroup

$$N = \exp(\mathbb{R}Y \oplus \mathbb{R}T)$$

of $\tilde{A}(\mathbb{R})$. Then the unitarily induced representation

$$U_\lambda = \underset{N\uparrow\tilde{A}(\mathbb{R})}{\text{Ind}} (\chi_\lambda | N) \qquad (\lambda \in \mathbb{R}^\times)$$

is an irreducible unitary linear representation of $\tilde{A}(\mathbb{R})$ the isomorphy class of which Kirillov corres-

ponds to the Coad$(\tilde{A}(\mathbb{R}))$-orbit $\lambda T^* + \mathfrak{z}^\perp$. Thus the isomorphy classes of the one-parameter family $(U_\lambda)_{\lambda \in \mathbb{R}^\times}$ form that part of the unitary dual of

$\tilde{A}(\mathbb{R})$ which consists of irreducible representations of dimension > 1. In other words, apart from the "degenerate" family of one-dimensional irreducible unitary linear representations of $\tilde{A}(\mathbb{R})$ parametrized by the pairs $(\alpha, \beta) \in \mathbb{R} \oplus \mathbb{R}$ there is the "generic" family $(U_\lambda)_{\lambda \in \mathbb{R}^\times}$ of irreducible unitary

linear representations of dimension > 1 parametrized by $\lambda \in \mathbb{R}^\times = \mathbb{R} - \{0\}$. Every irreducible unitary linear representation of $\tilde{A}(\mathbb{R})$ is unitarily isomorphic to one of the degenerate or the generic family.

To discover the complex Hilbert spaces of dimension > 1 on which the representations U_λ act

let us recall the geometric formulation of the inducing construction in the language of fiber bundles. Form the trivial complex line bundle

$$(\tilde{A}(\mathbb{R}) \times \mathbb{C}, \text{pr}_1, \tilde{A}(\mathbb{R}))$$

over $\tilde{A}(\mathbb{R})$ on which the subgroup N operates to the left by

$$\left(\begin{bmatrix} 0 \\ y_1 \end{bmatrix}, s_1\right) \left(\begin{bmatrix} x_2 \\ y_2 \end{bmatrix}, s_2, z\right) = \left(\begin{bmatrix} x_2 \\ y_1 + y_2 \end{bmatrix}, s_1 + s_2 + \tfrac{1}{2} y_1 x_2, \chi_\lambda(s_1) z\right)$$

and $\tilde{A}(\mathbb{R})$ acts to the right according to

$$(v_1,s_1,z)(v_2,s_2) = (v_1+v_2,s_1+s_2+\tfrac{1}{2}A(v_1,v_2),z).$$

The space E of the N-orbits in

$$(\tilde{A}(\mathbb{R}) \times \mathbb{C},\mathrm{pr}_1,\tilde{A}(\mathbb{R}))$$

is a complex line bundle over the subgroup

$$N\backslash\tilde{A}(\mathbb{R}) = \tilde{A}(\mathbb{R})/N = \{([\begin{smallmatrix}x\\0\end{smallmatrix}],0)\mid x \in \mathbb{R}\}$$

of $\tilde{A}(\mathbb{R})$ and U_λ acts on the complex Hilbert space

of cross-sections of E over $N\backslash\tilde{A}(\mathbb{R})$ which may be

identified with the space $L^2(\mathbb{R})$. Moreover, the
vector subspace of smooth cross-sections for U_λ

acting on $L^2(\mathbb{R})$ can be identified with the
Schwartz-Bruhat space $\mathscr{d}(\mathbb{R})$. Fix any function
$f \in \mathscr{d}(\mathbb{R})$. Since U_λ operates on f by right trans-

lations we get by the computations above in the
basic presentation (exponential coordinates) of
$\tilde{A}(\mathbb{R})$

$$U_\lambda(\begin{bmatrix}x\\y\end{bmatrix},s)f(t) = e(\lambda(s+ty+\tfrac{1}{2}xy))f(t+x) \qquad (\lambda \in \mathbb{R}^\times)$$

where $([\begin{smallmatrix}x\\y\end{smallmatrix}],s) \in \tilde{A}(\mathbb{R})$ and $t \in \mathbb{R}$ is identified with

the element $([\begin{smallmatrix}t\\0\end{smallmatrix}],0)$ of the subgroup $N\backslash\tilde{A}(\mathbb{R})$. At the

Lie algebra level these representations take the
form

$$U_\lambda(X) = \frac{d}{dt}, \quad U_\lambda(Y) = 2\pi i\lambda t, \quad U_\lambda(T) = 2\pi i\lambda.$$

Every 3-dimensional, two-step nilpotent, connected, simply connected, analytic group having
one-dimensional center is isomorphic to $\tilde{A}(\mathbb{R})$. For
the harmonic analysis on nilpotent Lie groups that
are not necessarily two-step nilpotent, the reader

is referred to the original paper by Kirillov [10].
Since $\tilde{A}(\mathbb{R})$ is the semi-direct product of N with
the closed subgroup K = exp($\mathbb{R}$X), the Mackey machine provides an alternative but less geometric
way to classify the isomorphy classes of the irreducible unitary linear representations of $\tilde{A}(\mathbb{R})$
which avoids Lie algebraic arguments, i.e., differential methods. For this approach which allows
to deduce the Stone-von Neumann theorem from the
imprimitivity theorem, see for instance the monograph of Warner [27].

3. THE RADAR CROSS-AMBIGUITY FUNCTION AND THE LINEAR SCHRÖDINGER MODEL

Let Planck's constant λ be standardized by setting
λ = 1. According to the Stone-von Neumann theorem
the representation

$$U:=U_1 = \underset{N\uparrow\tilde{A}(\mathbb{R})}{\text{Ind}} (\chi_1|N)$$

of $\tilde{A}(\mathbb{R})$ in the complex Hilbert space $L^2(\mathbb{R})$ is up
to unitary isomorphy the unique irreducible unita-
ry linear representation of $\tilde{A}(\mathbb{R})$ which subduces
the unitary central character $\chi_1|\tilde{Z}$ = e. The reali-
zation

$$U(\begin{bmatrix}x\\y\end{bmatrix},s)f(t) = e(s+ty+\tfrac{1}{2}xy)f(t+x) \qquad (f\in\mathscr{d}(\mathbb{R}))$$

is called the linear Schrödinger representation of
the real Heisenberg nilpotent group $\tilde{A}(\mathbb{R})$ in its
basic presentation. Let the Schwartz-Bruhat space

$\mathscr{d}(\mathbb{R})$ of smooth vectors for U acting on $L^2(\mathbb{R})$
be equipped with the scalar product inherited by

the complex Hilbert space $L^2(\mathbb{R})$. For any pair
$(f,g) \in \mathscr{d}(\mathbb{R}) \times \mathscr{d}(\mathbb{R})$ let f $\otimes$ $\bar{g}$ denote the dyadic
tensor product. It is well known that the sesqui-
linear mapping $(f,g)\longrightarrow f \otimes \bar{g}$ induces an isomorphism
of $\mathscr{d}(\mathbb{R}) \otimes \mathscr{d}(\mathbb{R})$ onto the complex vector space of
linear operators $\mathscr{d}(\mathbb{R}) \to \mathscr{d}(\mathbb{R})$ of finite rank
which extends to a Hilbert space isomorphism of

$L^2(\mathbb{R}) \underset{2}{\hat{\otimes}} L^2(\mathbb{R})$ to the space of Hilbert-Schmidt operators on $L^2(\mathbb{R})$. From an elementary computation we infer the following theorem which is of central importance in our applications.

Theorem 1.   The radar cross-ambiguity function with respect to the complex envelopes f and g in $\mathscr{S}(\mathbb{R})$ satisfies the identity

$$H(f,g;x,y) = tr_U(f \otimes \overline{g})\left(\begin{bmatrix} x \\ y \end{bmatrix}, 0\right)$$

for all $\begin{bmatrix} x \\ y \end{bmatrix} \in \mathbb{R} \oplus \mathbb{R}$ where tr_U denotes the trace functional relative to the linear Schrödinger representation U.

Using the terminology introduced by Howe [9], the group-representation theoretic characterization of Theorem 1 can be rephrased as follows: The radar cross-ambiguity function $H(f,g;.,.)$ is the restriction of the coefficient function $tr_U(f \otimes \overline{g})$ to the polarized cross-section

$$\left\{ \left(\begin{bmatrix} x \\ y \end{bmatrix}, 0\right) \Big| x \in \mathbb{R}, \ y \in \mathbb{R} \right\}$$

to the center $\tilde{Z}$ in $\tilde{A}(\mathbb{R})$. This cross-section gives rise to the decomposition

$$\{0\} \rightarrow \tilde{Z} \rightarrow \tilde{A}(\mathbb{R}) \rightarrow \mathbb{R} \oplus \mathbb{R} \rightarrow \{0\}.$$

Since there are other cross-sections to $\tilde{Z}$ in $\tilde{A}(\mathbb{R})$ yielding different coordinates and, moreover, many rather different looking ways of realizing the representation U of $\tilde{A}(\mathbb{R})$, Theorem 1 lies at the basis of various different expressions for $H(f,g;.,.)$ which may be useful in other situations. See Theorem 4 and Theorem 8 infra.

The preceding result connects the theory of radar waveform design with harmonic analysis on the Heisenberg nilpotent group. It provides a group-representation theoretic description of the signal cross-correlation function $H(f,g;.,.)$ simultaneously in both time shifts x and frequency

shifts y (cf. Gabor [8]), which is particularly
suited for the combined time-frequency analysis of
non-stationary signals. Moreover, it offers an in-
vestigation of the radar cross-ambiguity function
on the Lie group as well as on the Lie algebra level. The present paper is devoted to an elaborati-
on of this point. In particular, Theorem 1 will
enable us to determine the group of signal energy
preserving invariants of the radar ambiguity surface (Theorem 2 infra). Before giving the details in
the next section, let us formulate two direct consequences which represent important geometric properties of the ambiguity surface (cf. [19,20]).

Corollary 1.   For $f \in \mathscr{S}(\mathbb{R})$ the tensor product
function

$$\left(\begin{bmatrix} x \\ y \end{bmatrix}, s \right) \longrightarrow H(f;x,y)\, e(s)$$

is of positive type on the real Heisenberg nilpo-
tent group $\tilde{A}(\mathbb{R})$. In particular, the radar auto-
ambiguity function $H(f;.,.)$ with respect to f is
of positive type on the symplectic time-frequency
plane $(\mathbb{R} \oplus \mathbb{R};A)$ and satisfies the condition

$$\sup_{(x,y)\, \in\, \mathbb{R}\, \times\, \mathbb{R}} |H(f;x,y)| = H(f;0,0)$$

("peak condition").

        Let us fix a (bi-invariant) Haar measure on
$\tilde{A}(\mathbb{R})$ which in our coordinate systems is just Le-

besgue measure on the underlying manifold $\mathbb{R}^3$. A
typical coadjoint orbit in the Kirillov orbit pic-
ture for the unitary dual of a connected, simply
connected nilpotent Lie group is an affine algebraic variety homeomorphic either to a single
point or to an Euclidean space of even dimension.
The geometric fact mentioned above that the unique
$Coad(\tilde{A}(\mathbb{R}))$-orbit corresponding to the Schrödinger

representation U is a "flat" variety in $\mathscr{N}^*$, to wit the affine plane $T^{*} + \mathfrak{z}^{\perp}$, implies by a result of Brezin [5] and Moore-Wolf [11] that the unitary linear representation U of $\tilde{A}(\mathbb{R})$ is square inte-grable modulo the center $\tilde{Z}$. Of course, various less geometric arguments based for instance on the in-tegrated version of U also are possible to estab-lish the important fact that U is a member of the discrete series of $\tilde{A}(\mathbb{R})$. Since U has formal de-gree 1, an application of the Frobenius-Schur or-thogonality relations yields

<u>Corollary 2.</u>   For complex envelopes $f, g \in \mathscr{S}(\mathbb{R})$ the radar cross-ambiguity function $H(f, g; ., .)$ be-longs to the Schwartz-Bruhat space $\mathscr{S}(\mathbb{R} \times \mathbb{R})$.

More general, if f, g are elements of $L^{2}(\mathbb{R})$ then

we have $H(f, g; ., .) \in \mathscr{C}(\mathbb{R} \times \mathbb{R}) \cap L^{2}(\mathbb{R} \times \mathbb{R})$. If

f', g' also belong to the space $L^{2}(\mathbb{R})$, the identity

$$\int_{\mathbb{R} \times \mathbb{R}} \int H(f, g; x, y) \overline{H}(f', g'; x, y) \, dx \, dy = \langle f | f' \rangle . \langle g' | g \rangle$$

holds ("Moyal's identity"). In particular, we have the identity

$$\int_{\mathbb{R} \times \mathbb{R}} \int |H(f, g; x, y)|^{2} \, dx \, dy = ||f||^{2} . ||g||^{2}$$

("geometric form" of the radar uncertainty princip-le).

The preceding identity shows that the total ambiguity for signal waveforms is a constant de-termined by the total input energies of the radar signals involved and cannot be diminished by vary-ing their envelope functions.

<u>Remark.</u>   Since the linear Schrödinger representa-tion U of $\tilde{A}(\mathbb{R})$ is irreducible it follows from Co-rollary 1 that the sum $H(f_{1}; ., .) + H(f_{2}; ., .)$ of

any two radar auto-ambiguity functions with respect to signal envelope functions $f_k \neq 0$ out of

$\mathscr{A}(\mathbb{R})$ is never a radar auto-ambiguity function.
In particular, the mapping given by

$$\mathscr{A}(\mathbb{R}) \ni f \rightarrow H(f;.,.) \in \mathscr{A}(\mathbb{R} \times \mathbb{R})$$

is non-linear (cf. [19,20]).

Let us conclude this section by noticing that
our proof of the radar uncertainty principle works
in terms of the Lie group $\tilde{A}(\mathbb{R})$ whereas the Heisenberg uncertainty principle of quantum mechanics is
usually formulated in terms of local operators at
the Lie algebra level.

4. THE RADAR AMBIGUITY SURFACE

Consider the group of unitary intertwining operators

$$G = \{S \in U(L^2(\mathbb{R})) | S \circ U(v,s) = U(v',s) \circ S, v,v' \in \mathbb{R} \oplus \mathbb{R}, s \in \mathbb{R}\}.$$

Let the unitary group $U(L^2(\mathbb{R}))$ of $L^2(\mathbb{R})$ be
equipped with the strong operator topology and G
with the induced relative topology. Then G has the
structure of a Lie group and we have $v' = h(S)v$
where $h(S): \mathbb{R} \oplus \mathbb{R} \rightarrow \mathbb{R} \oplus \mathbb{R}$ is a homeomorphism
(cf. Borel-Wallach [3]). The mapping

$$h: G \rightarrow Sp(1,\mathbb{R})$$

is a surjective Lie group homomorphism and the irreducibility of U implies via Schur's lemma that

$$\{1\} \rightarrow \mathbb{1}id_{L^2(\mathbb{R})} \rightarrow G \xrightarrow{h} Sp(1,\mathbb{R}) \rightarrow \{1\}$$

is a short exact sequence.

On the other hand it is well known that the
fundamental group of $Sp(1,\mathbb{R}) = SL(2,\mathbb{R})$ is infinite cyclic so that there exists a unique connected
Lie group $Mp(1,\mathbb{R})$ which double covers $Sp(1,\mathbb{R})$.
The real Lie group $Mp(1,\mathbb{R})$ is called the metaplec-

tic group. Let $\varepsilon : \text{Mp}(1,\mathbb{R}) \ni \tilde{\sigma} \longrightarrow \sigma \in \text{Sp}(1,\mathbb{R})$ denote the covering homomorphism, i.e.,

$$\{0\} \to \mathbb{Z}_2 \to \text{Mp}(1,\mathbb{R}) \overset{\varepsilon}{\to} \text{Sp}(1,\mathbb{R}) \to \{1\}$$

is a short exact sequence. The metaplectic group $\text{Mp}(1,\mathbb{R})$ has a well-known infinite dimensional unitary linear representation $\tilde{\sigma} \longrightarrow T_{\tilde{\sigma}}$, the so-called

unitary oscillator representation which has certain formal resemblances to the spin representation of the double covering of $\text{SO}(1,\mathbb{R}) = \mathbb{1}$. Since $\text{Mp}(1,\mathbb{R})$ is a subgroup of G, to wit the commutator group of G, and the Lie group homomorphism h extends the covering projector ε, and the oscillator representation of $\text{Mp}(1,\mathbb{R})$ coincides with the realization

$\text{Mp}(1,\mathbb{R}) \to \underline{U}(L^2(\mathbb{R}))$, the Stone-von Neumann theorem applied to $\tilde{A}(\mathbb{R})$ takes the form of I. Segal's famous metaplectic formula

$$T_{\tilde{\sigma}} \circ U(\begin{bmatrix} x \\ y \end{bmatrix}, s) \circ T_{\tilde{\sigma}}^{-1} = U(\sigma\begin{bmatrix} x \\ y \end{bmatrix}, s) \qquad (\sigma \in \text{Sp}(1,\mathbb{R}))$$

for all elements $(\begin{bmatrix} x \\ y \end{bmatrix}, s) \in \tilde{A}(\mathbb{R})$. Notice that the

intertwining operators $T_{\tilde{\sigma}} \in \text{Mp}(1,\mathbb{R})$ can be compu-

ted explicitly in terms of polarizations of $\mathcal{W}$ (or equivalently in terms of Lagrangian lines in W = $\log \mathbb{R} \oplus \log \mathbb{R}$) and appropriate partial Fourier transforms (cf. [16]). In view of Theorem 1 supra we therefore have the following theorem which represents our main result.

Theorem 2.   Let f,g denote two complex envelope functions in the Schwartz-Bruhat space $\mathcal{S}(\mathbb{R})$ and let $\sigma \in \text{Sp}(1,\mathbb{R})$ be given. Then the identity

$$H(f,g;x,y) = H(T_{\tilde{\sigma}}f, T_{\tilde{\sigma}}g; \sigma(x,y))$$

holds for all $(x,y) \in \mathbb{R} \times \mathbb{R}$. In particular, the radar auto-ambiguity function $H(f;.,.)$ with respect to $f \neq 0$ is invariant under the cyclic subgroup of $\text{Sp}(1,\mathbb{R})$ generated by $\{\sigma\}$ if and only if there exists a number $\zeta_{\tilde{\sigma}} \in \mathbb{1}$ such that

$$T_{\tilde{\sigma}}f = \zeta_{\tilde{\sigma}}f$$

holds, i.e., if and only if $f \in \mathscr{S}(\mathbb{R})$ is an eigen-vector of the unitary operator $T_{\tilde{\sigma}}: L^2(\mathbb{R}) \to L^2(\mathbb{R})$.

If $f \in \mathscr{S}(\mathbb{R})$, $f' \in \mathscr{S}(\mathbb{R})$, and $\sigma \in Sp(1,\mathbb{R})$ are given such that

$$H(f;x,y) = H(f';\sigma(x,y))$$

holds for all $(x,y) \in \mathbb{R} \times \mathbb{R}$, there exists a number $\eta_{\tilde{\sigma}} \in \mathbb{1}$ such that

$$f' = \eta_{\tilde{\sigma}}T_{\tilde{\sigma}}f$$

holds.

Conversely, let S denote a unitary automor-phism of $L^2(\mathbb{R})$ which maps $\mathscr{S}(\mathbb{R})$ onto itself and satisfies for all $f \in \mathscr{S}(\mathbb{R})$ the following condition:

For any pair $(x,y) \in \mathbb{R} \times \mathbb{R}$ there exists a pair $(x',y') \in \mathbb{R} \times \mathbb{R}$ such that the identity

$$H(f;x,y) = H(Sf;x',y')$$

holds, i.e., the mapping $H(f;.,.)$ and $H(Sf;.,.)$ on $\mathbb{R} \times \mathbb{R}$ have the same range and therefore generate the same radar auto-ambiguity surface over the time-frequency plane $\mathbb{R} \times \mathbb{R}$.

Then there exists a unique mapping

$$\sigma \in Sp(1,\mathbb{R})$$

and a complex number $\eta_{\tilde{\sigma}} \in \mathbb{1}$ with the property

$$\sigma \begin{bmatrix} x \\ y \end{bmatrix} = \begin{bmatrix} x' \\ y' \end{bmatrix}, \qquad S = \eta_{\tilde{\sigma}}T_{\tilde{\sigma}}.$$

The preceding result shows that the symplec-tic group $Sp(1,\mathbb{R})$ forms the signal energy preserving invariants of the radar ambiguity surface. It sharpens and corrects the investigations by Reis [13].

5. $\mathbb{Z}_4$-SYMMETRIC RADAR AUTO-AMBIGUITY FUNCTIONS

The unitary automorphisms $T_{\tilde{\sigma}}$ of $L^2(\mathbb{R})$ can be computed explicitly in terms of polarizations of $\mathcal{N}$. However, in the present section we will restrict our consideration to signal envelope functions generating radar auto-ambiguity surfaces that are invariant under the cyclic group $\mathbb{Z}_4$ of powers of a quarterturn about the origin of the time-frequency plane $\mathbb{R} \oplus \mathbb{R}$.

Let $(W_m)_{m \geq 0}$ denote the sequence of standardized Hermite-Weber functions (harmonic oscillator wave functions) and let the orthogonal direct sum

$$L^2(\mathbb{R}) = \bigoplus_{0 \leq k \leq 3} \mathcal{H}_k$$

be the Wiener decomposition of the complex Hilbert space $L^2(\mathbb{R})$. Thus the Hilbert sums

$$\mathcal{H}_k = \widehat{\bigoplus_{m \in \mathbb{N}}} \mathbb{C}W_{4m+k}, \quad k \in \{0,1,2,3\}$$

denote the four eigenvector spaces of the Fourier transform $\mathcal{F}_{\mathbb{R}} : L^2(\mathbb{R}) \to L^2(\mathbb{R})$ associated with its eigenvalues $(-i)^k$ (of infinite multiplicities).

Theorem 3.   Let $f \neq 0$ be a given signal envelope function in $\mathcal{S}(\mathbb{R})$. The condition

$$H(f;x,y) = H(f;-y,x)$$

holds for all pairs (x,y) of the time-frequency plane $\mathbb{R} \oplus \mathbb{R}$ if and only if

$$f \in \mathcal{H}_k$$

for a (unique) number $k \in \{0,1,2,3\}$.

<u>Proof.</u>   The Fourier transform $\mathcal{F}_{\mathbb{R}} \in U(L^2(\mathbb{R}))$ belongs to G and maps $\mathscr{S}(\mathbb{R})$ onto itself. Its image under the Lie group epimorphism h: G → Sp(1,$\mathbb{R}$) is J. Thus $f \in \mathscr{S}(\mathbb{R})$ has to be an eigenfunction of $\mathcal{F}_{\mathbb{R}}$ by Theorem 2 supra. —

Obviously the Hermitean central symmetry

$$H(f;x,y) = \overline{H}(f; -x,-y)$$

holds for any $f \in \mathscr{S}(\mathbb{R})$ and all pairs $(x,y) \in \mathbb{R} \times \mathbb{R}$. In particular the radar auto-ambiguity functions invariant under the cyclic group $\mathbf{Z}_4$ of quarterturns

about the origin of $\mathbb{R} \oplus \mathbb{R}$ are real valued.

The following plots serve as an illustration
of the generated radar auto-ambiguity surfaces
over the time-frequency plane in the present case.
They show the same surface from different view-
points.

6. RADIALLY SYMMETRIC RADAR AUTO-AMBIGUITY FUNCTIONS

Let $\mathfrak{sp}(1,\mathbb{R})$ denote the Lie algebra of $Sp(1,\mathbb{R})$ given by the traceless matrices

$$\left\{ \begin{pmatrix} a & b \\ c & d \end{pmatrix} \middle| a + d = 0, \ a,b,c,d \in \mathbb{R} \right\}$$

and identify the Lie algebra of $Mp(1,\mathbb{R})$ with $\mathfrak{sp}(1,\mathbb{R}) = \mathfrak{sl}(2,\mathbb{R})$. Moreover, let Exp denote the exponential mapping of $\mathfrak{sp}(1,\mathbb{R})$ into $Mp(1,\mathbb{R})$ and consider J as an element of $\mathfrak{sp}(1,\mathbb{R})$. Then it is easy to check that

$$\tilde{\sigma} = \mathrm{Exp}(\varphi J)$$

has as its projection $h(\tilde{\sigma})$ the rotation $\sigma \in SO(2,\mathbb{R})$ given by the matrix

$$\begin{pmatrix} \cos\varphi & -\sin\varphi \\ \sin\varphi & \cos\varphi \end{pmatrix} (\varphi \in \mathbb{R})$$

and that $\mathrm{Exp}(\mathbb{R}\,J)$ is a Cartan subgroup of $Mp(1,\mathbb{R})$ and $h(\mathrm{Exp}(\mathbb{R}\,J))$ is a Cartan subgroup of $Sp(1,\mathbb{R})$. In order to determine the $SO(2,\mathbb{R})$-invariant radar auto-ambiguity surfaces over the time-frequency plane $\mathbb{R} \oplus \mathbb{R}$ we will use the coordinates $\begin{bmatrix} p \\ q \end{bmatrix}$

relative to the basis $\{X,Y\}$ introduced in Section 2 supra to coordinatize the vector subspace $W = \log \mathbb{R} \oplus \log \mathbb{R}$ of $\mathfrak{u} = W \oplus \mathfrak{z}$. Notice by the way that $\exp W$ forms the isotropic cross-section

$$\left\{ \left(\begin{bmatrix} x \\ y \end{bmatrix}, \tfrac{1}{2}xy \right) \middle| \ x \in \mathbb{R}, \ y \in \mathbb{R} \right\}$$

to the center $\tilde{Z}$ in the real Heisenberg nilpotent group $\tilde{A}(\mathbb{R})$ in its dual pairing presentation (canonical coordinates) (cf. Howe [9]) which gives rise to the decomposition

$$\{0\} \to \tilde{Z} \to \tilde{A}(\mathbb{R}) \to \exp W \to \{0\}.$$

The isotropic cross-section links the dual pairing presentation to the basic presentation of $\tilde{A}(\mathbb{R})$ and lies by a complexification of W at the basis of the Bargmann-Fock-Segal model (or complex wave model). The linear Schrödinger representation U takes with respect to the dual pairing presentation of $\tilde{A}(\mathbb{R})$ the form

$$U\left(\begin{bmatrix} x \\ y \end{bmatrix}, s\right) f(t) = e(s+ty)f(t+x) \quad (f \in \mathcal{S}(\mathbb{R})).$$

Let U also denote the differentiated version of the linear Schrödinger representation U of $\tilde{A}(\mathbb{R})$ acting on $L^2(\mathbb{R})$. Then we find by Theorem 1 supra and the relation between the polarized and isotropic cross-section lifted to Schwartz-Bruhat functions:

Theorem 4.   The radar cross-ambiguity function with respect to the complex signal envelopes f and g in $\mathcal{S}(\mathbb{R})$ satisfies the identity

$$H(f,g;p,q) = \text{tr}_{\exp U}(\overline{g} \otimes f) \begin{bmatrix} -p \\ -q \end{bmatrix}$$

for all $\begin{bmatrix} p \\ q \end{bmatrix} \in \mathbb{R} \oplus \mathbb{R}$.

Switching back to the linear Schrödinger representation U at the Lie algebra level we observe that the formulae

$$U(X) = \frac{d}{dt} , \quad U(Y) = 2\pi i t$$

hold which give rise to the well-known momentum operator and position operator of elementary quantum mechanics, respectively. Consequently we have

$$U(pX+qY) = p\frac{d}{dt} + 2\pi i q t$$

for all $(p,q) \in \mathbb{R} \times \mathbb{R}$. Put

$$T^- = -\frac{1}{2}(X-iY)$$

and

$$T^+ = -\frac{1}{2}(X+iY).$$

Then we get the linear differential operators

$$a^-: = U(T^-) = -\frac{1}{2}(\frac{d}{dt} + 2\pi t)$$

and

$$a^+: = U(T^+) = -\frac{1}{2}(\frac{d}{dt} - 2\pi t)$$

as the boson annihilation and creation operators, respectively (cf. Cartier [6]). It follows

$$U(pX + qY) = (-p+iq)a^+ - (p+iq)a^-$$

for all $(p,q) \in \mathbb{R} \times \mathbb{R}$. Moreover, it is easy to check by taking the derivative of

$$T_{Exp(-\varphi J)} U(\tau\begin{bmatrix}x\\y\end{bmatrix},0) T_{Exp(\varphi J)} = U(\tau\begin{bmatrix}x\cos\varphi - y\sin\varphi\\x\sin\varphi + y\cos\varphi\end{bmatrix},0)$$

with respect to τ at $\tau = 0$ that the following identities

$$T_{Exp(-2\pi\varphi J)} a^+ T_{Exp(2\pi\varphi J)} = e^{2\pi i\varphi}a^+$$

and

$$T_{Exp(-2\pi\varphi J)} a^- T_{Exp(2\pi\varphi J)} = e^{-2\pi i\varphi}a^-$$

hold on the Schwartz-Bruhat space $\mathscr{S}(\mathbb{R})$ of smooth vectors for U for all $\varphi \in \mathbb{R}$. If we set

$$W_0: t \longrightarrow \sqrt[4]{2}\ e^{-\pi t^2}$$

then the Gaussian $W_0 \in \mathscr{S}(\mathbb{R})$ is a unit vector in the complex Hilbert space $L^2(\mathbb{R})$ and we have

$$T_{Exp(-2\pi\varphi J)} W_0 = e^{\pi i\varphi}W_0 \qquad (\varphi \in \mathbb{R})$$

Obviously

$$a^-(W_0) = 0.$$

The irreducibility of the linear Schrödinger representation U of $\tilde{A}(\mathbb{R})$ implies the well-known fact that the family $(W_m)_{m \geq 0}$ of Hermite-Weber functions (harmonic oscillator wave functions) given by the formulae

$$W_m = \frac{1}{\sqrt{\pi^m m!}} \, (a^+)^m (W_0) \qquad (m \in \mathbb{N})$$

forms a Hilbert basis of $L^2(\mathbb{R})$ the elements of which are simultaneously eigenfunctions in $\mathscr{S}(\mathbb{R})$ of the subgroup $T_{\text{Exp}(\mathbb{R}J)}$ of $U(L^2(\mathbb{R}))$. Conversely, any Hilbert basis of $L^2(\mathbb{R})$ the elements of which belong to $\mathscr{S}(\mathbb{R})$ and are simultaneously eigenfunctions of the group $T_{\text{Exp}(\mathbb{R}J)}$ form a total family of eigenfunctions of the Hermite linear differential operator of second kind

$$2(a^- a^+ + a^+ a^-) = 2i \left(\frac{d}{dr}\right)_{r=0} T_{\text{Exp}(rJ)}$$

with domain $\mathscr{S}(\mathbb{R})$. This operator is essentially self-adjoint in $L^2(\mathbb{R})$, its closure admits a pure point spectrum given by $\{-2\pi(2m+1) \mid m \in \mathbb{N}\}$ and all of its eigenvalues are simple. Consequently we established the following result:

<u>Theorem 5.</u>   Let the signal envelope function $f \in \mathscr{S}(\mathbb{R})$ have L^2-norm $||f|| = 1$. The radar auto-ambiguity function $H(f;.,.)$ is radially symmetric, i.e., $SO(2,\mathbb{R})$-invariant on the time-frequency plane $\mathbb{R} \oplus \mathbb{R}$ if and only if

$$f = \zeta W_m$$

holds for certain numbers $m \in \mathbb{N}$ and $\zeta \in \mathbf{T}$. Thus f is apart from a phase factor a harmonic oscillator wave function W_m.

Now let us specialize Theorem 4 supra to the Hermite-Weber functions as signal envelope functions. We find in the coordinates of the vector space W that

$$H(W_m; p, q) = \operatorname{tr}_{\exp \; U}(W_m \otimes W_m) \begin{bmatrix} -p \\ -q \end{bmatrix}$$

$$= \langle W_m | e^{U(pX+qY)}(W_m) \rangle \qquad (m \in \mathbb{N})$$

$$= \frac{1}{\sqrt{\pi^m m!}} \langle W_m | e^{U(pX+qY)}(a^+)^m (W_0) \rangle$$

for all $\begin{bmatrix} p \\ q \end{bmatrix} \in \mathbb{R} \oplus \mathbb{R}$. Let $\tilde{\delta}$ denote the right regular representation of $\tilde{A}(\mathbb{R})$ on $L^2(\tilde{A}(\mathbb{R})/\tilde{Z})$ transferred to $L^2(W)$ in the natural way by the extension of the isomorphism $\mathscr{A}(\tilde{A}(\mathbb{R})/\tilde{Z}) \to \mathscr{A}(W)$ associated with the isotropic cross-section exp W to the center $\tilde{Z}$ in $\tilde{A}(\mathbb{R})$. Then we obtain for $H(W_m; \cdot, \cdot)$,

considered as a function on W, the expression

$$H(W_m; p, q) = \frac{1}{\sqrt{\pi^m m!}} \tilde{\delta}(z^+)^m \langle W_m | e^{U(pX+qY)}(W_0) \rangle.$$

In view of the identity

$$U(pX+qY)^m (W_0) =$$

$$\sum_{0 \leq l \leq [\frac{1}{2}m]} \frac{m!}{2^l l! (m-2l)!} (\pi(p+iq))^l (-p+iq)^{m-l}(a^+)^{m-2l}(W_0)$$

which can be established via a recurrence relation by induction on $m \geq 0$ (cf. Howe [9]) we obtain

$$e^{U(pX+qY)}(W_O) = \sum_{m\geq 0} \frac{1}{m!} U(pX+qY)^m(W_O)$$

$$= \sum_{k\geq 0} (\sum_{j\geq 0} \frac{1}{2^j j!} (-\pi(p^2+q^2))^j) \frac{1}{k!}(-p+iq)^k(a^+)^k(W_O)$$

$$= e^{-(\pi/2)(p^2+q^2)} e^{(-p+iq)a^+}(W_O)$$

for all $\begin{bmatrix} p \\ q \end{bmatrix} \in \mathbb{R} \oplus \mathbb{R}$. Consequently

$$<W_m|e^{U(pX+qY)}(W_O)> = \sqrt{\frac{\pi^m}{m!}}(-p+iq)^m e^{-(\pi/2)(p^2+q^2)}$$

and therefore

$$H(W_m;p,q) = \frac{1}{m!}\tilde{\delta}(z^+)^m(-p+iq)^m e^{-(\pi/2)(p^2+q^2)} \qquad (m \in \mathbb{N}).$$

Observe that

$$\tilde{\delta}(X) = \frac{\partial}{\partial p} - \pi iq, \quad \tilde{\delta}(Y) = \frac{\partial}{\partial q} + \pi ip$$

and therefore

$$\tilde{\delta}(z^+) = -\frac{1}{2}(\frac{\partial}{\partial p} + i\frac{\partial}{\partial q}) + \frac{\pi}{2}(p+iq)$$

$$= -\frac{1}{2}e^{(\pi/2)(p^2+q^2)}(\frac{\partial}{\partial p} + i\frac{\partial}{\partial q})e^{-(\pi/2)(p^2+q^2)}.$$

Then we get our final result (cf. Wilcox [29])

$$H(W_m;p,q) = e^{-(\pi/2)(p^2+q^2)} \times$$

$$\sum_{0\leq j\leq m} \binom{m}{j} \frac{(-\pi(p^2+q^2))^j}{j!}$$

$$= L_m(\pi(p^2+q^2)) \qquad (m \in \mathbb{N})$$

where $L_m = L_m^{(0)}$ ($m \in \mathbb{N}$) denotes the mth Laguerre-Weber function (of order 0) given by

$$L_m(x) = e^{-\frac{1}{2}x} \sum_{0 \leq j \leq m} \binom{m}{j} \frac{(-x)^j}{j!} \qquad (x \in \mathbb{R}).$$

Thus we have established the following

Theorem 6.   The radar autoambiguity function with respect to the harmonic oscillator wave functions $(W_m)_{m \geq 0}$ as signal envelope admits the form

$$H(W_m; p, q) = L_m(\pi(p^2 + q^2)) \qquad (m \in \mathbb{N})$$

for all $(p, q) \in \mathbb{R} \times \mathbb{R}$.

For some related computations see Peetre [12]. This paper however does not indicate any application to the theory of radar detection nor to the Wigner quasi-probability distribution function of quantum mechanics whereas Wilcox's paper cited above ignores completely the group-theoretical aspects of the radar cross-ambiguity function and applies the generating function of the Hermite polynomials.

The figures displayed below show the radar ambiguity surfaces associated with $H(W_m; ., .)$ in

the cases of degree m= 0, m= 6, and m=10, respectively.

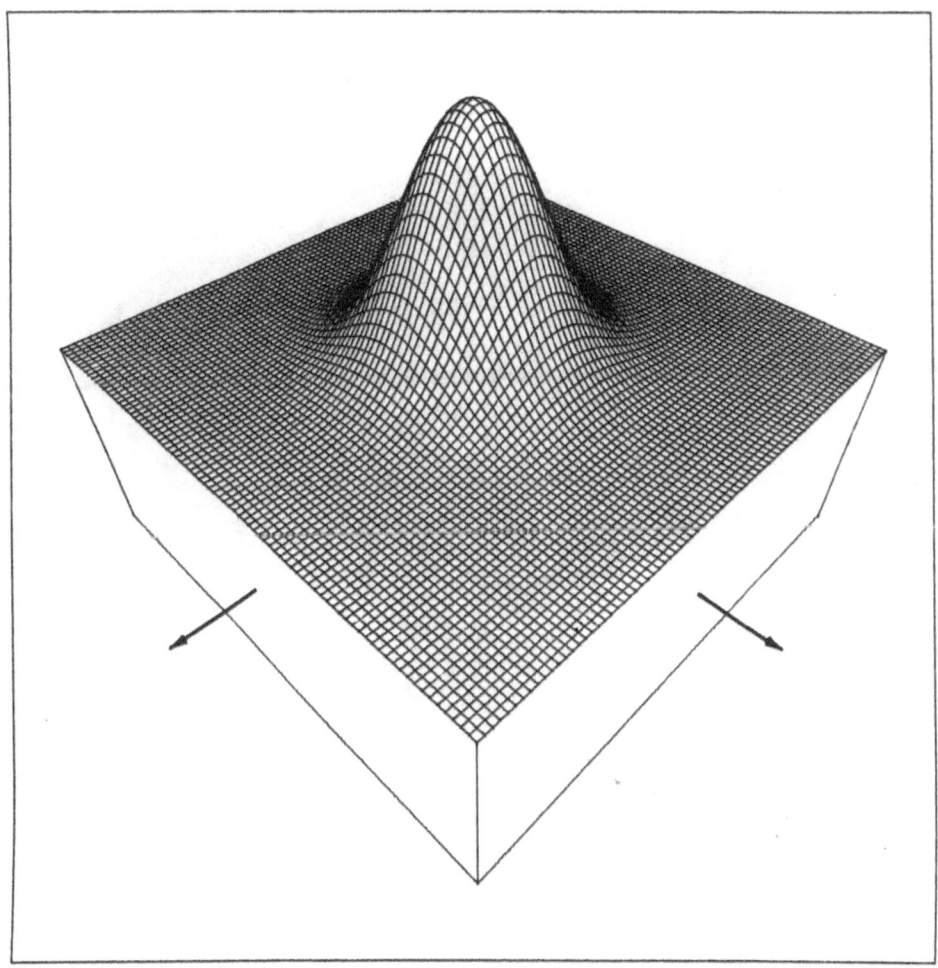

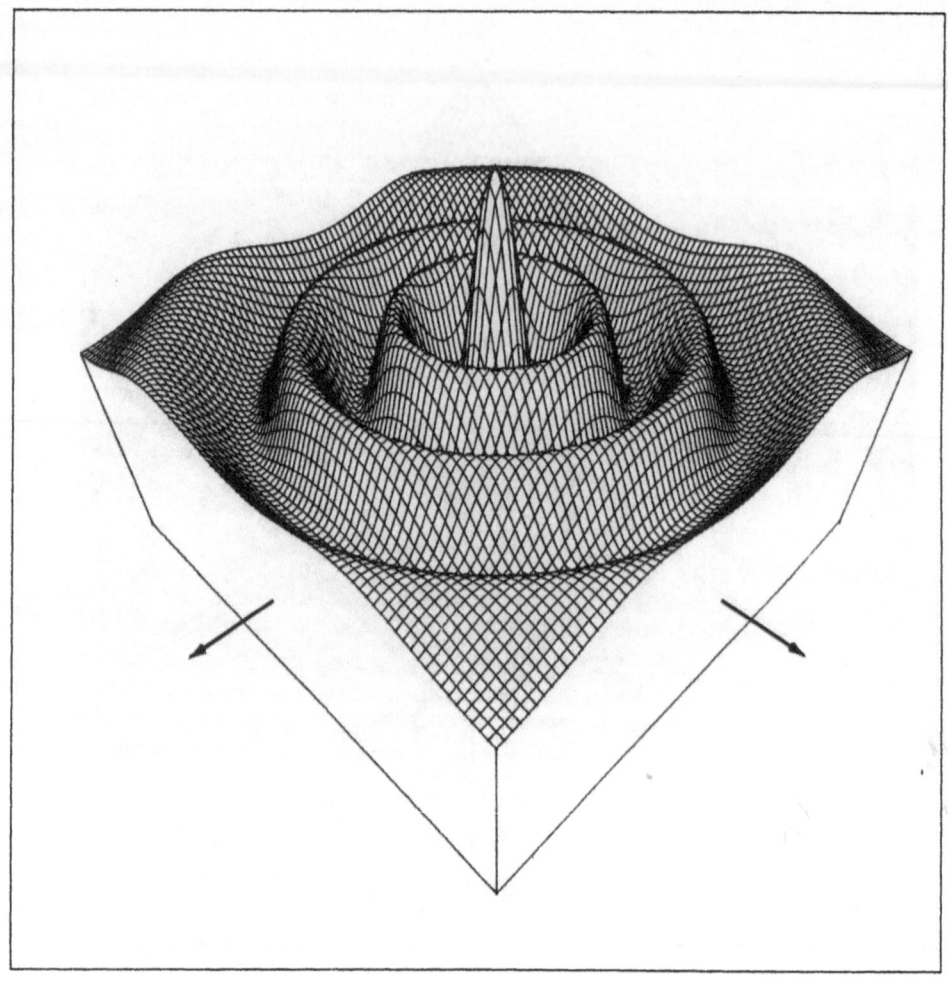

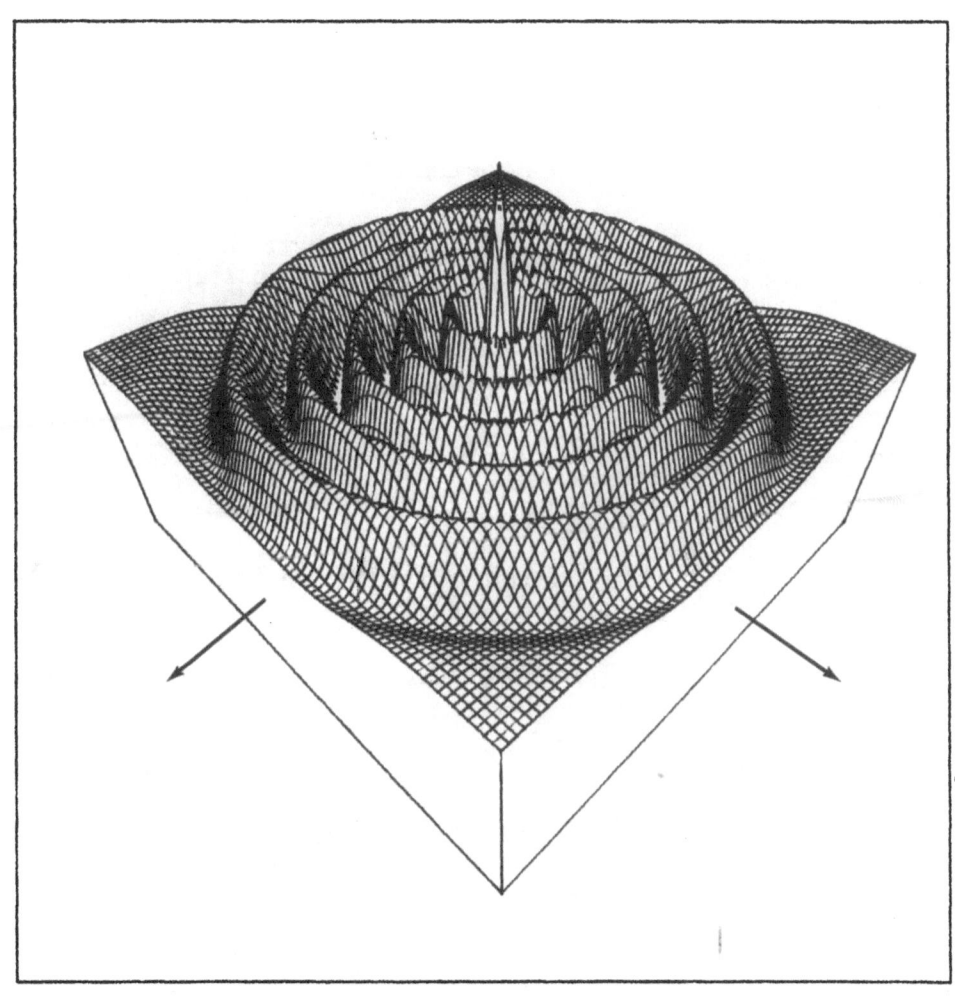

In the next section we shall need an explicit expression of the radar cross-ambiguity functions $H(W_m, W_n; ., .)$. Denoting by

$$L_n^{(\alpha)}(x) = e^{-\frac{1}{2}x} \sum_{0 \le j \le n} \binom{n+\alpha}{n-j} \frac{(-x)^j}{j!} \qquad (x \in \mathbb{R})$$

the nth Laguerre-Weber function of order $\alpha > -1$ so that $L_n^{(0)} = L_n$ ($n \in \mathbb{N}$) we obtain

Theorem 7. The radar cross-ambiguity function with respect to the harmonic oscillator wave functions $(W_m)_{m \ge 0}$ as signal envelopes admits the form

$$H(W_m, W_n; p, q) = \sqrt{\frac{n!}{m!}} \; (\sqrt{\pi}(p+iq))^{m-n} L_n^{(m-n)}(\pi(p^2+q^2)) \qquad (m \ge n),$$

$$H(W_m, W_n; p, q) = \sqrt{\frac{m!}{n!}} \; (\sqrt{\pi}(-p+iq))^{n-m} L_m^{(n-m)}(\pi(p^2+q^2)) \qquad (n \ge m),$$

for all $(p,q) \in \mathbb{R} \times \mathbb{R}$.

Proof. Similarly to the computations indicated above we get by Theorem 4 supra

$$H(W_m, W_n; p, q) = \mathrm{tr}_{\exp} U(W_n \otimes W_m) \begin{bmatrix} -p \\ -q \end{bmatrix}$$

$$= \langle W_n | e^{U(pX+qY)} (W_m) \rangle$$

$$= \frac{1}{\sqrt{\pi^m m!}} \langle W_n | e^{U(pX+qY)} (a^+)^m (W_0) \rangle \qquad (m,n \in \mathbb{N})$$

$$= \frac{1}{\sqrt{\pi^m m!}} \tilde{\delta}(Z^+)^m \langle W_n | e^{U(pX+qY)} (W_0) \rangle$$

$$= \frac{(-1)^m}{2^m \sqrt{\pi^{m-n} m! n!}} e^{(\pi/2)(p^2+q^2)} \times$$

$$(\frac{\partial}{\partial p} + i\frac{\partial}{\partial q})^m (-p+iq)^n e^{-\pi(p^2+q^2)}$$

for all $\begin{bmatrix} p \\ q \end{bmatrix} \in \mathbb{R} \oplus \mathbb{R}$. Supposing $m \geq n$ we see that

$$H(W_m, W_n; p, q) = \sqrt{\frac{n!}{\pi^{m-n} m!}} e^{-(\pi/2)(p^2+q^2)} \times$$

$$\sum_{0 \leq j \leq n} \binom{m}{j} \frac{1}{(n-j)!} (\pi(p+iq))^{m-j} (-p+iq)^{n-j}$$

$$= \sqrt{\frac{n!}{m!}} e^{-(\pi/2)(p^2+q^2)} (\sqrt{\pi}(p+iq))^{m-n} \times$$

$$\sum_{0 \leq j \leq n} \binom{m}{j} \frac{1}{(n-j)!} (-\pi(p^2+q^2))^{n-j}$$

holds. Inserting the expression

$$L_n^{(m-n)}(x) = e^{-\frac{1}{2}x} \sum_{0 \leq j \leq n} \binom{m}{j} \frac{(-x)^{n-j}}{(n-j)!} \qquad (m \geq n)$$

the result follows. -

As a special result we infer the following

Corollary.   For all quadratic lattice points $(\mu, \nu) \in \mathbb{Z} \times \mathbb{Z}$ the identities

$$H(W_m, W_n; \mu, \nu) = \sqrt{\frac{n!}{m!}} (\sqrt{\pi}(\mu+i\nu))^{m-n} L_n^{(m-n)} (\pi(\mu^2+\nu^2)) \qquad (m \geq n),$$

$$H(W_m, W_n; \mu, \nu) = \sqrt{\frac{m!}{n!}} (\sqrt{\pi}(-\mu+i\nu))^{n-m} L_m^{(n-m)} (\pi(\mu^2+\nu^2)) \qquad (n \geq m)$$

hold.

7. THE LATTICE MODEL OF THE LINEAR SCHRÖDINGER REPRESENTATION

The preceding identities for the evaluations of the radar cross-ambiguity function with respect to the harmonic oscillator wave functions at the points (μ, ν) of the quadratic lattice $\mathbb{Z}^2$ of the time-frequency plane $\mathbb{R}^2$ suggests another reali-

zation of the linear Schrödinger representation
U of $\tilde{A}(\mathbb{R})$.

Denote by D the image of the cubic integer
lattice $\mathbb{Z}^3$ under the embedding map

$$\mathbb{Z}^3 \ni (\mu,\nu,\sigma) \longrightarrow (\begin{bmatrix} \mu \\ \nu \end{bmatrix}, \sigma) \in \tilde{A}(\mathbb{R}).$$

Then D is a discrete and cocompact principal sub-
group of $\tilde{A}(\mathbb{R})$ and the homogeneous space $D\backslash\tilde{A}(\mathbb{R})$
of right cosets mod D forms a two-step compact
nilmanifold, the Heisenberg compact nilmanifold.
It is not difficult to see that this manifold is
a principal circle bundle over the compact two-

dimensional torus $\mathbb{T}^2$.

The compact nilmanifold $D\backslash\tilde{A}(\mathbb{R})$ admits a
unique probability measure invariant under right
translations. This measure on $D\backslash\tilde{A}(\mathbb{R})$ will be
assumed in all that follows. Decompose the com-

plex Hilbert space $L^2(D\backslash\tilde{A}(\mathbb{R}))$ into the Hilbert
sum

$$L^2(D\backslash\tilde{A}(\mathbb{R})) = \widehat{\bigoplus_{n \in \mathbb{Z}}} M_n$$

of Hilbert subspaces $(M_n)_{n \in \mathbb{Z}}$ which are comple-

tions of the vector subspaces

$$\{f \in \mathscr{C}^\infty(D\backslash\tilde{A}(\mathbb{R})) \mid f(v,s+s')=e(ns)f(v,s'),$$
$$v \in \mathbb{R} \oplus \mathbb{R}, s, s' \in \mathbb{R}\} \qquad (n \in \mathbb{Z})$$

of smooth left-$\mathbb{Z}^3$-periodic complex-valued functions
on $\tilde{A}(\mathbb{R})$. The complex Hilbert spaces $(M_n)_{n \in \mathbb{Z}}$ are

primary summands with respect to the right quasi-

regular representation δ of $\tilde{A}(\mathbb{R})$ on $L^2(D\backslash\tilde{A}(\mathbb{R}))$
and the subduced representation δ_n forms a unitary

linear representation of $\tilde{A}(\mathbb{R})$ in M_n which con-

tains exactly $|n|$ copies of the single irreducible unitary linear representation $U_n (n \in \mathbb{Z})$ of $\tilde{A}(\mathbb{R})$

when $n \neq 0$; cf. Auslander [1]. In particular, this multiplicity result shows that the unitary linear representation δ_1 of $\tilde{A}(\mathbb{R})$ acting on M_1 is irre-

ducible and therefore, by the theorem of Stone-von Neumann, unitarily isomorphic to the linear Schrödinger representation $U := U_1$ of $\tilde{A}(\mathbb{R})$. Indeed,

the Weil-Brezin isomorphism (cf. Brezin [4]) given by the formula

$$w_1 : \mathscr{A}(\mathbb{R}) \ni f \longrightarrow ((\begin{bmatrix} x \\ y \end{bmatrix}, s) \longrightarrow e(s) \times$$

$$\sum_{\mu \in \mathbb{Z}} f(\mu+x)e(\mu y)) \in M_1 \cap \mathscr{C}^\infty (D \backslash \tilde{A}(\mathbb{R}))$$

extends to a unitary isomorphism $L^2(\mathbb{R}) \to M_1$ of U

onto δ_1, to wit, w_1 is an intertwining operator

between the unitary linear representations U and δ_1 according to the formula

$$w_1 \circ U = \delta_1 \circ w_1.$$

Consequently, from Theorem 1 supra we conclude the following

Theorem 8.   The radar cross-ambiguity function satisfies

$$H(f,g;x,y) = \text{tr}_{\delta_1} (w_1(f) \otimes \overline{w_1(g)}) (\begin{bmatrix} x \\ y \end{bmatrix}, 0)$$

where f,g are elements of $\mathscr{A}(\mathbb{R})$ and $\begin{bmatrix} x \\ y \end{bmatrix} \in \mathbb{R} \oplus \mathbb{R}$.

In other words, $H(f,g;.,.)$ coincides with the restriction of the trace evaluation $\text{tr}_{\delta_1} (w_1(f) \otimes \overline{w_1(g)})$

to the polarized cross-section to $\tilde{Z}$ in $\tilde{A}(\mathbb{R})$.

An application of the Fourier inversion formula for doubly periodic infinitely differentiable functions yields the following result for the product function $w_1(f).\overline{w_1(g)}$.

Corollary. For any functions f,g in $\mathscr{S}(\mathbb{R})$ the absolutely convergent Fourier expansion

$$(w_1(f).\overline{w_1(g)})\left(\begin{bmatrix}x\\y\end{bmatrix},0\right) = \sum_{(\mu,\nu)\,\in\,\mathbb{Z}\times\mathbb{Z}} H(f,g;\mu,\nu)e(-\nu x+\mu y)$$

holds for all points $\begin{bmatrix}x\\y\end{bmatrix} \in \mathbb{R} \oplus \mathbb{R}$ with respect to the topology of uniform $\mathscr{C}^\infty$-convergence.

In view of this identity, the L^2-Parseval-Plancherel theorem for double Fourier series, to wit, the canonical unitary isomorphy between the complex Hilbert spaces $L^2(\mathbb{T}^2)$ and $L^2(\mathbb{Z}^2)$ yields the following relations for the evaluations of radar ambiguity functions at quadratic lattice points:

Theorem 9. For all pairs $(m,n) \in \mathbb{N} \times \mathbb{N}$ the identities

$$\sum_{(\mu,\nu)\in\mathbb{Z}\times\mathbb{Z}} H(W_m;\mu,\nu).H(W_n;\mu,\nu) = \sum_{(\mu,\nu)\in\mathbb{Z}\times\mathbb{Z}} |H(W_m,W_n;\mu,\nu)|^2$$

hold ("Poisson-Plancherel identities" for radially symmetric radar autoambiguity functions).

If we apply the Corollary of Theorem 7 supra then the following consequence of Theorem 9 becomes obvious.

Corollary. In the case $m\geq n$ we have the Poisson-Plancherel identities for Laguerre-Weber functions of different orders

$$\sum_{(\mu,\nu)\in\mathbb{Z}\times\mathbb{Z}} L_m(\pi(\mu^2+\nu^2))L_n(\pi(\mu^2+\nu^2)) = \frac{n!}{m!}\,\pi^{m-n}\sum_{(\mu,\nu)\in\mathbb{Z}\times\mathbb{Z}}(\mu^2+\nu^2)^{m-n}\times$$
$$(L_n^{(m-n)}(\pi(\mu^2+\nu^2)))^2.$$

Of course, the most simplest case occurs when m=1, n=0. Then we get

$$\sum_{(\mu,\nu)\in\mathbb{Z}\times\mathbb{Z}} e^{-\pi(\mu^2+\nu^2)}(1-\pi(\mu^2+\nu^2)) = \pi\sum_{(\mu,\nu)\in\mathbb{Z}\times\mathbb{Z}} e^{-\pi(\mu^2+\nu^2)}(\mu^2+\nu^2)$$

which reads in a slightly different form as follows:

$$2\pi\sum_{(\mu,\nu)\in\mathbb{Z}\times\mathbb{Z}}(\mu^2+\nu^2)e^{-\pi(\mu^2+\nu^2)} = \sum_{(\mu,\nu)\in\mathbb{Z}\times\mathbb{Z}} e^{-\pi(\mu^2+\nu^2)}$$

From this identity we derive by a short calculation the following well-known relation between holomorphic theta series

$$\frac{1}{4\pi}\sum_{\mu\in\mathbb{Z}} e^{-\pi\mu^2} = \sum_{\mu\in\mathbb{Z}} \mu^2 e^{-\pi\mu^2}$$

which is a consequence of the heat equation for the classical first order Jacobi theta functions, too (cf. Bellman [2]). It fits well in André Weil's [28] fundamental discovery in the 1960s that many features of theta series can be explained in terms of group-representation theoretic constructions involving the basic group which stands at the crossroads of quantum mechanics and radar analysis, to wit, the real Heisenberg nilpotent group $A(\mathbb{R})$.

Lehrstuhl für Mathematik I
der Universität Siegen
Hölderlinstraße 3
D-5900 Siegen
Federal Republic of Germany

REFERENCES

1.   Auslander, L.: Lecture notes on nil-theta
      functions. Regional Conference Series in
      Mathematics, No. 34. Providence, R.I.:
      American Mathematical Society 1977

2.   Bellman, R.: A brief introduction to theta
      functions. New York: Holt, Rinehart and
      Winston 1961

3.   Borel, A., Wallach, N.: Continuous cohomology,
      discrete subgroups, and representations of re-
      ductive groups. Annals of Mathematics Studies,
      No. 94. Princeton, N.J.: Princeton University
      Press 1980

4.   Brezin, J.: Harmonic analysis on nilmanifolds.
      Trans. Amer. Math. Soc. 150 (1970), 611-618

5.   Brezin, J.: Geometry and the method of Kirillov.
      In: Non-Commutative Harmonic Analysis, pp.
      13-25. Edited by J. Carmona, J. Dixmier, and
      M. Vergne. Lecture Notes in Mathematics, Vol.
      466. Berlin-Heidelberg-New York: Springer 1975

6.   Cartier, P.: Quantum mechanical commutation
      relations and theta functions. In: Algebraic
      Groups and Discontinuous Subgroups, pp. 361-
      383. Edited by A. Borel and G.D. Mostow. Pro-
      ceedings of Symposia in Pure Mathematics, Vol.
      IX. Providence, R.I.: American Mathematical
      Society 1966

7.   De Bruijn, N.G.: A theory of generalized func-
      tions, with applications to Wigner distribu-
      tion and Weyl correspondence. Nieuw. Arch.
      Wisk 21 (1973), 205-280

8.   Gabor, D.: Theory of communication. J. IEE 93
      (1946) 429-457

9.   Howe, R.: Quantum mechanics and partial dif-
      ferential equations. J. Funct. Anal. 38 (1980),
      188-254

10.   Kirillov, A.A.: Unitary representations of
      nilpotent Lie groups. Uspehi Mat. Nauk. 17
      (1962), 57-110. Russ. Math. Surveys 17 (1962),
      53-104

11.   Moore, C.C., Wolf, J.: Square integrable re-
      presentations of nilpotent Lie groups. Trans.
      Amer. Math. Soc. 185 (1973), 445-462

12.   Peetre, J.: The Weyl transform and Laguerre
      polynomials. Matematiche (Catania) 27 (1972),
      301-323

13.   Reis, F.B.: A linear transformation of the am-
      biguity function plane. IRE Trans. Information
      Theory IT-8 (1962), 59

14.   Schempp, W.: Radar reception and nilpotent har-
      monic analysis I. C.R. Math. Rep. Acad. Sci.
      Canada 4 (1982), 43-48

15.   Schempp, W.: Radar reception and nilpotent har-
      monic analysis II. C.R. Math. Rep. Acad. Sci.
      Canada 4 (1982), 139-144

16.   Schempp, W.: Radar reception and nilpotent har-
      monic analysis III. C.R. Math. Rep. Acad. Sci.
      Canada 4 (1982), 219-224

17.   Schempp, W.: Radar reception and nilpotent har-
      monic analysis IV. C.R. Math. Rep. Acad. Sci.
      Canada 4 (1982), 287-292

18.   Schempp, W.: Radar reception and nilpotent har-
      monic analysis V. C.R. Math. Rep. Acad. Sci.
      Canada 5 (1983), 121-126

19.   Schempp, W.: Radar ambiguity functions of po-
      sitive type. In: General Inequalities IV. Edi-
      ted by L. Losonczi and W. Walter. ISNM Series.
      Basel-Boston-Stuttgart: Birkhäuser (in print)

20.   Schempp, W.: Radar ambiguity functions and the
      linear Schrödinger representation. In: Func-
      tional Analysis and Approximation. Edited by

P.L. Butzer, B. Sz.-Nagy, and R.L. Stens. ISNM Series. Basel-Boston-Stuttgart: Birkhäuser (in print)

21. Schempp, W.: On the Wigner quasi-probability distribution function I. C.R. Math. Rep. Acad. Sci. Canada 4 (1982), 353-358

22. Schempp, W.: On the Wigner quasi-probability distribution function II. C.R. Math. Rep. Acad. Sci. Canada 5 (1983), 3-8

23. Schempp, W.: On the Wigner quasi-probability distribution function III. C.R. Math. Rep. Acad. Sci. Canada 5 (1983), 35-40

24. Schempp, W.: Radar ambiguity functions, the Heisenberg group, and holomorphic theta series (to appear)

25. Schempp, W.: The complex Laplace-Beltrami operator, nilpotent harmonic analysis, and holomorphic theta series (to appear)

26. Ville, J.: Théorie et applications de la notion de signal analytique. Câbles et Transmission 2 (1948), 61-74

27. Warner, G.: Harmonic analysis on semi-simple Lie groups I. Die Grundlehren der math. Wissenschaften, Bd. 188. Berlin-Heidelberg-New York 1972

28. Weil, A.: Sur certains groupes d'opérateurs unitaires. Acta Math. 111 (1964), 143-211. Also in: Collected papers, Vol. III, pp. 1-69. New York-Heidelberg-Berlin: Springer 1980

29. Wilcox, C.H.: The synthesis problem for radar ambiguity functions. MRC Technical Summary Report, No. 157. Madison, WI: The University of Wisconsin 1960

30. Woodward, P.M.: Probability and Information Theory, with Applications to Radar. London: Pergamon Press 1953, New York: McGraw Hill 1953

Louis Auslander

A FACTORIZATION THEOREM FOR THE FOURIER TRANSFORM
OF A SEPARABLE LOCALLY COMPACT ABELIAN GROUP

Introduction

By a separable locally compact Abelian group, abbre-
viated SLCA group, we will mean a locally compact Abelian
group whose topology comes from a separable metric. Let A
be a SLCA group and let Δ be a subgroup of A with a
discrete induced topology and such that A/Δ is compact.
Let $\hat{A}$ be the dual group of A and let

$$\mathcal{J} \colon L^2(A) \to L^2(\hat{A})$$

be the Fourier transform. In this paper we will show that
there is a natural factorization of $\mathcal{J}$ into the product of
three unitary operators. When one specializes this
factorization to the case where $A = \mathbf{Z}/r_1 r_2$, $\mathbf{Z}$ the integers

one obtains the Cooley-Tukey algorithm [5]; when A is a
finite Abelian group one obtains the results in [3], and
for more general groups one obtains the results of A. Weil
in [6,7].
    Our method is to functorially assign a separable
locally compact nilpotent group N to the pair $(A,\hat{A})$ and
use the representation theory of N to obtain the desired
factorization. Special cases of this approach by the
author and collaborators can be found in [1,2,3]. It
should be remarked that our results reduce a proof of
the Plancherel theorem for SLCA groups to the Plancherel
theorem for separable compact Abelian groups.
    The assumption of separability has been made,
because I know nothing of the representation theory of
non-separable groups.

§1. The group N and the Fourier Transform
    We will begin by summarizing the data of Abelian
harmonic analysis in a group theoretic form. Let A be
a SLCA group, let $\hat{A}$ be its dual group and let
$T = \{<a,\hat{a}>| \; a \in A, \; \hat{a} \in \hat{A}\}$. Then T is a closed subgroup of
the complex numbers of absolute value 1. Now consider the

261

R. A. Askey et al. (eds.), Special Functions: Group Theoretical Aspects and Applications, 261–269.
© 1984 by D. Reidel Publishing Company.

second cohomology group $H^2(A \oplus \hat{A}, T)$ with $A \oplus \hat{A}$ acting trivially on T. Then the mapping

$$(a,\hat{a}) \rightarrow <a,\hat{a}> \varepsilon\ T$$

is easily seen to be a 2-cocycle and so determines a unique group N that satisfies the exact sequence

$$1 \rightarrow T \rightarrow N \rightarrow A \oplus \hat{A} \rightarrow 0$$

with T the center of N. It is easily seen that N is a 2-step nilpotent group with commutator subgroup $[N,N] = T$.

A more explicit, if less functorial, presentation of the group N can be given as follows: As a set $N = A \times \hat{A} \times T$ and multiplication is given by

$$(a,\hat{a},s)(b,\hat{b},t) = (a+b,\ \hat{a}+\hat{b},\ st<a,\hat{b}>)$$

N has been called the Heisenberg group of A. The duality theorem suggests we also consider the group M that is the Heisenberg group of $\hat{A}$. Then as a set $M = \hat{A} \times \hat{\hat{A}} \times T = \hat{A} \times A \times T$. Our original functorial definition of N implies that N is isomorphic to M. In our presentation this iso-morphism is easily seen to be given by the mapping

(1) $D(a,\hat{a},t) = (\hat{a},a,t^{-1}<a,\hat{a}>)$

One of the most important objects associated with the pair A and $\hat{A}$ is the Fourier transform $\mathcal{J}$. $\mathcal{J}$ is a unitary operator

(2) $\mathcal{J}: L^2(A) \rightarrow L^2(\hat{A})$

which is formally given by the formula

(3) $(\mathcal{J}f)(\hat{a}) = \int_A f(a) <a,\hat{a}>\ da$

where da is a Haar measure on A. For $f(a) \varepsilon\ L^2(A)$ define

(4) $(S(b)f)(a) = f(a+b) \quad a,b\ \varepsilon\ A$

(5) $(M(\hat{a})f)(a) = <a,\hat{a}>\ f(a) \quad a\ \varepsilon\ A, a\ \varepsilon\ A$

(Clearly S(b), b ε A, is a continuous unitary represen-tation of A and M($\hat{a}$), $\hat{a}$ ε $\hat{A}$, is a continuous unitary representation of $\hat{A}$). It is a classical result that

(6) $\mathcal{J}S(b) = \hat{M}(-b)\mathcal{J}$

(7) $\mathcal{J}M(\hat{a}) = \hat{S}(\hat{a})\mathcal{J}$

where

(8) $(\hat{M}(-b)\hat{f})(\hat{a}) = <-b,\hat{a}>\hat{f}(\hat{a})$ $f \in L^2(\hat{A})$

(9) $(\hat{S}(\hat{a})\hat{f})(\hat{b}) = f(a+b)$ $\hat{f} \in L^2(\hat{A})$

From our point of view the groups N and M are center stage and so the most important objects become irreducible unitary representations of N and M.

Let G be a separable locally compact group and ρ_1 and ρ_2 be irreducible unitary representations of G on Hilbert spaces L_1 and L_2, respectively. A unitary operator $U: L_1 \rightarrow L_2$ is called an intertwining operator for ρ_1 and ρ_2 provided

(10) $U\rho_1 = \rho_2 U$

We will use (L,ρ) to denote that ρ is a unitary representation of G in L and

$$(L_1,\rho_1) \xrightarrow{U} (L_2,\rho_2)$$

to denote that U intertwines ρ_1 and ρ_2.

One of the crucial results of the theory then states that if ρ_1 or ρ_2 is irreducible and $(L_1,\rho_1) \xrightarrow{U_1} (L_2,\rho_2)$ then $U = cU_1$ where $|c| = 1$, $c \in \mathbf{C}$. In this context formulas 6 and 7 will be interpreted as showing that $\mathcal{J}$ is an intertwining operator between irreducible representations of N on $L^2(A)$ and $L^2(\hat{A})$.

We will now carry out this in some detail. One verifies easily that

$$S(b)M(\hat{a}) = <b,\hat{a}>M(\hat{a})S(b)$$

and, since in N

$$(b,0,1)(0,\hat{a},1) = (0,0,<b,a>)(0,\hat{a},1)(b,0,1)$$

it follows that the mapping

$$(\rho(b,\hat{a},t)f)(a) = t<a,\hat{a}>f(a+b)   f \in L^2(A)$$

is a faithful unitary representation of N on $L^2(A)$. One of the fundamental results of Abelian harmonic analysis is that ρ is an irreducible representation of N.

If we define

$$(\mu(\hat{a},a,t)\hat{f})(\hat{b}) = t^{-1}<\hat{b},-a>\hat{f}(\hat{b}+\hat{a}) \qquad f \varepsilon L^2(\hat{A})$$

one verifies by a computation, similar to the above, that μ is a unitary representation of M on $L^2(\hat{A})$. The crucial result then becomes

$$\mathcal{J}\rho = \mu \circ D\mathcal{J}$$

where D is the isomorphism defined by formula 1.

We have thus proven

Theorem 1. $\mathcal{J}$ is, up to scalar multiple, uniquely determined by the property

$$(L^2(A),\rho) \xrightarrow{\mathcal{J}} (L^2(\hat{A}), \mu \circ D).$$

§2.   The Weil-Brezin Mapping

Let Δ be a discrete subgroup of A such that A/Δ is compact. Let $\Delta^* \subset \hat{A}$ be the annihilator of Δ. Then $\hat{A}/\Delta^*$ is the dual group of Δ and hence is compact. By a slight abuse of notation we will use $<\delta,\hat{a}>$, $\delta \varepsilon \Delta$, $\hat{a} \varepsilon \hat{A}$ to denote the pairing of Δ and $\hat{A}/\Delta^*$. Further Δ^* is isomorphic to (A/Δ) and so is discrete. Again, by a slight abuse of notation we will use $<a,\delta^*>$, $a \varepsilon A$, $\delta^* \varepsilon \Delta^*$ to denote the pairing of A/Δ and Δ^*.

Now in N consider the subset

$$\Omega = \{(\delta,\delta^*,1) \mid \delta \varepsilon \Delta, \delta^* \varepsilon \Delta^*\}$$

It is easily verified that Ω is a discrete Abelian subgroup of N and $\Omega\backslash N$ is compact. Haar measure on N is left and right invariant and we will normalize it so that $\Omega\backslash N$ has total measure 1 in the induced measure. Form the Hilbert space $L^2(\Omega\backslash N)$. It will be most convenient to view functions in $L^2(\Omega\backslash N)$ as functions f on N such that

$$f(\omega n) = f(n) \qquad n \varepsilon N, \omega \varepsilon \Omega$$

We define a unitary representative U of N on $L^2(\Omega\backslash N)$ as follows: For $f \varepsilon L^2(\Omega\backslash N)$ let

$$(U(m)f)(n) = f(nm) \qquad m,n \varepsilon N$$

Clearly $f(\omega nm) = f(nm)$ and the fact that Haar measure on

N is left and right invariant implies that $U(m)f \in L^2(\Omega\backslash N)$
and that U is a unitary operator.

Because T is the center of N, the subspace of $L^2(\Omega\backslash N)$
defined by

$$H_1 = \{f \in L^2(\Omega\backslash N) \mid f(nt) = t\, f(u),\ t \in T,\ n \in N\}$$

is a closed subspace of $L^2(\Omega\backslash N)$ invariant under U. Let
$U_1 = U|H_1$. Then U_1 is a unitary representation of N on H_1.

We must now go through a similar construction for M.
Let $\Phi \subseteq M$ be defined by

$$\phi = \{(\delta^*,\delta,1) \mid \delta^* \in \Delta^*,\ \delta \in \Delta\}.$$

Again form $L^2(\Phi\backslash M)$ only this time consider the subspace
of $L^2(\phi\backslash M)$ defined by

$$H_{-1} = \{f \in L^2(\Phi\backslash M) \mid f(mt) = t^{-1}f(m),\ m \in M,\ t \in T\}.$$

Let U_{-1} be the action of M on $L^2(\phi\backslash M)$ restricted to H_{-1}.
Then U_{-1} is a unitary representation of M.

Our next task is to define intertwining operators
W_1 and W_{-1} such that

$$(L^2(A),\delta) \xrightarrow{\ W_1\ } (H_1,U_1)$$

$$(L^2(\hat{A}),\mu) \xrightarrow{\ W_{-1}\ } (H_{-1},U_{-1})$$

We will first give a "formal" definition and verify
"formal" properties and only later worry about questions
of rigor.

For $f \in L^2(A)$ define

$$W_1(f)(a,\hat{a},t) = t \sum_{\delta \in \Delta} f(a+\delta)\ \langle\delta,\hat{a}\rangle$$

Now

$$W_1(f)((\delta,\hat{\delta},1)(a,\hat{a},t)) = W_1(\delta)(a+\delta,\ \hat{a}+\hat{\delta},\ t\langle\delta,\hat{a}\rangle)$$

$$= t\langle\delta,\hat{a}\rangle \sum f(a+\delta+\delta_1)\langle\delta_1,\hat{a}+\hat{\delta}\rangle$$

$$= t \sum f(a+(\delta+\delta_1))\langle\delta+\delta_1,\hat{a}\rangle$$

$$= W_1(f)(a,\hat{a},t).$$

Hence $(W_1 f)(\omega n) = (W_1 f)(n)$, $\quad n \in N$, $\omega \in \Omega$.

We will next verify the intertwining property

$$W_1(f)((a,\hat{a},t)(b,\hat{b},s)) = W_1(f)(a+b,\ \hat{a}+\hat{b},\ ts<a,\hat{b}>)$$

$$= ts<a,\hat{b}> \sum f(a+b+\delta)<\delta,\hat{a}+\hat{b}>$$

$$= t \sum \{s<a+\delta,\hat{b}>f(a+\delta+b)\}<\delta,\hat{a}>.$$

But $\delta(b,\hat{b},s)(f)(a) = s<a,\hat{b}>f(a+b)$. Hence

$$W_1(\delta(b,\hat{b},s)f)(a,\hat{a},t) = t \sum \{s<a+\delta,\hat{b}>f(a+\delta+b)\} \times <\delta,\hat{a}>.$$

This proves that formally W_1 has all the desired properties.

We must now worry about the analytic validity of these formulas.

In $L^2(A)$ consider the dense subspace of continuous functions with compact support, $C_c(A)$.

Assertion I. We will first show that for $f \in C_c(A)$, $W_1(f) \in H_1$ and that $||W_1(f)|| = ||f||$. This implies that W_1 extends to an isometry of $L^2(A)$ into H_1.

Assertion II. We will show that $W_1(L^2(A))$ is dense in H_1. The fact that W_1 is an isometry then implies that W_1 is a surjection and completes the proof the W_1 is a unitary operator.

Proof of Assertion I. First note that the mapping M_1 given by $(M_1 f)(a,\hat{a},t) = f(a,\hat{a},1)$ is a unitary operator $M_1: H_1 \to L^2(A/\Delta \times \hat{A}/\Delta^*)$. Thus we need only show that

$$W: f \to \sum_{\delta \in \Delta} f(a+\delta)<\delta,\hat{a}>$$

is norm preserving.

To do this, let $\overline{X}$ be a measurable subset of A such that every element $a \in A$ can be written in one and only one way as

$$a = x + \delta \qquad x \in \overline{X}, \ \delta \in \Delta.$$

Let C be the support of $f \in C_c(A)$ and let $C_\delta = C \cap (\overline{X} + \delta)$, $\delta \in \Delta$. Then

$$= \underset{\Delta_f}{\cup} C_\delta$$

where Δ_f is a finite subset of Δ and the sets in the union are disjoint. Let $dx = da\,|\overline{X}$. Then

$$\int_A |f|^2 \, da = \sum_{\delta \in \Delta_f} \int_{\overline{X}} |f(x+\delta)|^2 \, dx$$

$$= \int_{\overline{X}} \sum_{\delta \in \Delta_f} |f(x+\delta)|^2 \, dx$$

Now for x fixed, $f(x+\delta)$ is an L^2 function on Δ and so, by the Plancherel theorem for discrete groups, we have (using our notational conventions) that

$$\sum |f(x+\delta)|^2 = \int_{\hat{A}/\Delta^*} |f(x+\delta)<\delta,\hat{a}>|^2 \, d\hat{a}$$

where $d\hat{a}$ is normalized Haar measure on $\hat{A}/\Delta^*$ and

$$\int_A |f|^2 \, da = \int_{A/\Delta} \int_{\hat{A}/\Delta^*} |\sum f(x+\delta)<\delta+\hat{a}>|^2 \, d\hat{a} \, da$$

where da is normalized Haar measure on A/Δ. This proves assertion I.

$\underline{\text{Assertion II}}$. $<\delta^*,a> <\delta,\hat{a}>$, $\delta \in \Delta$, $\delta^* \in \Delta^*$ are a basis of $L^2(A/\Delta \oplus \hat{A}/\Delta^*)$. But if f equals the restriction of $<\delta^*,a>$ to $\overline{X} + \delta$, we have easily that $W_1(f) = <\delta^*,a> <\delta,\hat{a}>$. This proves assertion II.

$\underline{\text{Theorem 2}}$.   The Weil-Brezin mapping W_1 satisfies

$$(L^2(A),\rho) \xrightarrow{\;W_1\;} (H_1,U_1)$$

We will now define W_{-1} in a similar manner, but we will only carry out the formal computations.

Define, for $\hat{f} \in L^2(\hat{A})$

$$W_{-1}(\hat{f})(\hat{a},a,t) = t^{-1} \sum_{\Delta} \hat{f}(\hat{a}+\delta^*)<\delta^*,-a>$$

Because $(\delta^*,\delta,1)(\hat{a},a,t) = (a+\delta^*, a+\delta, <\delta^*,a>t)$ we have

$$W_{-1}(f)((\delta^*,\delta,1)(\hat{a},a,t)) = t^{-1}<\delta^*,-a>\sum_{\Delta^*} \hat{f}(\hat{a}+\delta^*+\delta^{\#})<\delta^{\#},-a>$$

$$= W_{-1}(\hat{f}).$$

To verify the intertwining property note that

$$W_{-1}(t)((\hat{a},a,s)(\hat{b},b,t)) = W_{-1}(\hat{f})(\hat{a}+\hat{b},\ a+b,\ st<\hat{a},b>)$$

$$= s^{-1}t^{-1}<\hat{a},b>^{-1}\sum_{\delta^{\#}\varepsilon\Delta^*} \hat{f}(\hat{a}+\hat{b}+\delta^{\#})<\delta^{\#},-(a+b)>$$

$$= s^{-1}\sum(t^{-1}\hat{f}(\hat{a}+\hat{b}+\delta^{\#})<a,-b><\delta^{\#},-b><\delta^{\#},-a>$$

$$= W_{-1}(\mu(\hat{b},b,t)(\hat{f})).$$

This proves

　　Theorem 3.　The unitary operator W_{-1} satisfies the diagram

$$(L^2(\hat{A}),\mu) \xrightarrow{W_{-1}} (H_{-1},U_{-1})$$

and

$$(L^2(\hat{A}),\mu \circ D) \xrightarrow{W_{-1}} (H_{-1},U_{-1} \circ D)$$

　　The last assertion is the only thing that has not been verified and it is obvious.

　　Notice $(L^2(\hat{A}),\mu)$ is a unitary representation of M and $(L^2(\hat{A}),\mu \circ D)$ is a unitary representation of N.

§3. Factoring the Fourier Transform

　　For $F(a,\hat{a},t) \varepsilon H_1$ define $F^{\#} \varepsilon H_{-1}$ by $M(F) = F^{\#}$ where $F^{\#}(\hat{a},a,t) = F(D^{-1}(\hat{a},a,t))$.

　　Theorem 4.　With all notation as above

$$(H_1,U_1) \xrightarrow{M} (H_{-1},U_{-1} \circ D)$$

　　Proof.　Because $D(a,\hat{a},t) = (\hat{a},a,t^{-1}<a,\hat{a}>)$

$$D^{-1}(\hat{a},a,t) = (a,\hat{a},t^{-1}<a,\hat{a}>)$$

Hence $M(F)(\hat{a},a,t) = t^{-1}<a,\hat{a}>F(a,\hat{a},1)$. This easily implies $M(F) \varepsilon H_{-1}$ and that M is a unitary operator. Because

$$U_{-1}(D(MF) = F^{\#}((\hat{a},a,t)D(b,\hat{b},c))$$

$$= F(D^{-1}((\hat{a},a,t)D(b,\hat{b},c))$$

$$= F(D^{-1}(\hat{a},a,t)(b,\hat{b},c))$$

$$= MU_1F$$

we have that M is an intertwining operator. This completes
the proof of our main result.

<u>Theorem 5.</u> $\mathcal{J} = c(W_{-1})^{-1}MW_1$ $|c| = 1$.

These mappings have been explicitly worked out in the
cases where A is the reals or a finite Abelian group in
[1,2,3]. Notice that W_1 and $(W_{-1})^{-1}$ are "essentially"
tensor products of Fourier transforms and M is a simple
multiplication operator.

References

[1] L. Auslander and R. Tolimieri, Abelian harmonic analysis,
    theta functions and function algebras on a nilmanifold,
    <u>Lecture Notes in Math.</u> vol. 436, Springer-Verlag 1975

[2] L. Auslander and R. Tolimieri, Is computing into the
    finite Fourier transform pure or applied mathematics?
    <u>Bull. (New Series) of the A.M.S.</u> 1(1979) 847-897

[3] L. Auslander, R. Tolimieri and S. Winograd, Hecke's
    theorem in quadratic reciprocity, finite nilpotent
    groups and the Cooley-Tukey algorithm. <u>Advances in Math.</u>
    43(1982) 122-172

[4] L. Brezin, Harmonic analysis in nilmanifolds, <u>Trans.</u>
    <u>Amer. Math. Soc.</u> 150(1970) 611-618

[5] J.W. Colley and J.V. Tukey, An algorithm for the machine
    calculation of complex Fourier series, <u>Math. Comput.</u>
    19(1965) 297-301

[6] A. Weil, L'Integration dans les groupes topologue.
    Hermann and Cie

[7] A. Weil, Sur certaines groupes d'operateurs unitaires,
    <u>Acta Math.</u> 111(1964) 143-211

F. Alberto Grünbaum

BAND AND TIME LIMITING, RECURSION RELATIONS AND SOME NONLINEAR EVOLUTION EQUATIONS

TABLE OF CONTENTS

1. Band and time limiting

Let Af the Finite Fourier Transform of a function $f(x) \in L^2(R)$, given by

$$(Af)(\lambda) = \int_{-T}^{T} e^{i\lambda x} f(x)dx, \quad \lambda \in [-\Omega, \Omega]$$

One can consider A as the result of first chopping f to the interval $[-T,T]$, then taking its Fourier transform and then chopping again, this time to the interval $[-\Omega, \Omega]$. This explains the title of this section.

A number of applied areas require a careful analysis of the singular value decomposition of this operator as a function of T and Ω. A prototype of such an application is discussed in the beautiful series of papers [1,2,3,4,5]. For some other applications, extensions, etc., see [6,7,8,9,10,11].

The central issue is the effective computation of the eigenvectors of the map A^*A. One easily sees that

R. A. Askey et al. (eds.), Special Functions: Group Theoretical Aspects and Applications, 271–286.
© 1984 by D. Reidel Publishing Company.

$$(A^*Af)(x) = \int_{-T}^{T} \frac{\sin \Omega(x-y)}{x-y} \; f(y)dy, \; x \in [-T,T]$$

We see that A^*A is given by a finite convolution integral operator with kernel

$$k(x-y) = \int_{-\Omega}^{\Omega} e^{i\lambda x} \, e^{-i\lambda y} \; d\lambda \tag{1}$$

Look now at a different example. Instead of doing Fourier analysis based on the eigenfunctions of

$$- \frac{d^2}{dx^2}$$

consider the eigenfunctions of

$$- \frac{d^2}{dx^2} + x^2$$

and decompose L^2 in terms of these eigenfunctions, the familiar Hermite functions.

If we denote these functions by $h_n(x)$, it is clear that there are two natural analogs of (1), namely the $N \times N$ matrix.

$$G_{ij}(T) = \int_{-\infty}^{T} h_i(\xi) \, h_j(\xi) \; d\xi \tag{2}$$

or the integral operator in $(-\infty,x)$ with kernel given by

$$\sum_{k=0}^{N} h_k(x) \, h_k(y) \tag{3}$$

2. Commuting differential operator

One has the following remarkable

Theorem. The operator

$$Df \equiv ((T^2 - x^2)f')' - \Omega^2 x^2 f$$

commutes with A^*A introduced in the previous section.

Comment: This result has been found and proved several times, by different people and in very different contexts. See [1,12,13,14].

In an effort to extend this result to other geometrical setups which arise in several applications we were forced to look into the real meaning of this theorem. So far, we have not succeeded in this enterprise, but in the process we have found a number of interesting properties and new questions. The interested reader can consult [15].

For a very small sample we have the following

Theorem. For each N,T there exists a tridiagonal matrix with single spectrum which commutes with (2) and a second order differential operator which commutes with (3).

Comment: This property holds true if we start with any of the "classical orthogonal polynomials": Jacobi, Laguerre, Bessel, Hahn, Krawtchouk, Askey-Wilson, ... On the other hand the property fails on any example that we have considered thus far and which does not come from this list [16].

One should remark that this property, referred to as the COMMUTING PROPERTY for the rest of the paper is not to be expected in general. If $k(\xi)$ in A^*A is replaced by some other convolution kernel the commuting property fails [7]. Similarly if the operator

$$- \frac{d^2}{dx^2} + x^2$$

above is replaced by an arbitrary second order operator of the form

$$- \frac{d^2}{dx^2} + V(x)$$

one does not have this property except for exceptional choices of $V(x)$.

Most of this lecture is an attempt to find some of these $V(x)$, as well as to study some related questions.

3. Differential equation in the spectral parameter

A look at several of the examples where the commuting property holds led us to consider the following pair of problems.

a) Find "potentials" $V(x)$ such that for some family of eigenfunctions $\phi(x,\lambda)$ of the Schroedinger equation

$$\left[- \frac{d^2}{dx^2} + V(x) \right] \phi(x,\lambda) = \lambda^2 \phi(x,\lambda)$$

we have an equation of the form

$$\sum_{i=0}^{M} a_i(\lambda) \frac{\partial^i \phi}{\partial \lambda^i} = \Theta(x)\phi(x,\lambda).$$

b) Find potentials $V(x)$ such that for some family of eigenfunctions $\phi_n(x)$ we get an equation

$$\sum_{i=-M}^{M} a_i(n)\phi_{n+i}(x) = \Theta(x)\phi_n(x).$$

The initial work on problem a) was done in collaboration with H. Duistermaat [17]. Here we show a couple of examples relating to problem a). More examples are

treated in section 4. For examples pertaining to problem
b) see section 5.

Take $V(x,t) = \dfrac{6x(x^3 - 2t)}{(x^3 + t)^2}$, then if

$$\phi(\lambda,x) = \frac{(\lambda^2 x^3 + 3i\lambda x^2 - 3x + \lambda^2 t)}{\lambda^2(x^3 + t)} e^{i\lambda x}$$

we've

$$\left[-\frac{d^2}{dx^2} + V(x,t) \right] \phi(\lambda,x) = \lambda^2 \phi(\lambda,x)$$

and

$$\left[\left(\frac{d}{d\lambda} - \frac{2}{\lambda} \right)\left(\frac{d}{d\lambda} + \frac{2}{\lambda} \right) \right]^2 \phi - 4it \frac{d\phi}{d\lambda} = (x^4 + 4tx)\phi$$

For a different example, take

$$V(x,t) = \frac{15x^4 - 18tx^2 - t^2}{4x^6 + 8tx^4 + 4t^3 x^2}$$

with eigenfunctions

$$\phi(x,k) = -\sqrt{x} \; J_2(xk) + \frac{2t}{(t + x^2)kx} \sqrt{x} \; J_1(xk)$$

One then has

$$\left[\left(-\partial_K^2 + \frac{15}{4k^2} \right)^2 + 2t\left(-\partial_K^2 - \frac{1}{4k^2} \right) \right]\phi = (x^4 + 2tx^2)\phi$$

Observe that the function $V(x,t)$ in the first
example is a solution of the Korteweg-deVries equation
[18, 19,20,21,22,23,24].

A bit more on this in section 4.

4. The characters of GL(N)

As is well known any representation of GL(N) is given by a set of integers

$$(n_1, n_2, \ldots, n_N) \qquad n_i \text{ non decreasing}$$

Then if $e_1, e_2, \ldots, e_N$ denote the eigenvalues of the element g in GL(N), Weyl's character formula gives

$$ch(n_1, n_2, \ldots, n_N) = \det(e_j^{i-1+n_j})/\det(e_j^{i-1}).$$

This is a polynomial in the eigenvalues which can be expressed in terms of the traces of powers of g, the so called "power sums"

$$s_j = e_1^j + e_2^j + \ldots + e_N^j.$$

Below we give some of the first few examples of these characters corresponding to the so called "triangular representations of GL(N)", given by integers

$$(0, 0, \ldots, 0, 1, 2, 3, \ldots, n-2, n-1, n)$$

The identity representation corresponds to $(0,0,0, \ldots, 0)$ and has character equal to

$$1$$

The N dimensional representation given by GL(N) itself has character

$$s_1$$

and corresponds to the integers $(0,0,0, \ldots, 1)$.

The representation with integers $(0,0,0, \ldots, 1,2)$ has character

$$(s_1^3 - s_3)/3.$$

In the case of the representation given by $(\emptyset,\emptyset, \ldots,\emptyset,1,2,3)$ the character is given-except for a multiplicative constant in this and each of the examples below-by

$$9\, s_1 s_5 - 5\, s_3^2 - 5\, s_1^3 s_3 + s_1^6 .$$

For the representation corresponding to $(\emptyset,\emptyset,\emptyset, \ldots, 1,2,3,4)$ we have

$$225\, s_3 s_7 - 225\, s_1^3 s_7 - 189\, s_5^2 + 315\, s_1^2 s_3 s_5$$

$$+ 63\, s_1^5 s_5 - 175\, s_1 s_3^3 - 15\, s_1^7 s_3 + s_1^{10}$$

See Adler and Moser [21] as modified by Airault and Ablowitz in [22].

The example below, corresponding to the integers $(\emptyset,\emptyset,\emptyset, \ldots, 1,2,3,4,5)$, is the first one where the recipe in [22] cannot be taken literally, but rather the variables

$$s_j^*$$

introduced in [21] should be brought into play. The character is given by

$$99225\, s_1 s_5 s_9^* - 55125\, s_3^2 s_9^* - 55125\, s_1^3 s_3 s_9^*$$

$$+ 11025\, s_1^6 s_9^* - 91125\, s_1 s_7^2 + 85050\, s_3 s_5 s_7$$

$$+ 42525\, s_1^3 s_5 s_7 + 70875\, s_1^2 s_3^2 s_7 - 14175\, s_1^5 s_3 s_7$$

$$- 2025\, s_1^8 s_7 - 35721\, s_5^3 - 59535\, s_1^2 s_3 s_5^2 - 11907\, s_1^5 s_5^2$$

$$- 11025\ s_1 s_3^3 s_5 + 33075\ s_1^4 s_3^2 s_5 + 945\ s_1^7 s_3 s_5$$

$$+ 252\ s_1^{10} s_5 + 6125\ s_3^5 - 12250\ s_1^3 s_3^4 - 1225\ s_1^6 s_3^3$$

$$+ 175\ s_1^9 s_3^2 - 35\ s_1^{12} s_3 + s_1^{15}.$$

Observe that in the expression above the coefficient of s_1^6 is given by

$$3^2 5^2 7^2 s_9^* - 5^2 7^2 s_3^3$$

therefore if one takes this coefficient to be , see [22]

$$3^2 5^2 7^2 s_9$$

one is creating a function with a different dependence on the previous variables.

The explicit computation of the Schur functions, that is, the computation of the characters for the next few representations gets involved rather fast.

For the partition $(0, \ldots, 1,2,3,4,5,6)$ the number of terms is 59, for the next one the number becomes 159 and for the partition $(0, \ldots, 1,2,3,4,5,6,7,8)$ one has the healthy value of 483 terms. The author would be happy to provide the explicit expressions of these Schur functions to any one in need.

The polynomials mentioned above were introduced by Adler and Moser in connection with the rational solutions of the Korteweg-deVries equation. The relation with the characters of some representations of GL(N) or equivalently the Schur functions for some representations of the symmetric group were apparently found by Sato.

We have shown that the rational solutions of the Korteweg-deVries equation have eigenfunctions which give

examples of question a) of section 3. The simplest
example was displayed in section 3. The general case
will appear elsewhere. See also [15].

5. Recursion relations and pure soliton potentials

In this section we treat a special class of so
called pure N-soliton potentials $V(x)$ for the
Schroedinger equation

$$\frac{d^2\phi}{dx^2} + V\phi = k^2\phi \tag{4}$$

Such a pure N-soliton potential is prescribed by its
eigenvalues and normalization constants. We have

<u>Theorem.</u> If V is given by

$$V(x) = \frac{2d^2}{dx^2} \log \tau(x) \tag{5}$$

with

$$\tau(x) = \det\left\{\delta_{ij} + c_j^2 \frac{e^{(K_i+K_j)x}}{K_i + K_j}\right\}$$

and

$$K_j = \frac{j}{2} \qquad j = 0, \ldots, N-1$$

$$K_N = \frac{N^2 - N + 4}{4}$$

then the eigenfunctions of (4) with

$$\phi(x,k) \sim e^{kx} \qquad x \to \infty$$

satisfy the finite order recursion relation

$$\sum_{-M}^{M} b_i(k)\phi(x,k+i) = \Theta(x)\phi(x,k) \tag{6}$$

with $b_i(k)$ and $\Theta(x)$ properly chosen and $M = \dfrac{N^2 - N + 2}{2}$.

Comments: This is, of course, not the only choice of potentials leading to (6).

A classical example not covered by the theorem is given by the choice

$$K_j = j \qquad j = 1, \ldots, N \tag{7}$$

and the very special choice of "normalization constants"

$$c_j = 2j(-1)^{j-1} \prod_{i \neq j} \frac{i + j}{i - j} \tag{8}$$

In this case one finds

$$V(x) = N(N+1) \operatorname{sech}^2(x)$$

and the eigenfunctions $\phi(x,k)$ are given in terms of the well known hypergeometric functions by the expression

$$\phi(x,k) = e^{-kx} {}_2F_1\left(-N, N+1, 1+k, \frac{e^{-x}}{e^x + e^{-x}}\right)$$

In this example we have a 3 term recursion relation which holds only if we insist on the choice given in (8) for the "normalization constants".

The whole point of the theorem above is that the choice given there picks up a complete isospectral manifold of potentials $V(x)$ enjoying property (6).

More details about this theorem appear in [25].

Here we present only the example of two solitons which gives a five term recursion relation.

We get

$$K_1 = 1/2 \qquad K_2 = 3/2$$

$$b_{-2}(k) = -\frac{1}{2} \qquad b_{-1}(k) = \frac{c_1^2(2k-1)^2}{(2k-3)(2k+1)} \qquad b_0(k) = 0$$

$$b_1(k) = \frac{c_2^2(2k-3)(2k+5)}{3(2k-1)(2k+3)}$$

$$b_2(k) = \frac{c_1^2 c_2^2(2k-3)(2k-1)(2k+5)(2k+7)}{24(2k+1)^2(2k+3)^2}$$

and

$$\Theta(x) = \frac{c_1^2 c_2^2}{24} e^{2x} + \frac{c_2^2}{3} e^x - c_1^2 e^{-x} - \frac{e^{-2x}}{2}$$

6. The commuting property for deformations from the potential $N(N+1) \operatorname{sech}^2(x)$

In this section we consider the analog of the commuting property for potentials obtained by deforming away from the celebrated

$$V(x) = N(N+1) \operatorname{sech}^2 x$$

This potential is obtained when one uses the expression given by (5) and the choices (7) and (8) in the previous section.

We consider fixing the choice (7) and letting the normalization constants c_i roam around freely. For any such potential we consider the eigenfunctions

$\phi(x,k)$ which behave like e^{kx} at $x = +\infty$.

Forming the Gramm matrix of the N normalized L^2 eigenfunctions

$$G_{ij}(x) = c_i c_j \int_{-\infty}^{x} \phi_i(\xi)\phi_j(\xi)d\xi \qquad 1 \leq i,j \leq N$$

we obtain a matrix valued function which satisfies

$$G(-\infty) = 0 \qquad G(+\infty) = I$$

We consider then the problem of finding a tridiagonal matrix with simple spectrum which would commute with $G(x)$.

We have found the following

Theorem. If the normalization constants c_i are chosen as follows

$$c_j(t,s) = c_j e^t e^{js} \qquad j = 1, \ldots, N \qquad (9)$$

with c_j given as in (8) and t,s arbitrary then one has a tridiagonal matrix $T(x)$ such that

$$T(x)G(x) = G(x)T(x)$$

Comments: This family of examples is special in that chopping to an $M \times M$ matrix with $M < N$ destroys the commuting property. Details and proofs will appear later.

The choice

$$t = 0, \quad s = 0$$

gives $V(x) = N(N+1) \operatorname{sech}^2 x$. The interpretation of the parameter s in (9) is trivial: it amounts to a shift of the variable x. The interpretation of the remaining

free parameter t is rather intriguing. Notice that it
does not correspond to the Korteweg-deVries or any of the
other familiar flows. It can of course be written as a
linear superposition of all these flows at the level of
the tangent space.

For example if N = 2 we get

$$c_1 = \sqrt{6} \qquad\qquad c_2 = \sqrt{12}$$

Then we get

$$c_1(t,0) = e^t c_1 \qquad c_2(t,0) = e^t c_2$$

$$V(x,t) = 2 \frac{d^2}{dx^2} \log \tau(x,t)$$

with

$$\tau(x,t) = 1 + 3e^{2x+t} + 3e^{4x+t} + e^{6x+2t}$$

One can then check that

$$V_t = \frac{7}{12} V_x + \frac{1}{48} (V_{xxx} + 6VV_x)$$

The second term on the right hand side is the
Korteweg-deVries infinitesimal generation acting on V,
but we cannot do away with some translation.

In fact we do not know of any nontrivial potential
V(x) such that the property discussed above holds not
only at t = 0 but also as V(x) evolves under KdV.

References

1. Slepian, D., Pollak, H.P. Prolate spheroidal wave
    functions, Fourier analysis and uncertainty I.
    Bell System Tech. Journal 40, No. 1 (1961) 43-64.

2. Landau, H.J., Pollak, H.O. Prolate spheroidal wave
    functions, Fourier analysis and uncertainty II.
    Bell Systems Tech. Journal 40, No. 1, (1961) 65-84.

3. Landau, H.J., Pollak, H.O. Prolate spheroidal wave
    functions, Fourier analysis and uncertainty III.
    Bell Systems Tech. Journal 41, No. 4 (1962)
    1295-1336.

4. Slepian, D. Prolate spheroidal wave functions,
    Fourier analysis and uncertainty IV. Bell Systems
    Tech. Journal 43, No. 6 (1964) 3009-3058.

5. D. Slepian. 1978. Prolate spheroidal wave functions.
    Fourier analysis and uncertainty. Bell Systems
    Tech. Journal 57, No. 5, 2371-1430.

6. F.A. Grünbaum. A study of Fourier space methods for
    "limited angle" image reconstruction. Numerical
    Functional Analysis and Optimization 2(1), 31-42.

7. F.A. Grünbaum, L. Longhi, M. Perlstadt. Differential
    operators commuting with finite convolution integral
    operators: some nonabelian examples. SIAM J. Applied
    Math., 42, 5, (1982) 941-955.

8. F.A. Grünbaum, 'Eigenvectors of a Toeplitz matrix:
    discrete version of the prolate spheroidal wave
    functions', SIAM J. Alg. Disc. Math. 2(2), (1981)
    136-141.

9. F.A. Grünbaum. 'Toeplitz matrices commuting with a
    tridiagonal matrix', Linear Algebra and its Applica-
    tions 40, (1981) 25-36.

10. F.A. Grünbaum. 'A remark on Hilbert's matrix'.
    Linear Algebra and its Applications, 43, (1982), 119-
    124.

11. F.A. Grünbaum. 'The eigenvectors of the discrete Fourier transform: a version of the Hermite functions', J. Math Anal. Applic. 88, (1982), 355-363.

12. E.T. Whittaker. 1915, Proc. London Math. Soc. (2) 14, 260-268.

13. E.L. Ince. 1922, On the connection between linear diff. systems and integral equations. Proc. of the Royal Society of Edinburg (42), pp. 43-53.

14. M.L. Mehta. Random matrices, Academic press, N.Y., 1967.

15. F.A. Grünbaum. The limited angle problem in tomography and some related mathematical problems. Internat. Colloq. Luminy (France), May 1982, North Holland.

16. F.A. Grünbaum. To appear, 'A new property of reproducting kernels for classical orthogonal polynomials', J. Math. Anal. Applic. 95, (1983).

17. J. Duistermaat and F.A. Grünbaum. In preparation, Differential equations in the eigenvalue parameter.

18. V. Bargmann. On the connection between phase shifts and scattering potential. Rev. Mod. Phys. 21, 1949, 489-493.

19. H. Airault, H. McKean, and J. Moser. 1977, Rational and elliptic solutions of the Korteweg-deVries equation and a related many body problem. Communications in Pure and Applied Math. (30), 95-148.

20. D. Chudnovski and G. Chudnovski. 1977, Pole expansions for nonlinear partial differental equations. Nuovo Cimento, 40B, 339-353.

21. M. Adler and J. Moser. 1978. On a class of polynomials connected with the Korteweg-deVries equation. Communications in Mathematical physics (61), 1-30.

22.   M. Ablowitz and H. Airault. 1981. Perturbations
      finies et forme particuliere de certaines solutions
      de l'equation de Korteweg-deVries. C.R. Acad.
      Sci. Paris, t. 292, 279-281.

23.   P. Lax. 1968. Integrals of nonlinear equations of
      evolutions and solitary waves. Communication on
      Pure and Aoolies Mathematics 21, 467-490.

24.   I. Gelfand and Dikii. 1976. Fractional powers of
      operators and Hamiltonian systems. Funkts. Anal.
      Prilozhen, 10, 4, 13-39.

25.   F.A. Grünbaum. Recursion relations and a class of
      isospectral manifolds for Schroedinger's equation,
      to appear in Nonlinear Waves, E. Debnath, Editor.

J.J. Seidel

HARMONICS AND COMBINATORICS *)

1. INTRODUCTION

The geometry of the sphere in $\mathbb{R}^d$, which provides a setting
for various combinatorial configurations, is governed by
spherical harmonics. Bounds for the cardinality of such configurations may be derived by use of a linear programming
method. This applies to equiangular lines, root systems,
and Newton numbers. The methods may be extended to the hyperbolic case, as well as to the discrete case. The present
paper aims to survey these harmonic methods and some of their
results. The paper does not consider complex, quaternionic,
octave, and other more general situations, which for instance
appear in [5] and [8].
Section 2 considers the orthogonal decomposition of polynomials on the unit sphere in $\mathbb{R}^d$ into spherical harmonic components. The zonal spherical harmonics are introduced in
Section 3, leading to the spherical addition formula and the
Gegenbauer polynomials. This applies to spherical codes in
Section 4, in particular to the three examples mentioned
above. Section 5 relates the theory to spherical designs and
cubature formulae. For the case of hyperbolic geometry Section 6 considers inner products, harmonic polynomials, and
the recently discovered addition formula. In Section 7 we
mention the theory of discrete harmonics, which is the Sym(d)
analogon of the O(d)-invariant spherical harmonics. Likewise,
in Section 8 the analogies between discrete and spherical
designs are indicated.

2. SPHERICAL HARMONICS

Let Pol(k) denote the real linear space of the polynomials
in d variables of degree $\leq k$ restricted to the unit space Ω_d
in $\mathbb{R}^d$. Let Hom(k) and Harm(k) denote the subspaces of the
homogeneous, and of the homogeneous harmonic polynomials of
degree k, respectively. A polynomial is called harmonic if
it is in the kernel of the Laplace operator

R. A. Askey et al. (eds.), Special Functions: Group Theoretical Aspects and Applications, 287–303.
© 1984 by D. Reidel Publishing Company.

$$\Delta = \frac{\partial^2}{\partial x_1^2} + \ldots + \frac{\partial^2}{\partial x_d^2} \; .$$

We use the following positive definite inner product:

$$\langle f,g \rangle = \frac{1}{|\Omega|} \int_\Omega f(\xi)g(\xi)d\omega(\xi) \; , \quad \text{for } f,g \in \text{Pol}(k) \; .$$

<u>Theorem 2.1.</u> $\text{Pol}(k) = \text{Harm}(k) \perp \text{Harm}(k-1) \perp \ldots \perp \text{Harm}(1) \perp \text{Harm}(0)$.

This well-known orthogonal decomposition [13] of $\text{Pol}(k)$ is a direct consequence of the following isomorphisms, whose proofs we briefly indicate.

$$\text{Pol}(k) \cong \text{Hom}(k) \perp \text{Hom}(k-1) \; .$$

Indeed, on the sphere we have $(x,x) = 1$, hence $\text{Hom}(k-2i)$ may be viewed as $(x,x)^i \text{Hom}(k-2i)$ which is contained in $\text{Hom}(k)$. Likewise, $\text{Hom}(k-2i-1) \cong (x,x)^i \text{Hom}(k-2i-1) \subset \text{Hom}(k-1)$. Orthogonality holds since the integral of an odd function over Ω vanishes.

$$\text{Hom}(k) \cong \text{Harm}(k) \perp \text{Hom}(k-2) \; .$$

Indeed, $\text{Harm}(k)$ is the kernel and $\text{Hom}(k-2)$ is the image of the Laplace operator, when applied to $\text{Hom}(k)$. This decomposition is orthogonal [13].

<u>Theorem 2.2.</u> $\dim \text{Pol}(k) = \binom{d + k - 1}{d - 1} + \binom{d + k - 2}{d - 1}$,

$$\dim \text{Hom}(k) = \binom{d + k - 1}{d - 1} \; ,$$

$$\dim \text{Harm}(k) = \binom{d + k - 1}{d - 1} - \binom{d + k - 3}{d - 1} \; .$$

This follows from Theorem 2.1 and from the isomorphism of $\text{Hom}(k)$ to the linear space spanned by the monomials

$$x_1^{k_1} x_2^{k_2} \ldots x_d^{k_d} \; , \quad k_1 + \ldots + k_d = k \; ,$$

whose number is easily calculated.

The elements of Harm(k) are called the spherical harmonics of degree k for the sphere Ω in $\mathbb{R}^d$. With the right topology we have

$$L^2(\Omega) \cong \sum_{i=0}^{\infty} \perp \text{Harm}(i)$$

and generalizations, cf. [13].

Example. For d = 2 the linear space Harm(k) has the orthogonal basis $\{\sqrt{2}\cos k\varphi, \sqrt{2}\sin k\varphi\}$, consisting of polynomials of degree k in $\cos\varphi$ and $\sin\varphi$.

3. ZONAL SPHERICAL HARMONICS

H := Harm(k) is a linear space provided with a nondegenerate inner product $< , >$. For any linear functional $\ell(h)$ defined on H there exists a unique $\tilde{\ell} \in H$ such that

$$\ell(h) = <\tilde{\ell}, h> , \quad \text{for } h \in H .$$

Now fix $\xi \in \Omega$ and define a linear functional on Harm(k) by

$$h \longrightarrow h(\xi) , \quad \text{for } h \in \text{Harm}(k) .$$

By the property mentioned above there exists a unique $\tilde{\xi} \in \text{Harm}(k)$ such that

$$<\tilde{\xi}, h> = h(\xi) , \quad \text{for } h \in \text{Harm}(k) .$$

This polynomial $\tilde{\xi}$, also written $Q_k(\xi, \cdot)$, is called the kth zonal spherical harmonic with pole ξ. The name zonal is explained as follows. Let O(d) denote the orthogonal group of $\mathbb{R}^d$.

Theorem 3.1. $Q_k(\sigma\xi, \sigma\eta) = Q_k(\xi, \eta)$, for $\sigma \in O(d)$.

Proof. Let $h \in \text{Harm}(k)$, $\sigma \in O(d)$, $\xi \in \Omega$. Put $\zeta = \sigma\eta$ in

$$\int_{\Omega} h(\eta) Q_k(\sigma\xi, \sigma\eta) \, d\omega(\eta) =$$

$$= \int_{\Omega} h(\sigma^{-1}\zeta) \, Q_k(\sigma\xi, \zeta) \, d\omega(\zeta) = h(\sigma^{-1}\sigma\xi) = h(\xi) .$$

The result follows from the uniqueness of $Q_k(\xi,\cdot)$ in

$$\langle h, Q_k(\xi,\cdot)\rangle = h(\xi) . \qquad \Box$$

<u>Corollary 3.2.</u> $Q_k(\xi,\cdot)$ is constant on the parallels $\perp \xi$.

Indeed, take $\sigma \in O(d-1)$, $\sigma\xi = \xi$, then $Q_k(\xi,\eta) = Q_k(\xi,\sigma\eta)$.

<u>Corollary 3.3.</u> $Q_k(\xi,\eta)$ depends on (ξ,η) only, that is,

$$Q_k(\xi,\eta) = Q_k(z), \text{ with } z = (\xi,\eta).$$

We shall identify $Q_k(z)$ as a Gegenbauer polynomial by use of the following addition formula.

<u>Theorem 3.4.</u>

$$Q_k(\xi,\eta) = \langle \tilde{\xi}, \tilde{\eta}\rangle = \sum_{i=1}^{\mu_k} f_{k,i}(\xi)\, f_{k,i}(\eta) ,$$

where $f_{k,1},\ldots,f_{k,\mu_k}$ denotes an orthonormal basis

of $\text{Harm}(k)$.

<u>Proof.</u>

$$\langle \tilde{\xi}, \tilde{\eta}\rangle = \tilde{\xi}(\eta) = \sum_{i=1}^{\mu_k} \langle f_{k,i}, \tilde{\xi}\rangle\, f_{k,i}(\eta) =$$

$$= \sum_{i=1}^{\mu_k} f_{k,i}(\xi)\, f_{k,i}(\eta) . \qquad \Box$$

<u>Corollary 3.5.</u> $Q_k(1) = \dim \text{Harm}(k)$.

<u>Proof.</u>

$$Q_k(1) = \sum_{i=1}^{\mu_k} f_{k,i}^2(\xi) = \sum_{i=1}^{\mu_k} \frac{1}{|\Omega|} \int_\Omega f_{k,i}^2(\xi)\, d\omega(\xi) =$$

$$= \sum_{i=1}^{\mu_k} 1 = \mu_k . \qquad \Box$$

The addition formula is the case $k = \ell$ of the formula

$$\int_{\Omega} Q_k(\xi,\zeta)\, Q_\ell(\zeta,\eta)\, d\omega(\zeta) = \delta_{k,\ell}\, |\Omega|\, Q_k(\xi,\eta) \ ,$$

which also holds for $k \neq \ell$, since then Harm(k) $\perp$ Harm(ℓ). Putting $\xi = \eta$ and $(\xi,\zeta) = z$ we find

$$|\Omega_{d-1}| \int_{-1}^{1} Q_k(z)\, Q_\ell(z)\, (1-z^2)^{\frac{1}{2}(d-3)}\, dz =$$

$$= \delta_{k,\ell}\, |\Omega_d|\, Q_k(1) \ .$$

This means that $Q_k(z)$, $k = 0,1,2,\ldots$, $-1 \leq z \leq 1$, constitute a family of polynomials in one variable z which is orthogonal with respect to the weight function

$(1-z^2)^{\frac{1}{2}(d-3)}$. The first few of these so-called Gegenbauer polynomials are

$$Q_0(z) = 1 \ , \quad Q_1(z) = dz \ , \quad Q_2(z) = \tfrac{1}{2}(d+2)(dz^2 - 1) \ ,$$

$$Q_3(z) = \tfrac{1}{6}\, d(d+4)((d+2)z^3 - 3z) \ ,$$

$$Q_4(z) = \tfrac{1}{24}\, d(d+6)((d+2)(d+4)z^4 - 6(d+2)z^2 + 3) \ .$$

The Gegenbauer polynomial $Q_k(z)$ has degree k, and is even (odd) for k even (odd). Hence any polynomial F(z) of degree μ has an expansion

$$F(z) = \sum_{k=0}^{\mu} f_k\, Q_k(z)$$

with unique Gegenbauer coefficients f_k.

4. SPHERICAL CODES

Let A denote a subset of the interval $[-1,1[$. A spherical A-code in $\mathbb{R}^d$ is a finite subset X of the d-dimensional unit sphere Ω such that

$$(\underline{x},\underline{y}) \in A \ , \quad \text{for all } x,y \in X, \quad x \neq y \ .$$

Theorem 4.1. If $|A| = s$ then for any A-code X:

$$|X| \leq \binom{d+s-1}{d-1} + \binom{d+s-2}{d-1} .$$

Proof. For each $y \in X$ define the polynomial F_y by

$$F_y(\xi) := \prod_{\alpha \in A} \frac{(y,\xi) - \alpha}{1 - \alpha} , \quad \xi \in \Omega .$$

These are $|X|$ polynomials of degree $\leq s$. Since

$$F_y(x) = \delta_{y,x} \quad \text{for } x,y \in X ,$$

these polynomials are independent. Hence $|X| \leq \dim \mathrm{Pol}(s)$, and application of Theorem 2.2 yields the result, which is called the absolute bound.   □

Sometimes better bounds may be achieved by use of the following "linear programming method". We look for an appropriate polynomial $F(z)$ with degree μ and Gegenbauer coefficients $f_0, f_1, \ldots, f_\mu$. We compare the following two expressions for a spherical A-code X:

$$\sum_{x,y \in X} F((x,y)) = |X| \, F(1) + \sum_{\alpha \in A} \mathrm{freq}(\alpha) \, F(\alpha) ,$$

$$\sum_{x,y \in X} F((x,y)) = f_0 |X|^2 + \sum_{k=1}^{\mu} f_k \sum_{x,y \in X} Q_k(x,y) .$$

Here

$$\mathrm{freq}(\alpha) := \#\{(x,y) \in X \times X \mid (x,y) = \alpha\} \geq 0 ,$$

$$\sum_{x,y \in X} Q_k(x,y) = \sum_i \left(\sum_{x \in X} f_{k,i}(x) \right)^2 \geq 0 .$$

The following theorem is a direct consequence of these considerations.

Theorem 4.2. Let $F(z)$ have nonnegative Gegenbauer coefficients, let $f_0 > 0$, and let $F(\alpha) \leq 0$ for all $\alpha \in A$. Then the cardinality of any A-code X satisfies

$$|X| \leq F(1) / f_0 .$$

We illustrate this theorem and its generalizations by the following examples.

Example 4.3. Equiangular lines [9]

Take $A = \{\alpha, -\alpha\}$ and take

$$F(z) = \frac{z^2 - \alpha^2}{1 - \alpha^2} = \frac{1 - d\alpha^2}{d(1 - \alpha^2)} Q_0(z) + \frac{2}{d(d + 2)(1 - \alpha^2)} Q_2(z) .$$

Application of Theorem 4.2. yields

$$|X| \leq \frac{d(1 - \alpha^2)}{1 - d\alpha^2} , \quad \text{for } \alpha^2 < \frac{1}{d} .$$

For $\alpha = \frac{1}{3}$ the bounds read

d	3	4	5	6	7	8
$\|X\|$	4	6	10	16	28	28

and examples exist meeting these bounds.

Example 4.4. Root systems [3]

Take $A = \{0, \frac{1}{2}, -\frac{1}{2}\}$ and take $F(z) = z(z^2 - \frac{1}{4})$. Application of Theorem 4.2 yields, for $d < 10$,

$$|X| \leq \frac{3d(d + 2)}{10 - d} ,$$

d	5	6	7	8	9
$\|X\|$	21	36	63	120	120

The root systems E_6, E_7, E_8 meet the bound in dimensions 6, 7, 8, respectively.

Example 4.5. Newton numbers [11]

Let τ_d denote the maximum number of nonoverlapping unit
spheres that can touch a given unit sphere in $\mathbb{R}^d$. Until recently only τ_1, τ_2, τ_3 were known. Odlyzko and Sloane [11],
using the linear programming method, determined τ_8 and τ_{24}
and improved existing bounds for τ_d. We illustrate this for
the case d = 8. We take

$$A = \{\alpha \in \mathbb{R} \mid -1 \le \alpha \le \tfrac{1}{2}\} ,$$

and

$$F(z) = (z+1)(z+\tfrac{1}{2})^2 z^2 (z-\tfrac{1}{2}) .$$

The Gegenbauer coefficients turn out to be nonnegative, and
$f_0 = 3/320$. Application of Theorem 4.2 yields $\tau_8 \le 240$. On
the other hand, the root system E_8 provides an example of 240
spheres. Therefore, $\tau_8 = 240$.

5. SPHERICAL DESIGNS

We briefly indicate the relevance of harmonic methods for
spherical designs, cf. [6]. A finite subset X of the unit
sphere is a spherical design of strength t if

$$\frac{1}{|X|} \sum_{x \in X} f(x) = \frac{1}{|\Omega|} \int_\Omega f(\xi) \, d\omega(\xi) , \quad \text{for all } f \in \mathrm{Pol}(t).$$

Equivalent definitions are the following:

For k = 1,2,...,t   ,   the kth moments of X equal the kth
                              moments of Ω.

For k = 1,2,...,t   , $\sum_{x \in X} h(x) = 0$, for all $h \in \mathrm{Harm}(k)$.

The last condition implies $\sum_{x \in X} Q_k(x,y) = 0$, for k = 1,2,...,t.

This leads to the linear programming method explained in
Section 4 of which Theorem 5.1 is a direct consequence.
Theorems 5.2 and 5.3 are somewhat more complicated, yet they
follow from similar techniques.

Theorem 5.1. Let X be a spherical A-code and a spherical
        t-design. Let F(z) have Gegenbauer coefficients $f_0 > 0$
        and $f_k \leq 0$ for $k > t$, and let $F(\alpha) \geq 0$ for all $\alpha \in A$.
        Then

$$|X| \geq F(1) / f_0.$$

Theorem 5.2. Let X be an A-code and a t-design, with $|A| = s$,
        $t = 2e$. Then $\dim \text{Pol}(e) \leq |X| \leq \dim \text{Pol}(s)$, hence $t \leq 2s$.

Moreover, if equality somewhere, then everywhere.

Theorem 5.3. Let X be a $(2e + 1)$-design which is antipodal
        (i.e. $X = -X$) and has s inner products $\neq \pm 1$. Then

$$2 \dim \text{Pol}(e) \leq |X| \leq 2 \dim \text{Hom}(s), \quad \text{hence } e \leq s .$$

Moreover, if equality somewhere, then everywhere.

Small examples for equality in 5.2 and 5.3 are the regular
pentagon ($d = 2$, $t = 4$, $|X| = 5$), and the icosahedron
($d = 3$, $t = 5$, $|X| = 12$). For the remaining examples we refer
to [6]:

$$(d, t, |X|) = (6, 4, 27), (23, 4, 275), (8, 7, 240), (24, 11, 2\binom{28}{5})) .$$

Spherical harmonics may be used in the construction of cubature formulae of strength t. The problem is to find finite
point sets

$$\{x_1, \ldots, x_N\} \subset \Omega \subset \mathbb{R}^d$$

and coefficients $a_1, \ldots, a_N$ such that

$$\int_\Omega f(\xi) \, d\omega(\xi) = \sum_{r=1}^N a_r f(x_r)$$

for all $f \in \text{Pol}(t)$. By use of the orthogonal decomposition
of $\text{Pol}(t)$ the problem of constructing cubature formulae of
arbitrary strength t can be solved. Indeed, in terms of the
harmonic components $f_0, f_1, \ldots, f_t$ of any $f \in \text{Pol}(t)$ the condition above reads

$$f_0 = \sum_{r=1}^{N} a_r \sum_{k=0}^{t} f_k(x_r) ,$$

for all $f_k \in$ Harm(k), for k = 0,1,...,t , that is,

$$1 = \sum_{r=1}^{N} a_r , \quad 0 = \sum_{r=1}^{N} a_r f_k(x_r) .$$

By taking bases $\{f_{k,i}\}$ for Harm(k), and points $x_1,...,x_N$
such that the matrix $[f_{k,i}(x_r)]$ has maximal rank, we can calculate the coefficients $a_1,...,a_N$. This solves the problem,
but not in an efficient way. Better cubature formulae may be
obtained by using suitable orbits of a finite subgroup G of
the orthogonal group $O(d)$. We refer to [7] for this application of the theory of harmonic invariants of G.

Spherical t-designs may be viewed as the special case of a
cubature formula in which all coefficients $a_1,...,a_N$ are equal.
It is interesting to notice that very recently Seymour and
Zaslavsky have given a (nonconstructive) proof for the existence of spherical t-designs in $\mathbb{R}^d$ for any t and any d,
cf. [12].

6. HYPERBOLIC HARMONICS

In $\mathbb{R}^d$ we introduce the following indefinite inner product:

$$[x,y] = -x_0 y_0 + x_1 y_1 + ... + x_d y_d , \text{ for } x,y \in \mathbb{R}^{d+1} .$$

The vector x is inside, on, outside the "light cone" whenever [x,x] is negative, zero, positive, respectively. We consider the one- and the two-sheeted unit hyperboloïds

$$\Omega^+ = \{x \in \mathbb{R}^{d+1} \mid [x,x] = +1\} ,$$

$$\Omega^- = \{x \in \mathbb{R}^{d+1} \mid [x,x] = -1\} .$$

Ω^- is disconnected; one part has $x_0 > 0$, the other part has
$x_0 < 0$.

In order to develop a theory of harmonics we introduce an
inner product for polynomials in terms of differential operators (thus avoiding integration over noncompact spheres).

Let S^* denote the polynomial algebra of the vector space, that is, the functions on $\mathbb{R}^{d+1}$ which are represented by polynomials in $d + 1$ variables.

<u>Definition.</u> $\langle f,g \rangle = (f(\partial)g)(0)$, for $f,g \in S$, where

$$\partial = \left(-\frac{\partial}{\partial x_0} , \frac{\partial}{\partial x_1} , \dots , \frac{\partial}{\partial x_d} \right) .$$

Thus the polynomial $f(x_0,x_1,\dots,x_d)$ is converted into a differential operator by replacing x_0 by $-\partial / \partial x_0$, x_1 by $\partial / \partial x_1,\dots,x_d$ by $\partial / \partial x_d$. This operator is applied to $g(x_0,x_1,\dots,x_d)$, and the value at $x_0 = x_1 = \dots = x_d = 0$ is taken.

For instance

$$f = x_0^{k_0} x_1^{k_1} \dots x_d^{k_d} \quad \text{has} \quad \langle f,f \rangle = (-1)^{k_0} \prod_{i=0}^{d} k_i !$$

The inner product thus defined is symmetric ($\langle f,g \rangle = \langle g,f \rangle$) and invariant (w.r.t. the orthogonal group of $[x,y]$). We refer to [1] for details. The analogous definition for the positive definite case coincides with the definition of Section 2 in terms of integrals, up to a scalar multiple (depending on k), on the space Harm(k).

The differential operator corresponding to the quadratic $[x,x]$ is the hyperbolic Laplace operator

$$\Box = -\frac{\partial^2}{\partial x_0^2} + \frac{\partial^2}{\partial x_1^2} + \dots + \frac{\partial^2}{\partial x_d^2} .$$

Let $\text{hom}_{d+1}(k)$ denote the linear space of the polynomials in S^* which are homogeneous of degree k. Let

$$\text{harm}_{d+1}(k) := \text{hom}_{d+1}(k) \cap \ker \Box .$$

Then as before we have

$$\text{hom}_{d+1}(k) \cong \text{harm}_{d+1}(k) \perp \text{hom}_{d+1}(k-2) ,$$

$$\dim \text{harm}_{d+1}(k) = \binom{d+k}{d} - \binom{d+k-2}{d} = \binom{d+k-1}{d-1} + \binom{d+k-2}{d-1} .$$

The second expression also represents a decomposition, cf. the next lemma.

Lemma 6.1. $\operatorname{harm}_{d+1}(k) \cong \operatorname{hom}_d(k) \perp \operatorname{hom}_d(k-1)$.

Proof. We separate the coordinate x_0 in $x = (x_0; x')$ and in $f \in \operatorname{hom}_{d+1}(k)$ as follows:

$$f(x) = \sum_{i=0}^{k} x_0^i f_i(x') \ , \ f_i \in \operatorname{hom}_d(k-i) \ .$$

$$\Box f = \sum_{i=0}^{k-2} x_0^i \Delta f_i - \sum_{i=2}^{k} i(i-1) x_0^{i-2} f_i \ ,$$

where Δ is the ordinary Laplace operator in the variables $x_1, \ldots, x_d$. The condition $\Box f = 0$ determines $f_2, \ldots, f_k$ in terms of the arbitrary polynomials $f_0 \in \operatorname{hom}_d(k)$ and $f_1 \in \operatorname{hom}_d(k-1)$ as follows:

$$f_{2j} = \frac{(-1)^j}{(2j)!} \Delta^j f_0 \ , \ f_{2j+1} = \frac{(-1)^j}{(2j+1)!} \Delta^j f_1 \ ,$$

$$f(x) = \sum_{j=0}^{\lfloor \frac{1}{2}k \rfloor} x_0^{2j} f_{2j}(x') + \sum_{j=0}^{\lfloor \frac{1}{2}k-1 \rfloor} x_0^{2j+1} f_{2j+1}(x') \ . \qquad \Box$$

This proves the decomposition into the two subspaces, which are orthogonal with respect to $< , >$. Any f restricted to the first subspace is an even function of x_0, hence has positive $<f,f>$. Any f restricted to the second subspace is an odd function of x_0, hence has negative $<f,f>$. This proves the following theorem.

Theorem 6.2. $\operatorname{harm}_{d+1}(k)$, provided with the inner product

$< , >$, is nondegenerate and has signature

$$\mu_k = \binom{d+k-1}{d-1}, \ \nu_k = \binom{d+k-2}{d-1} \ .$$

This leads to the hyperbolic addition formula, cf. [1]. For simplicity we take x and y such that $[x,x] = [y,y] = 1$ or -1.

Theorem 6.3.

$$Q_k([x,y]) = \langle \tilde{x}, \tilde{y} \rangle =$$

$$= \sum_{i=1}^{\mu_k} f_{k,i}(x) f_{k,i}(y) - \sum_{j=1}^{\nu_k} g_{k,j}(x) g_{k,j}(y) .$$

Here $\{f_{k,i}\}$ denotes an orthonormal basis for $hom_d(k)$, $\{g_{k,j}\}$ for $hom_d(k-1)$, the signature (μ_k, ν_k) is as in Theorem 6.2, and the $Q_k(z)$ turn out to be the same Gegenbauer polynomials as in Section 3.

The techniques referred to in this section are useful for combinatorics. For instance, the addition formula yields a proof for the following result ([2],[1])

Theorem 6.4. An s-distance set in hyperbolic d-space has at most $\binom{d+s}{d}$ points.

7. DISCRETE HARMONICS

We summarize the theory of Delsarte [4]. Given $d \geq 2r > 0$, the discrete sphere in $\mathbb{R}^d$ is defined to be the set of all r-subsets (blocks) of a d-set:

$$\Omega := \{x \in \mathbb{R}^d \mid x_1^2 + \dots + x_d^2 = r , \quad x_i \in \{0,1\}\} .$$

The automorphism group of Ω is the symmetric group Sym(d) on d symbols.
For the functions defined on Ω we use the inner product

$$\langle f,g \rangle = \sum_{x \in \Omega} f(x) g(x) .$$

Hom(k) is the linear space of the homogeneous polynomials, of degree ≤ 1 in each of the d variables, of total degree k, restricted to Ω. Hom(k) is spanned by the restrictions of

$$x_1^{k_1} \dots x_d^{k_d} , \quad k_i \in \{0,1\} , \quad \sum_{i=1}^{d} k_i = k ,$$

hence has dimension $\binom{d}{k}$. As an invariant differential operator we use

$$\Delta := \frac{\partial}{\partial x_1} + \ldots + \frac{\partial}{\partial x_d} : \text{Hom}(k) \longrightarrow \text{Hom}(k-1) .$$

Example. $(x_1 - x_2)(x_3 - x_4) \ldots (x_{2k-1} - x_{2k}) \in \text{Ker } \Delta$.

We lift Hom(k) to Hom(r) by use of the symmetric function $s(r-k)$ of degree $r-k$ in the remaining $d-k$ variables:

$$\psi : \text{Hom}(k) \longrightarrow \text{Hom}(r) : f_k \longrightarrow f_k s(r-k) .$$

Defining the discrete harmonic polynomials by

$$\text{Harm}(i) := \psi(\text{Ker } \Delta \cap \text{Hom}(i)) ,$$

we arrive at the following orthogonal decomposition.

Theorem 7.1.

$$\text{Hom}(t) = \text{Harm}(t) \perp \text{Harm}(t-1) \perp \ldots \perp \text{Harm}(1) \perp \text{Harm}(0).$$

Corollary 7.2.

$$\mu_t := \dim \text{Harm}(t) = \binom{d}{t} \doteq \binom{d}{t-1} .$$

Now the theory of discrete harmonics may be developed along the lines of Section 3. For any $\xi \in \Omega$ the zonal harmonic $Q_k(\xi, \cdot)$ is introduced, and we readily arrive at the addition formula for discrete harmonics:

Theorem 7.3.

$$Q_t((\xi, \eta)) = \sum_{i=1}^{\mu_t} f_{t,i}(\xi) f_{t,i}(\eta) .$$

Here $f_{t,1}, \ldots, f_{t,\mu_t}$ denote an orthonormal basis for Harm(t).

The $Q_t(z)$ belongs to the Hahn polynomials, a family of polynomials in the discrete variable $z \in \{0,1,\ldots,r\}$ which is orthogonal w.r.t. the weight function

$$w(z) = \binom{r}{z}\binom{d-r}{r-z} ,$$

that is, the number of $\eta \in \Omega$ such that $(\xi, \eta) = z$ for any $\xi \in \Omega$. Any polynomial $F(z)$ in the discrete variable z has a unique Hahn expansion. As in Section 4, a key tool will be the linear programming method, that is, to find an appropriate $F(z)$, to express in two ways $\sum\limits_{x,y \in X} F((x,y))$ for a subset X of Ω, and to take into account the constraints, so as to derive useful inequalities for the size of $|X|$.

8. DISCRETE t-DESIGNS

A t-design t-(d,r,λ) is a collection X of r-subsets of a d-set such that each t-subset is in a constant number λ of blocks. This definition, when viewed in terms of the discrete sphere Ω, is anlogous to the definition of a spherical t-design in Section 5. Indeed, in both definitions we require

$$\sum_{x \in X^{\sigma}} f(x) \ , \quad \text{for } f \in \text{Hom}(t) \ ,$$

to be constant with respect to the elements σ of a group. In Section 5 this is the orthogonal group $O(d)$. In the present section this is the symmetric group $\text{Sym}(d)$.

Example. The 5-design property of the Steiner system 5-$(24,8,1)$ is expressed in terms of the set X of the blocks of size 8 by

$$1 = \sum_{x \in X} x_{\sigma(1)} \ x_{\sigma(1)} \ x_{\sigma(3)} \ x_{\sigma(4)} \ x_{\sigma(5)} \ ,$$

for all permutations σ of the 24 variables x_i.

Using the orthogonal decomposition of $\text{Hom}(t)$ we find the following characterization of t-designs:

Theorem 8.1. A set of blocks X forms a t-design whenever

$$\sum_{x \in X} h(x) = 0, \text{ for all } h \in \sum_{i=1}^{t} \text{Harm}(i).$$

The linear programming method yields the following generalization of Fisher's inequality, due to Wilson.

Theorem 8.2. $|X| \geq \binom{d}{e}$ for any 2e-design X.

In the proof the method of investigating $\sum\limits_{x,y \in X} F((x,y))$ is applied with

$$F(z) = (Q_0(z) + \ldots + Q_e(z))^2 ,$$

and it turns out that

$$F(r) = \binom{d}{e}^2 , \qquad f_0 = \binom{d}{e} .$$

In the case of equality we have $F((x,y)) = 0$ for all $x \neq y$.

As an example we mention the $253 = \binom{23}{2}$ blocks of the design $4-(23,7,1)$.

REFERENCES

1. E. Bannai, A. Blokhuis, P. Delsarte, J.J. Seidel, An addition formula for hyperbolic space, Journal Combin. Theory A, 36 (1984), to appear.
2. A. Blokhuis, An upper bound for the cardinality of s-distance sets in E^d and H^d, Memo 1982-08, Technol. Univ. Eindhoven.
3. P.J. Cameron, J.M. Goethals, J.J. Seidel, E.E. Shult, Line graphs, root systems and elliptic geometry, J. Algebra 43 (1976), 305-327.
4. P. Delsarte, Hahn polynomials, discrete harmonics, and t-designs, Siam J. Appl. Math. 34 (1978), 157-166.
5. P. Delsarte, J.M. Goethals, J.J. Seidel, Bounds for systems of lines and Jacobi polynomials, Philips Research Repts 30 (1975), 91-105 (Bouwkamp volume).
6. P. Delsarte, J.M. Goethals, J.J. Seidel, Spherical codes and designs, Geometricae Dedicata 6 (1977), 363-388.
7. J.M. Goethals, J.J. Seidel, Cubature formulae, polytopes and spherical desings, pp. 203-210 in The Geometric Vein, Coxeter Festschrift, ed. C. Davis a.o., Springer 1982.
8. S. Hoggar, t-designs in projective spaces, Europ. J. Combinat. 3 (1982), 233-254.
9. P.W.H. Lemmens, J.J. Seidel, Equiangular lines, J. Algebra 24 (1973), 494-512.
10. C. Müller, Spherical harmonics, Lecture Notes in Mathem. 17 (1966), Springer.

11. A.M. Odlyzko, N.J.A. Sloane, New bounds on the number of unit spheres that can touch a unit sphere in n dimensions, J. Combin. Theory A, 26 (1979), 210-214.
12. P.D. Seymour, T. Zaslavsky, Averaging sets: a generalization of mean values and spherical designs, manuscript.
13. E.M. Stein, G. Weiss, Introduction to Fourier analysis in Euclidean spaces, Princeton Univ. Press 1971.

*) This paper was published earlier in Proc. Seminar on Combinatorics and Applns in honour of S.S. Shrikhande, Calcutta 1982, eds. K.S. Vijayan, N.M. Singhi, published by the Indian Statistical Institute.

Subject Index

305

Mathematics and Its Applications